A BIRD-FINDING GUIDE TO COSTA RICA

A BIRD-FINDING GUIDE TO

COSTA RICA

Barrett Lawson

Comstock Publishing Associates
a division of
Cornell University Press
ITHACA, NEW YORK

First published 2009 by Cornell University Press
First printing, Cornell Paperbacks, 2009

Printed in the United States of America

Library of Congress Cataloging-in-Publication Data

Lawson, Barrett, 1982–
A bird-finding guide to Costa Rica / Barrett Lawson.
p. cm.
Includes bibliographical references and index.
ISBN 978-0-8014-7584-9 (pbk. : alk. paper)
1. Bird watching—Costa Rica—Guidebooks. 2. Birding sites—Costa Rica—Guidebooks. 3. Birds—Costa Rica. 4. Costa Rica—Guidebooks. I. Title.
QL687.C8L39 2009
598.072'347286—dc22
2009016843

Cornell University Press strives to use environmentally responsible suppliers and materials to the fullest extent possible in the publishing of its books. Such materials include vegetable-based, low-VOC inks and acid-free papers that are recycled, totally chlorine-free, or partly composed of nonwood fibers. For further information, visit our website at www.cornellpress.cornell.edu.

Paperback printing 10 9 8 7 6 5 4 3 2 1

This book is dedicated to my father,
and all the wonderful birding
memories we share.

Contents

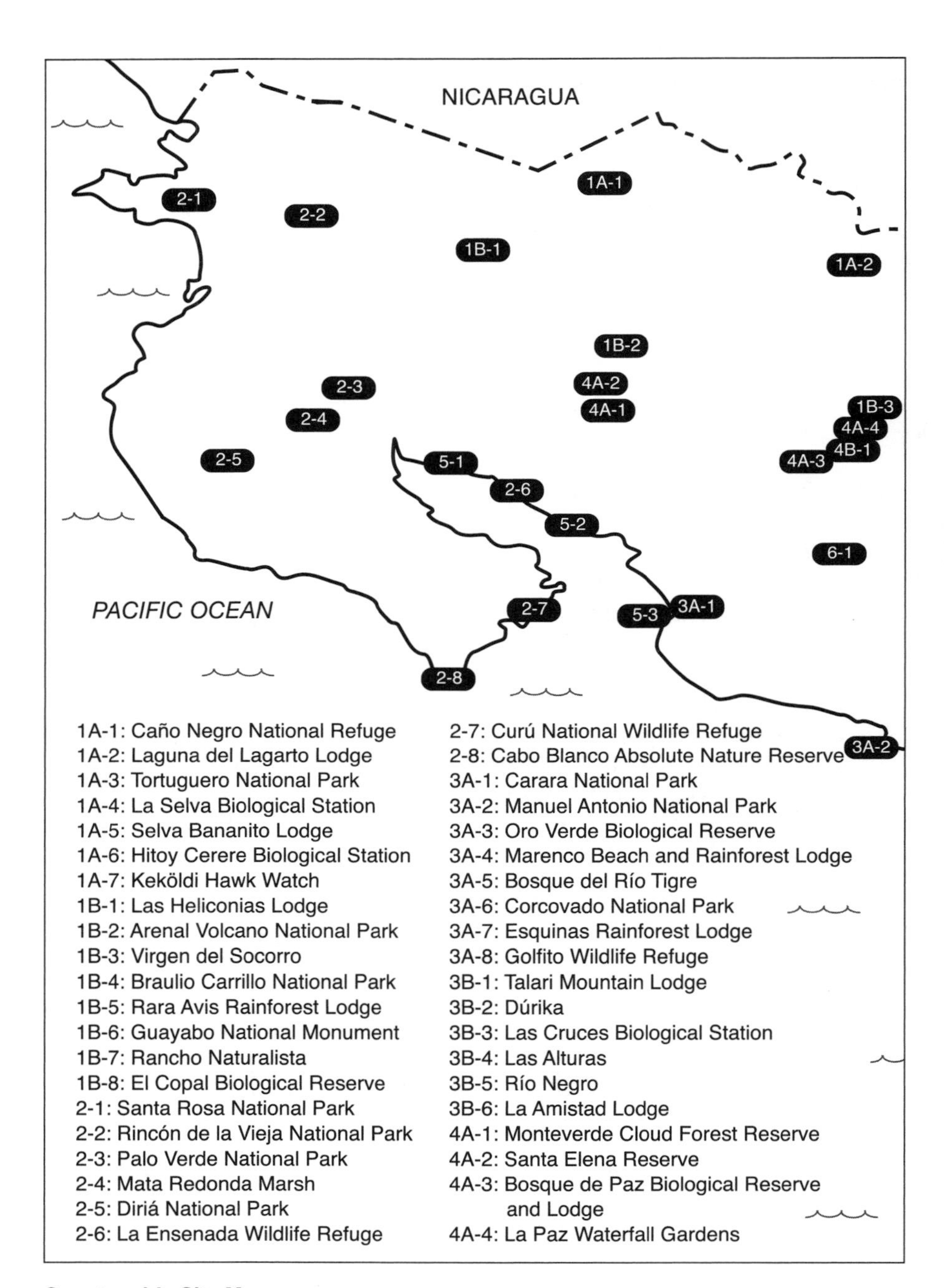

Countrywide Site Map

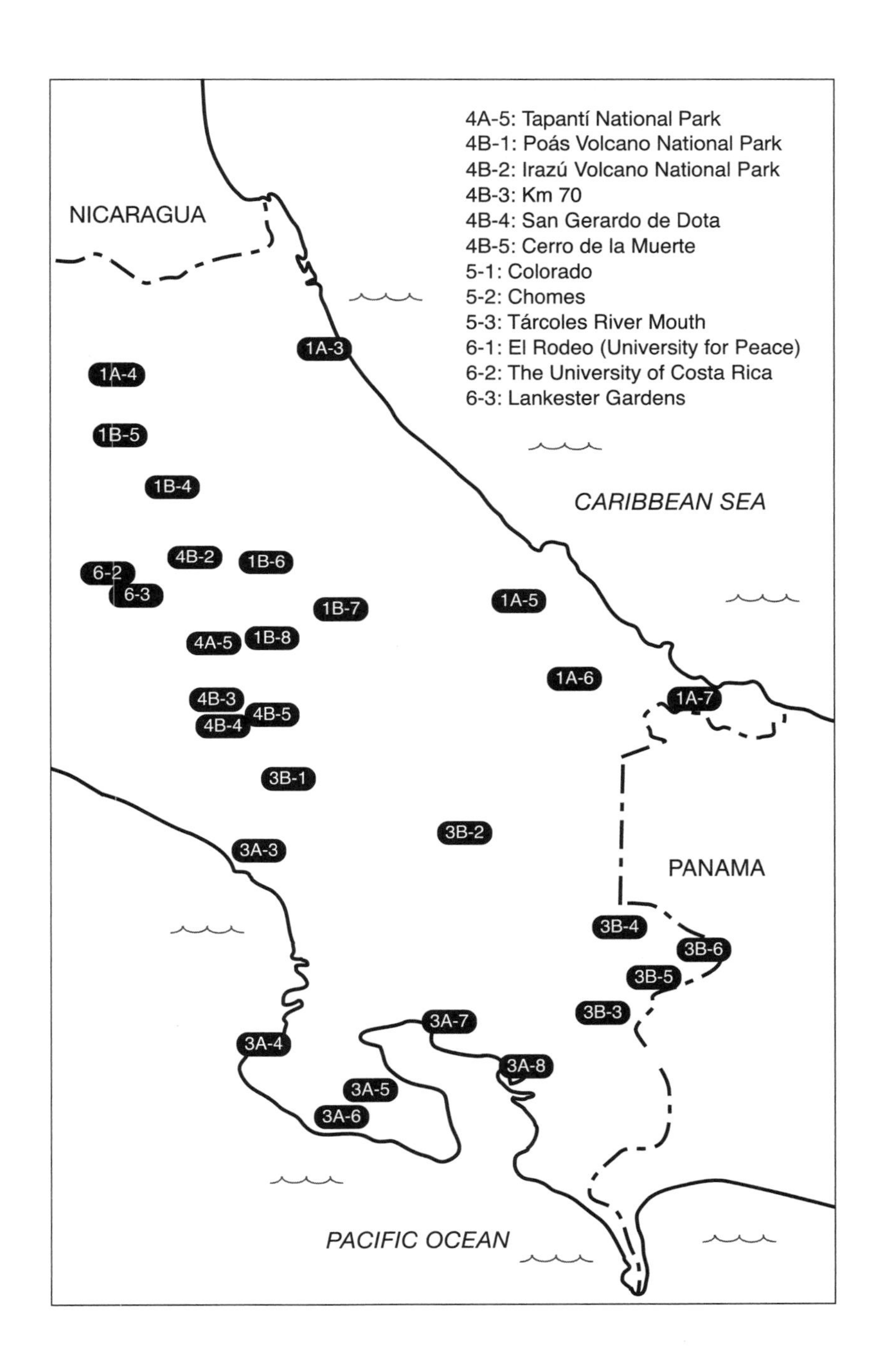
4A-5: Tapantí National Park
4B-1: Poás Volcano National Park
4B-2: Irazú Volcano National Park
4B-3: Km 70
4B-4: San Gerardo de Dota
4B-5: Cerro de la Muerte
5-1: Colorado
5-2: Chomes
5-3: Tárcoles River Mouth
6-1: El Rodeo (University for Peace)
6-2: The University of Costa Rica
6-3: Lankester Gardens
NICARAGUA
CARIBBEAN SEA
PANAMA
PACIFIC OCEAN
1A-3
1A-4
1B-5
1B-4
4B-2
1B-6
6-2
6-3
1B-7
1A-5
4A-5
1B-8
1A-6
1A-7
4B-3
4B-5
4B-4
3B-1
3B-2
3A-3
3B-4
3B-6
3B-5
3B-3
3A-7
3A-4
3A-8
3A-5
3A-6

Preface

I first arrived in Costa Rica as a student, filled with excitement at the idea of experiencing so many new bird species. At the time, I was a competent and confident birder with a strong handle on the birds of the Eastern United States. Nonetheless, when I stepped out onto the grounds of La Selva Biological Station, I was totally overwhelmed. I found a host of flycatchers, tanagers, hummingbirds, and woodcreepers, none of which I had seen before. My only reference was a field guide, which described over 850 species and felt like it weighed a ton. By the end of the day I had managed to identify only a few of the more colorful species of tanagers, toucans, and trogons. I was completely lost.

Not long after, the allure of the tropics drew me back, and I moved to Costa Rica, determined to learn everything I could about the country's birdlife. Now, after years of tromping through Costa Rican forests, I have developed an extensive knowledge of the birdlife found within. I have also found that the places where one can search for birds in Costa Rica are as diverse, exhilarating, and confusing as the birds themselves. Lush rainforest, mystical cloud forest, windswept *páramo*, parched dry forest, and everything in between can be found within this small country. Yet information about many of these exciting birding destinations was dispersed and hard to find, a fact I found puzzling and strove to rectify with the writing of this guide.

This bird-finding guide describes more than 50 of Costa Rica's most important birding destinations, making all of these exciting locations easily accessible to the general public. Its goal is to make birding in Costa Rica an enjoyable and more manageable adventure by listing which species are likely to be seen at each site, explaining where to search out the rarities, and providing accurate driving directions throughout the country. This book, along with a standard travel guide and a good country map, are all you will need to plan and execute your perfect birding expedition in Costa Rica. Whether you are a novice or an experienced tropical birder, part of an organized tour or exploring on your own, the specific information that you need to improve your Costa Rican birding experience has been made easy to find and use in this guide. This is the book I wish I had had on that first overwhelming day at La Selva Biological Station, and I sincerely hope that it will help you unlock the splendor and excitement of birding in Costa Rica.

Acknowledgments

I would like to start by thanking my parents, Robert and Janet Lawson, for their invaluable support throughout the writing of this book. Not only did they spend countless hours editing my drafts from afar, but their thoughtful insights also helped me to develop many of the unique elements of this guide. More important, their moral support was a pillar of stability during the ups and downs of the writing process.

I would like to extend my deep appreciation to my editor at Cornell University Press, Heidi Lovette. Her keen insights have improved this guide tremendously, and her guidance has been invaluable. My sincere thanks also go out to Julio Sánchez, my chief ornithological consultant, who has shared with me his lifetime of experience with Costa Rican birds; to Jim Zook and Richard Garrigues, for offering their input on the "Target Birds" lists; to José Alberto Pérez Arrieta (Cope), for providing me with the beautiful illustrations that grace this guide; and to Robert Dean, for his astute suggestions during the review process.

Many other people have assisted and supported me in a variety of ways, and I am grateful to each of them. These people include Candace Akins, Rodolfo Alvarado, Mary Babcock, Lance Barnett, Esteban Biamonte, Luis Campos, Carlos Castro, Randel Chavez, Kathy Clayton-Seymour, Michael Coogan, Erika Deinert, Melvin Fernández, Abraham Gallo, Mauricio Garcia, Leonardo Garrigues, Charlie Gómez, Ericka Gómez, Jose Huertas, Wilbur Jarquin, Liz Jones, Julia Lawson, Sawyer Lawson, Justo López, Daryl Loth, Daniel Martínez, Deedra McClearn, John McCuen, Alison Olivieri and the rest of the San Vito Bird Club, Jackie Pascucci, Alex Páez Balma, Pablo Porras, Vinicio Porras, César Sánchez, Luis Sandoval, George Serrano, George Serrano Jr., Susan Specter, Julie Tilden, Mark Wainwright, George Whipple, and Zak Zahawi.

Finally, I would like to extend my gratitude to Strickland Wheelock for cultivating my love of birding as a youth, and for helping me to develop the skills needed to complete this guide; to Alina López for starting me on my Costa Rican adventures; to the entire López-Hidalgo family for showing me their limitless hospitality and kindness; to Christopher Salter and Marcela Ramirez for their friendship; and to Tatiana Ácon for her moral support and companionship.

A BIRD-FINDING GUIDE TO COSTA RICA

Introduction

The **Jabiru** is a massive stork that can sometimes be found wading through freshwater marshes. It is best searched for at Palo Verde National Park (2-3) and Mata Redonda Marsh (2-4) during the dry season, when parched conditions concentrate the small population into these few remaining wetlands.

Why Visit Costa Rica?

Costa Rica is one of the world's top ecotourist destinations, seeing 1.9 million visitors in 2007 alone. Its beautiful weather, stunning scenery, and proximity to North America all help to entice a wide variety of tourists to Costa Rica's spectacular countryside. Major draws include fun outdoor activities such as nature and volcano viewing, whitewater rafting, hiking, fishing, surfing, zip line canopy tours, lazing on the beach, and, of course, birding.

Birders have found Costa Rica an especially alluring destination, and the odds are good that a number of birders you know have already visited the country. But what makes this little nation so popular among birders? Why not visit Honduras, Ecuador, or Brazil? Costa Rica's reputation as a premier birding destination has not come about by chance. If you look closely, you will find many conditions within the country that coalesce to create an ideal birding mecca.

"Biodiversity" is often the first word that comes to mind when describing Costa Rica's allure to a naturalist. This tiny country contains an incredible diversity of ecosystems thanks to its varied geography and climate. Tropical dry forest, wet lowland forest, mountain cloud forest, and just about everything in between can be found in Costa Rica. Each of these ecosystems is home to an exciting and unique set of organisms, which leads to the amazing biodiversity found within the country as a whole. Incredibly, Costa Rica boasts a bird list over 880 species long, nearly as many as occur in all of North America. To put this in global terms, about 9% of the earth's bird species have been found in Costa Rica, a country smaller than the state of West Virginia and occupying only 0.03% of the world's land mass! It stands to reason that Costa Rica attracts so much attention from the international birding community.

Staggering biodiversity is not the only attribute that birders find attractive about this country. Costa Rica has one of Latin America's most stable governments, enjoying freedom and democracy for over 50 years. Even during the 1980s, when most of the region was experiencing political unrest, Costa Rica was a base of stability. It is also one of the few countries in the world without a standing army, which it abolished in 1949. Because of its political stability, birders can feel safe while exploring remote parts of the country or traipsing through untamed forests.

Costa Rica's lack of a military has allowed the government to spend more money developing excellent education and healthcare systems. As a result the general population is well educated, with a literacy rate of 96%, and quite healthy. Most Costa Ricans, or "Ticos," as they affectionately call themselves, share the peaceful values of their government. They are eager to share the beauty of their country, and take great pride in its international reputation as an ecological gem. Ticos are also generally warm and welcoming people who go out of their way to help visitors. It is the kindness of its people that makes Costa Rica such a hospitable destination.

The Costa Rican government also has very progressive policies regarding the conservation of its land and resources. This has gone hand in hand with the country's developing ecotourism industry. Costa Rica currently maintains 27 national parks as well as many biological reserves, national wildlife refuges, national forests, and a number of other protected lands. According to the 2008 Environmental Performance Index, Costa Rica ranked fifth in the world and first in the Americas for its environmental policies. One example of the government's proactive stance on conservation is a law passed in 1996 that pays Costa Ricans roughly $20 per acre per year for conserving forest. While this is not comparable to what the landowners could make by logging, it is a nice incentive to conserve. Thanks to both public and private efforts, over 50% of Costa Rica's total area is considered protected, including some of the largest tracts of land in all of Central America. It is no wonder there are so many exciting places to look for birds!

As international tourism has grown in Costa Rica, a large and impressive infrastructure has developed to accommodate visitors. This has many positive implications for birders. Most travelers will touch down in the small but modern airport near San José. From there, transportation options range from public bus to small airplanes. Luxury hotels, hostels, and everything in between can be found throughout the country, as well as excellent restaurants where the food is delicious and the drinking water is safe. Highly adventurous birders looking for a wild, indigenous experience can head to remote national parks and locally run reserves that international travelers rarely visit. No matter what you are looking for, Costa Rica has it all.

An Overview of Costa Rica's Geology, Geography, and Climate

Geology

The land now known as Costa Rica is not very old in geologic time. Four million years ago North and South America were not connected, and ocean existed between what is now northern Nicaragua and Colombia. Ever since the breakup of Gondwanaland 130 million years ago, evolution had been acting independently on these two separate continents, and each possessed very different flora and fauna. Then, about three million years ago, the Cocos Plate, a small tectonic plate in the Pacific, began moving northeast. Running up against the Caribbean Plate, it slowly rose up out of the sea to create the land that is now Nicaragua, Costa Rica, and Panama. The Americas were connected.

The creation of the land bridge between North and South America was an extremely important biological event. It allowed species to disperse from one continent to the other. From the north came thrushes, sparrows, cats, deer, raccoons, rodents, oaks, walnuts, and mistletoe. From the south came hummingbirds, antbirds, parrots, sloths, armadillos, opossums, orchids, bromeliads, and heliconias.

All these organisms met and interacted with each other for the first time in the land that is now Costa Rica and Panama. There is no doubt that Costa Rica's historical position as the mixing point for the two continents helps to explain the staggering diversity that is found today.

Geography

Modern Costa Rica is only 19,600 square miles, about twice the size of Vermont, and home to just over four million people. It sits at 10 degrees north of the equator, bordering Nicaragua to the north and Panama to the south. To the east is the Caribbean Sea, and to the west the Pacific Ocean.

Costa Rica's most dramatic physical features are its four major mountain ranges and seven active volcanoes. Running down the center of the country northwest to southeast, the ranges are the Cordillera de Guanacaste, Cordillera de Tilarán, Cordillera Central, and Cordillera de Talamanca. The mountains get progressively

Costa Rican Geography. The insert shows Costa Rica's seven provinces.

larger and older as they extend southward. The highest peak, Mt. Chirripó, is found in the Cordillera de Talamanca, where it rises an impressive 3,820 meters (12,530 feet) although it is only 50 km (30 miles) from the Pacific Ocean and 70 km (45 miles) from the Caribbean Sea. This helps to put in perspective the severity of the slopes of these mountains, and is why Costa Rica has earned the nickname "the Switzerland of Central America."

The Central Valley is another important feature in the Costa Rican landscape. It sits roughly in the middle of the country, at an elevation of 700 to 1,400 meters, surrounded by the Talamanca and Central mountain ranges. This relatively flat area is ideal for human settlement and today is home for approximately half the population of the country.

Costa Rica's two coastlines, the Pacific and the Caribbean, are very different. The Pacific is the larger and generally the more interesting of the two, as well as the more popular among tourists. Habitats found along this coast are varied, from rocky outcroppings to beautiful sand beaches and dense mangroves. The Caribbean Coast, in contrast, is relatively monotonous and has only a few popular beach destinations along its southern end. Two large hook-shaped peninsulas, the Nicoya and the Osa, are important features on the Pacific coastline. The Nicoya Peninsula is the larger of the two, located on the northern part of the coast, and home to some of the most popular beaches in the country. The Osa Peninsula is found near the Panama border and is famous for Corcovado National Park, one of the most remote and untamed areas in all of Costa Rica.

Politically, Costa Rica is broken into seven provinces, each with a capital city of the same name (except Guanacaste, where the capital is Liberia, and Limón, where the capital is Puerto Limón). Four of these provinces, Alajuela, Heredia, Cartago, and San José, have their capitals in the Central Valley, the capital of San José being the national capital. One confusing by-product of the Costa Rican province system is that city names are often repeated in other provinces. For example, there is a Puerto Viejo in the province of Heredia as well as one in the province of Limón. This is important to remember as you travel through the country.

Climate

Costa Rica's climate is a complex system, functioning very differently from that in the temperate zone. Because it is located so close to the equator, Costa Rica's angle to the sun is fairly constant throughout the year, leading to relatively stable temperatures. As a result, the effects of summer, winter, fall, and spring are not obviously visible.

Instead of four seasons, Costa Rica has two: rainy and dry. The dry season runs roughly from December through April, while the rainy season fills up the rest of the year (except for a short period of dry weather called the *veranillo,* which usually occurs sometime in August or September). Interestingly, locals

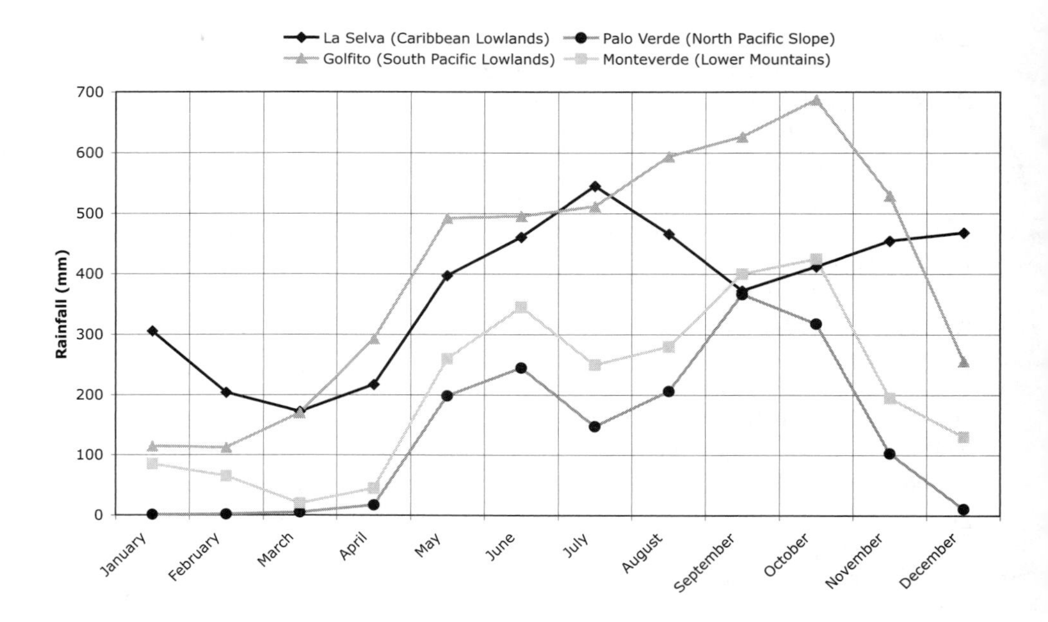

refer to their seasons as *verano* and *invierno*—"summer" and "winter." However, this terminology can cause confusion since the timing of the seasons runs roughly opposite to what is found in the northern temperate zones (even though Costa Rica is still a Northern Hemisphere country). For this reason, it is best to forget this terminology and just think in terms of rainy and dry seasons.

These two seasons operate somewhat differently, depending on where you are in the country. If you visit Costa Rica during the dry season, don't be foolish and leave your rain gear at home! For most of Costa Rica, "dry" means "not as wet." The exception, however, is the region from the Guanacaste Province through the western part of the Central Valley. In this area, "dry" actually means "dry," and you can almost count on there being no rain. On the other hand, the rainy season is not nearly as bad as many people think. In many places the days start off sunny and beautiful. Then, around noon, they start clouding over, and there is usually an afternoon downpour. It can rain extremely hard for an hour or two, but it often clears up after that. Many people are reluctant to book their vacation during the rainy season, but this can be a very good time to visit the country.

While Costa Rica is a tropical country, do not assume that everywhere you go will be hot. In the lowlands temperatures often reach 35°C (95°F), but in the mountains they can be quite chilly. The chart below shows how elevation affects average temperatures. Keep in mind that this chart shows averages, and temperatures at individual locations at any given elevation could fluctuate a good deal. The best policy is to be prepared for a range of weather conditions.

Temperature Variations Due to Elevation

Altitude (m)	Average Daily Maximum	Average Daily Minimum
0	32°C (90°F)	22°C (72°F)
1,000	27°C (80°F)	18°C (64°F)
2,000	18°C (64°F)	11°C (52°F)
3,400	10°C (50°F)	4°C (40°F)

Birding in the Neotropics

Most birders using this book will have logged the majority of their field time in the temperate zone of either North America or Europe. In the neotropics, and Costa Rica in particular, you will find many differences in bird behavior, diversity, and distribution. Below I discuss some of the most important and unique aspects of the neotropics with regard to its birdlife. If you find these topics interesting, I highly recommend getting a copy of Steven Hilty's excellent book, *Birds of Tropical America*.

Forest Structure and Microhabitats

Tropical forests are extremely complex ecosystems, and within the forests there are a number of microhabitats that are important to birders. It is common to break down a tropical forest into three distinct layers: understory, interior, and canopy. The understory runs from the forest floor to a few meters above the ground. Depending on the age of the forest, and the amount of light reaching the understory, the vegetation in this area can be sparse or quite dense. Birds found almost exclusively in the forest understory include antthrushes, antpittas, tinamous, and many species of wrens. While these birds may be quite close as you walk through the forest, they are often wary and difficult to spot.

The canopy level of the forest refers to the crowns of the tall trees. It is an area of high sunlight, and high diversity, as many species of plants and animals live their entire lives in the canopy. It is also, however, a difficult area to access, often reaching 45 meters above your head. Many species of birds spend the majority of their time in the forest canopy, among them the Scarlet-rumped Cacique, Green Shrike-Vireo, and Brown-capped Tyrannulet. Identifying birds in the canopy is often frustrating because you are looking through multiple layers of foliage at the stomach of a bird that is far away and often backlit. Many of these species are easier to see near forest edges, where they come a bit lower.

The forest interior is everything in between the understory and canopy. It is generally more open than the previous two strata and a relatively easy place to spot birds. Ruddy-tailed Flycatcher, White-throated Shrike-Tanager, and Wedge-billed Woodcreeper are just a few of the species often found in the forest interior.

The forest gap is a very interesting and important phenomenon in the tropical forest life cycle, and it also attracts a unique group of birds. A gap is created when an old canopy tree falls, often crushing other trees on its way down. Suddenly precious sunlight is available in abundance on the forest floor, and plant growth accelerates. This high-sunlight environment often favors fast-growing pioneer species that are not found in the low-light interior of a forest. The plant life in a forest gap is often very dense and tangled, as plants race to take advantage of the light. Many different species of birds, including the Fasciated Antshrike, Thicket Antpitta, Dusky Antbird, and Black-throated Wren, favor gap habitats over the dark interior forest.

The forest edge is similar to a gap. Here, because the forest has been cut on one side, more light is able to penetrate, leading to dense plant growth of many fast-growing, light-loving species. The birds that frequent gaps are also likely inhabitants of forest edges, as are many canopy species, which will sometimes descend to lower heights here.

Forest Age

Not all Costa Rican forests show the clear structure and microhabitats described in the section above. In general, the older the forest, the more developed and defined are its structure and microhabitats. Primary forest is the oldest tropical forest type, being at least a few hundred years old. Over time, old trees die, new ones grow, but the forest structure, as a whole, remains relatively static. Access to sunlight is the principal limiting factor for plant growth, and forest gaps play a crucial role in plant recruitment. True primary forests are not very common in Costa Rica. Every ecotourist destination seems to offer a "primary forest," a catch phrase in the industry, though many are not.

New-growth forest is a young forest in its first 20 to 40 years of reestablishment after being cut or burned. Here you will find almost exclusively light-loving, fast-growing pioneer plants such as heliconia, cecropia, balsa, and bamboo. The habitat is often dense, although diversity is low, and stratification into the microhabitats of understory, canopy, and interior is hardly noticeable. Many bird species that prefer forest edge and gap habitats also feel at home in a new-growth forest.

Secondary forest is the name given to anything in between new-growth and primary forest and includes a large variety of forest types. Secondary forests usually have developed a closed canopy, although they contain a high percentage of fast-growing pioneer plants and are not as complex as primary forests in terms of microhabitats.

The age of a forest has a direct influence on which species of birds reside within. Many species, including Tawny-faced Gnatwren and Olive-backed Quail-Dove, occur only in primary or very old secondary forests, while others, such as Yellow-bellied Elaenia and Black-headed Saltator, prefer new growth. It is also worth noting that while primary forests are indeed the most species-diverse habitats, and hold great potential for rarity, the birds here are often secretive and difficult to see. New-growth habitats are generally easier to bird because they are frequented by many bold, active foraging species. The best course of action in the tropics, as anywhere, is to spend time birding a wide range of habitats.

Rarity

One of the most striking differences between bird ecology in the temperate zone and in the tropics is the prevalence of rarity, or low density of a species' population over a large section of forest. There are a number of factors that account for tropical rarity. First, tropical forests contain almost five times more large birds than temperate forests. Large birds require more food than smaller birds, and many, like hawks and owls, are high up on the food chain. This means that they require large territories to be able to meet their nutritional needs, and thus they are necessarily spread thin in their distribution.

A more important cause of tropical rarity stems from the fine partitioning of resources within the forest by tropical birds. Many species have evolved to take advantage of very small and precise niches in their environment. Those that rely on specific habitats, such as tree-fall gaps, streams, and heliconia stands, are only as common as the habitat to which they are tied. Other tropical specialists might forage in very particular ways and thus are able to take advantage of only a small percentage of the food items available in an area. For example, the Striped Woodhaunter makes a living searching for insects inside tank bromeliads. To find enough food, this specialist must occupy a territory far larger than that of the average temperate bird, which will forage opportunistically, taking prey wherever it can be found. This fine partitioning of resources forces tropical birds to defend territories that are generally 10 times larger than those of their temperate counterparts, and thereby leads to lower-density populations.

For birders, tropical rarity can be bittersweet. If you have only a brief time to spend in the tropics, you may be disappointed because it is simply impossible to see everything that is out there. Yet it means that you can continue to bird an area day after day and still feel the excitement of finding new, unusual species each time you go out. In the end, I believe that rarity is one of tropical America's most alluring traits. No matter how much time you spend birding in the tropics, there will always be new birds to see, and reason to return.

Diversity

"Diversity" is a word that comes up almost immediately in any discussion of tropical ecosystems. Neotropical wet forests are the most diverse habitats in the world, supporting an astounding variety of living organisms, from fungi and bacteria to plants, birds, and mammals. Regarding avifauna, some small sections of rainforest in Costa Rica hold close to 200 species of breeding birds, a number four to five times larger than that in most temperate forests! How is this possible? Scientists have been searching for the secret to tropical diversity for generations, and while they understand some factors that influence bird diversity, such as evolutionary time, geographical area, and climate, they have not yet put the entire puzzle together.

To start, tropical ecosystems provide birds with a number of important food sources that are not available consistently, or in sufficient quantity, outside of the tropics. Fruits, army ant swarms, large insects, and dead leaf clusters are examples of important tropical food sources that either have no temperate counterpart or are not consistently available. A great variety of tropical birds are supported by these unique food sources. White-necked Puffbird and Great Jacamar are large-insect specialists. Ocellated Antbird and Bicolored Antbird always feed with army ants. Great Green Macaws eat unripe fruits, and Buff-throated Foliage-gleaner and Checker-throated Antwren constantly search for prey in dead leaf clusters. All of these species have no ecological counterparts in temperate forests.

Even within groups of birds that feed on a similar food source, insect gleaners, for example, there is more diversity in the tropics than in temperate climates. This is partly because the complexity and diversity of microhabitats within a tropical forest allow many birds to specialize more precisely. One insect gleaner may focus its attention on the underside of leaves, another on the tops, and a third on leaves out at the tips of branches. This fine partitioning of food sources allows for high species diversity within an environment.

Mixed-Species Flocks

For most tropical birds the risk of predation is a serious one. Many species of bird-eating hawks lurk in the forests, waiting for their opportunity to swoop in on an unsuspecting victim. The threat is especially intense for birds that hunt insects and spend the majority of their time carefully surveying leaves, trunks, and bromeliads rather than scanning for danger. Some of these species have developed a behavioral strategy, known as the "mixed-species flock," to improve their defense against predators. Mixed-species flocks are groups of principally insectivorous birds from many different species that feed together in an effort to avoid predation. Within a mixed-species flock you are likely to find only one or two individuals of each species. This reduces competition for food between members of the flock because each species has its own narrow foraging niche. Species A may specialize in searching dead leaves for insect larvae, species B may prefer to look for prey in bromeliads, species C checks along the trunks of trees, while species D sallies for small insects in the air. None of these birds compete directly with each other, so there is little cost to foraging together. There is, however, an important added benefit. With so many birds foraging in the same area, it is very difficult for a predator to approach unnoticed, and as soon as one member spots danger, it will sound the alarm and everyone will dart to safety.

Most mixed flocks are made up of five or six key species, such as Checker-throated Antwren, White-flanked Antwren, and Streak-crowned Antvireo, which feed exclusively with the flock for almost their entire lives. Each of these core species defends an identical territory, which becomes the territory of the flock. Each day they forage together, moving through their collective territory, while many other species join them opportunistically. Some will feed with the flock for a number of hours; others may stop in for only a few minutes.

Birding a tropical mixed-species flock can be both exhilarating and frustrating. When a large flock moves through, the entire forest seems to be filled with birds. Yet they often forage so actively and move so quickly that getting a good look at anything is impossible. Other times the mixed-species flock will be smaller, less obtrusive, and moving at a manageable pace. These birds may actually afford you good looks. Whatever the dynamics of the flock you encounter, be sure to bird it as thoroughly as possible. Since there are often many more species in a flock than you first suspect, mixed-species flocks often provide the most productive birding during a walk through a tropical forest.

Army Ants

Witnessing a large army ant raid is one of the most exciting experiences for naturalists and birders in the neotropics. There are a number of species of army ant, but *Eciton burchelli*, a dark, moderately sized ant with long legs, is the one most likely to be encountered. These ants live in lowland and lower middle-elevation forests and are often found in great swarms that can include as many as half a million individuals. The army has a mobile base of operations called a bivouac, which is nothing more than a massive clump of worker ants, within which are found the larvae and a queen. The bivouac is usually located under a log, in a pile of brush, or within the crotch of a tree, although its location changes frequently to ensure fresh prey for the swarm.

Every morning the ants begin their raid by streaming out of the bivouac and forming a large raiding column, which plunders through the forest floor and undergrowth in search of insects and small vertebrates. The front of the column is a dense mat of ants a few meters wide that swarms over everything, inspecting each nook and cranny for prey. When the ants find a prey item, they attack in great numbers, killing and dismembering it so it can be carried back to the bivouac. Behind the foraging front are trails of ants that run between the bivouac and the front, and you will frequently encounter these while walking through a neotropical forest.

For birders, army ants are of special interest because the raiding swarms are often accompanied by a host of birds that feed along with the ants. At first glance you might think that the birds are eating the ants themselves, but this is not the case. Because of their high formic-acid content, ants are a very unpalatable prey. Instead, birds at an army ant swarm feed on the unfortunate insects desperately trying to escape the ants. In their rush to avoid being eaten by the ants, the insects become easy prey for the hungry birds lurking above.

A number of antbird species are especially regular participants at army ant raids. While the entire family bears the name "antbirds," only a small percentage frequently feed with army ants. Some of these species have specialized so completely that they are unlikely to survive without the benefits of ant swarms. Biologists have given these species the title "obligate ant followers." In Costa Rica the most common antbirds to find at an army ant raid are Oscillated Antbird, Bicolored Antbird, and Spotted Antbird. Northern Barred-Woodcreeper is another common participant in the feeding frenzy, as well as woodcreepers in the *Dendrocincla* genus. While these species are the most consistent participants at army ant raids, many others will opportunistically show up, especially at very large raids.

Fruit Eaters

One major difference in lifestyle between temperate and tropical birds is the importance of fruit in the diet. In the temperate zone fruit is not abundant throughout the year, and an explosion of insect prey during the spring and summer months supplies a very easy food source. For these reasons, it is almost impossible to find birds in the temperate zone that specialize solely on fruit. In the tropics, however, fruit takes on a stronger significance because of its year-round availability. Many species, and indeed many tropical families of birds, are almost exclusively fruit specialists. Trogons, manakins, many tanagers, and even some flycatchers eat almost exclusively fruit. While fruiting trees in the tropics are far more common than in the temperate zone, it is still a patchy and ephemeral food source marked by staggered fruiting cycles. Fruit specialists will often congregate in large numbers around an especially productive fruit tree in season.

For birders, finding a popular fruiting tree is like striking gold. Fruiting trees can attract an amazing range of species, from the fruit specialists to opportunistic omnivores. It is not uncommon for more than 50 species to visit a single fruiting tree over the course of a day or two! If you are lucky enough to find a very active fruit tree, hang around and see who shows up, or return to the tree later for a second look. You are almost guaranteed to find a hubbub of activity.

Migration

Most people think of bird migration as a phenomenon that involves the temperate zone, but migration plays an important part in the lives of strictly tropical birds as well. In fact, Costa Rica has three different types of migrants: northern, southern, and elevational. Most North American visitors will be quite familiar with the northern migrants. These are the wood warblers, thrushes, tanagers, and shorebirds that spend their summer months nesting in the temperate zone, only to retreat to the tropics with the onset of winter. Most northern migrants begin to return to Costa Rica during September, and leave by April, although the exact timing for each species varies.

Costa Rica also sees a few southern migrants, those that move up from South America to nest. These species follow a lifestyle similar to that of northern migrants, except they use Costa Rica as a breeding ground instead of a wintering ground. They usually arrive in January and stay until September. Examples include Yellow-green Vireo, Piratic Flycatcher, and Sulphur-bellied Flycatcher.

Elevational migration is the least understood by most visiting birders, yet it is quite common in Costa Rica. This occurs when populations of birds move up or down the mountainsides in search of patchy and ephemeral food sources. Most people's image of bird migration involves birds traveling incredible distances, the most dramatic case being the Arctic Tern, which migrates between the North and South Pole. In contrast, elevational migrants may find it necessary to travel only a few miles before they change environments significantly enough to encounter new sources of food. While this form of migration may not be as impressive as intercontinental migration, it is a very important part of bird ecology in Costa Rica.

Most elevational migrants are either fruit or nectar specialists, like hummingbirds, tanagers, manakins, and cotingas. In Costa Rica the abundance of fruit or flowers in any given location changes throughout the year. As availability declines in one habitat, the birds that feed on the food sources there begin to travel upslope or downslope in search of more lucrative habitats. Most insectivorous birds, on the other hand, are not elevational migrants, as their food source stays relatively constant throughout the year, allowing them to maintain a fixed range.

An important difference between elevational migration and latitudinal migration is timing. The temperate zone sees a flood of birds migrating in during the spring and moving out in the fall. While some migrants may arrive slightly earlier or later than others, the timing of the event is relatively uniform. Elevational migrants, on the other hand, are not nearly as well synchronized. While much of the movement occurs with the changing of the wet and dry seasons, there are no firm patterns. The best policy for a visiting birder is to be conversant with the birds most likely to be present but not to rule out other possibilities.

Leks

Lek displays are another interesting behavior exhibited by many tropical birds. A lek is an area where males of the same species gather together to display for females who come to the lek for mate selection and fertilization. The female chooses one male with whom to copulate, after which she goes off to raise her family alone. The male returns to the lek to continue displaying. Among the more common tropical families that display this behavior are manakins, cotingas, and hummingbirds. While lek behavior is most frequently found in the tropics, there are a few temperate birds, such as Sage Grouse and Ruff, that also exhibit it.

Many lek displays are quite extraordinary to watch. The male Red-capped Manakin, for instance, performs for a female along a small horizontal branch. He first jumps up from his perch into the air while emitting a staccato chirp. Landing back on the branch, he lifts his tail slightly, exposing his bright yellow thighs. The final element of the dance is a moonwalk, where the bird uses small, rapid steps to seemingly slide down the branch toward the female.

Many leks are a very confined area. At an Orange-collared Manakin lek, you can often see 10 or more individuals at a time bouncing from branch to branch while producing wing snaps and buzzing notes. Others, known as "exploded leks," are more dispersed. Males may display as much as 100 meters apart, but generally within earshot of one another.

How This Book Works

Region and Subregion Breakdown

The sites in this book are grouped into six regions: Caribbean Slope, North Pacific Slope, South Pacific Slope, Mountains, Coastline, and Central Valley. Three of these, Caribbean, South Pacific, and Mountains, are split into subregions. Rather than basing my groupings on the proximity of one site to another, I use regional distinctions based on general patterns of avian distribution, which reflect factors like climate, elevation, and natural barriers. Understanding these broad geographical patterns is the crucial first step to mastering bird identification in Costa Rica. The first three regions, Caribbean Slope, North Pacific Slope, and South Pacific Slope, form the foundation of my regional breakdown. The Continental Divide creates the border between the Caribbean and Pacific slopes, and an imaginary line, roughly along the Río Tárcoles, separates the North Pacific and South Pacific slopes. Each of these three regions has a distinct avifauna. The North Pacific Slope is covered with tropical dry forest, which is home to many unique species not found elsewhere in the country. The Caribbean and South Pacific slopes are covered in wet forest and are fairly similar in both climate and habitat. However, they are separated by a very large and imposing mountain range, the Cordillera de Talamanca, which has kept many species isolated on one side or the other.

Beyond these three major regions of the country, elevation is the next most important factor that affects basic avian distribution in Costa Rica. The elevation of any given location has a huge influence on its climate (most notably temperature) and on the ecosystems that exist there. In reality, a shift in elevation from lowlands to high mountains is very similar to a latitudinal shift from tropical to temperate. As a result, you will find widely differing bird populations at different elevations within the same geographic region. For example, while all of

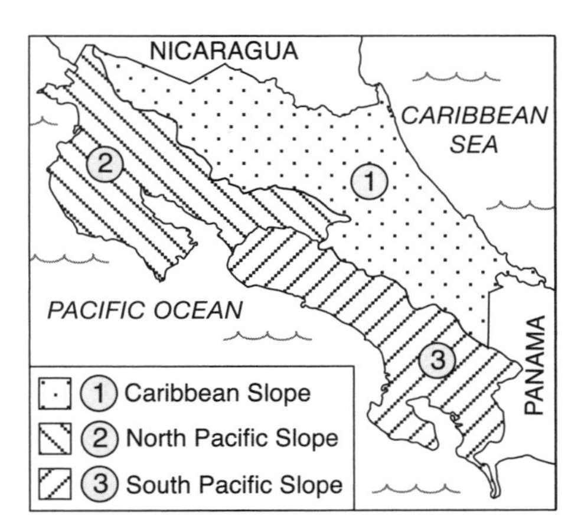

Costa Rica's Three Principal Regions

the following species are found on Costa Rica's Caribbean Slope, the Bronzy Hermit is generally found below 400 meters, the Rufous-browed Tyrannulet between 500 and 1,300 meters, and the Timberline Wren only above 2,700 meters. Thus, two sites on the Caribbean Slope, even if seemingly close together on a map, will produce very different species lists if they are at different elevations.

The effects of elevation are strong enough to warrant the creation of a fourth region, Mountains. The birdlife found in Costa Rica's highlands is totally distinct from that found downslope, and the Mountains region includes sites where high-elevation specialists can be found. In reality, the Mountains region has been cut out of the previous three regions, taking all sites above roughly 1,200 meters and placing them in their own region. By grouping all of the sites above 1,200 meters from across the country, I no longer distinguish between sites found on the Caribbean Slope and those found on the Pacific Slope, because at these elevations there is very little difference in avifauna between one side of the Continental Divide and the other.

Three of the four regions described so far, Caribbean Slope, South Pacific Slope, and Mountains, are further divided into two subregions that are also based on the effects of elevation. Creating elevational subregions is a way of more finely grouping sites that share similar birdlife. In the cases of the Caribbean and South Pacific slopes, the first subregion is called Lowlands and includes all sites up to about 500 meters. The second subregion, Middle-Elevations, includes everything between 500 meters and about 1,200 meters, the start of the Mountains region. The two subregions within the Mountains region are Lower Mountains, which runs roughly from 1,200 meters to about 2,000 meters, and Upper Mountains, which continues upward from this point.

The book's final two regions are much smaller than the four previously discussed. The Coastline includes sites along the coast where the primary birding focus is to find shorebirds, gulls, terns, and mangrove specialists. The Central Valley isolates a few sites located within Costa Rica's Central Valley, even though

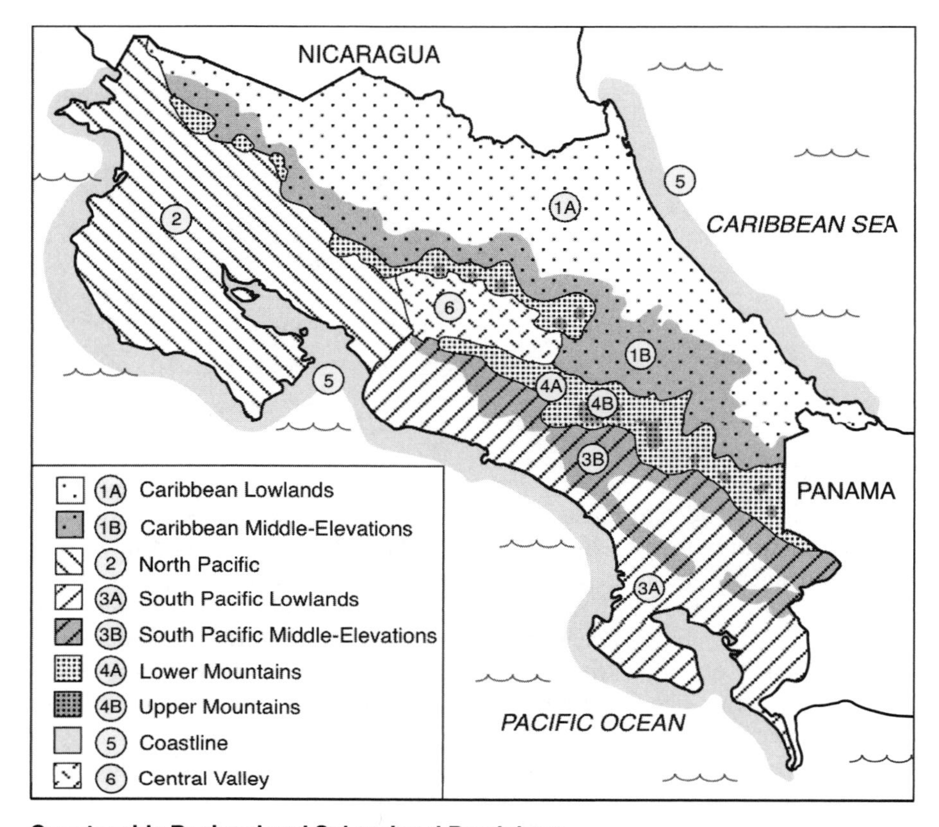

Countrywide Regional and Subregional Breakdown

this area is not ecologically cohesive. The creation of this region was mainly for the convenience of travelers who may find themselves in this populous area.

Bird Lists

Throughout this book you will find four recurring types of bird lists. Two of these, "Common Birds to Know" and "Regional Specialties," pertain to regional (or subregional) areas. The other two, "Species to Expect" and "Target Birds," pertain to specific birding sites. Most site descriptions contain both a "Target Birds" list and a "Species to Expect" list. However, in some cases, where a site is of lesser importance or very similar to a nearby site, these lists have been omitted. Finally, at the end of the book is a list entitled "Costa Rican Checklist with Select Site Lists" that presents a complete bird list for eight representative sites.

For convenience, I have included annotations in my lists that refer you to illustrations and species descriptions in two major field guides. The first number after each bird name is given in parentheses and references the appropriate page in Garrigues and Dean's guide, *The Birds of Costa Rica*. The second annotation is a set of two numbers separated by a colon, the first referring to plate number and the second to picture number in Stiles and Skutch's *Guide to the Birds of Costa Rica*. For example, the entry "White-lined Tanager" is followed by "46:16," as its picture is #16 on Plate 46. Also, please note that endemic species are marked with an asterisk.

Regional Bird Lists

Regional Specialties: At the beginning of each chapter, you will find a list of birds whose range is restricted to the region being discussed. This will help you to know which birds are especially important to look for in that particular part of the country. These lists have been compiled liberally, intending to get you thinking about the general regionality of bird distribution rather than every little exception. For example, even though the Olivaceous Piculet can be found at one site on the Caribbean Slope, it is still included on the South Pacific's Regional Specialties list because it is widespread and much easier to find in this region.

Common Birds to Know: These lists are geared toward less experienced tropical birders and present the 15 to 30 most abundant species within each region or subregion. If you want to do any studying before your trip, these are the species to start with.

Site-Specific Bird Lists

Target Birds: This list includes species of special note that you should be looking for or hoping to see at a given site. These could be flashy, popular species such as Resplendent Quetzal and Scarlet Macaw, or rare species that seem to show up regularly in the area. Keep in mind that inclusion on this list does not necessarily mean that a bird is easily found at the site. What it does mean is that your chances of seeing the bird are greater there than at most other locations throughout the country.

Species to Expect: This list includes many of the birds that are regularly found at a site, and is intended to be an example of the species that three or four days of birding could produce. While this is not a complete list, you can expect that 80% to 90% of the species you encounter at the site will be included on it. The species in bold type are those that are *especially* common, and hence good birds to study before birding the area.

Costa Rican Checklist with Select Site Lists: For eight select sites, I have compiled a complete list of birds. Each of these eight sites was carefully chosen as most representative of the entire region or subregion that it falls within. Referring to this list will give you a good point of reference for any other sites within that region. These lists are presented in a comprehensive chart at the end of the book and include basic abundance ratings.

Trail Difficulty Ratings

At the beginning of each site description, I rate the site's trail difficulty on a scale of 1 to 3, with 3 being the most difficult. This rating is based on both footing and incline. A rating of 1–2 indicates that there are both easy and moderate trails on site. Any trails that are extremely difficult are noted in the text. I recommend taking these ratings seriously, especially if you are not in peak physical condition. The Costa Rican landscape is steep and muddy, and the climate is hot, which makes birding here physically demanding.

Maps

Most people find maps a useful counterpart to written directions, and many maps are provided to help you find your way around. You will find local maps within most of the site descriptions. While these maps are not precisely to scale, they are clear and easy to read. This book does not provide trail maps for locations that offer maps on site. The majority of eco-lodges and national parks offer their own trail maps, and it is always a good idea to ask for a map when you arrive on site. Large-scale regional maps can be found at the beginning of each chapter, and there are several countrywide maps, including a useful road map found at the end of this chapter as well as a countrywide site map located after the table of contents.

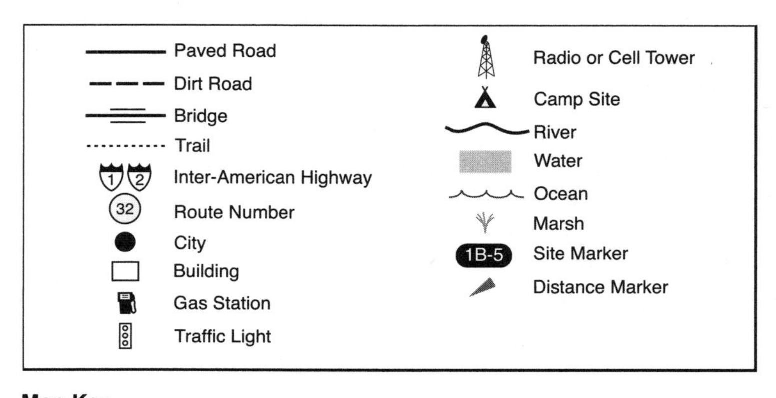

Map Key

Website

I maintain a corresponding website, www.birdingcr.com, which is worth checking before your trip to Costa Rica. This website includes links to hotels, lodges, national parks, and private reserves as well as to other Costa Rican birding websites. Other features include a photo gallery and updated information on changes that may have occurred since the printing of this book.

Logistics

Health and Vaccinations

Costa Rica is a tropical country and has a number of tropical diseases for which visitors from temperate regions may not be immunized. In general, these diseases are rare and you need not concern yourself too much, although it is always better to be safe than sorry. Call your doctor six to eight weeks prior to your trip to obtain complete and current information on vaccinations you might need and medications you may want to bring.

Entering the Country

For U.S., Canadian, and most European citizens, entering Costa Rica is quite easy. No visa is required, and all you will need to show is your passport. If you are planning an extended stay, you should be aware that entering in this fashion requires that you leave for at least 3 days every 90 days.

Money

Because the Costa Rican tourism industry is so strong, U.S. dollars are accepted at many tourist locations, including national parks and most hotels. Major credit cards like VISA and MasterCard are also accepted at most establishments and make for an easy alternative to carrying lots of cash. The use of traveler's checks is waning, and it is difficult and expensive to cash them. I recommend using ATMs instead. Almost all of the ATMs in the country will allow you to withdraw cash using a VISA card, and most of the machines run by private banks will recognize an ATM card from major U.S. banks, letting you withdraw cash directly from your account. Keep in mind, however, that these ATMs have a fairly low withdrawal limit.

In addition to U.S. dollars, you should carry with you some Costa Rican currency: the colón. (The exchange rate is always changing, so check online for current information.) Colones can be obtained at ATMs or by exchanging dollars at a bank, although you will need to show a passport for the latter transaction. Many local establishments like restaurants, shops, and taxis will want to be paid in colones. Aside from a variety of coins, you will generally see four denominations of bills: 1,000, 2,000, 5,000, and 10,000 colones. Try to have smaller bills handy because you often run into situations where people cannot break a 10,000-colón bill.

Language

Spanish is the national language of Costa Rica, but the country's booming tourism industry has inspired many Ticos to learn English. At almost every established ecotourism destination you will find people who are able to speak English. Nonetheless, some of the more remote sites described in this guide will require limited Spanish, and I recommend that you carry a basic phrase book or dictionary. Even if you lack confidence in your Spanish pronunciation, I encourage you to try speaking with local residents. Ticos are very friendly and more than willing to struggle to communicate with you. Although this might be somewhat uncomfortable, it will give you a more authentic Costa Rican experience.

Driving

Driving in Costa Rica is both memorable and challenging for visitors. While the rules-of-the-road are similar to those in the United States, mountainous terrain, poorly paved roads, and aggressive drivers can make driving a stressful experience. Below you will find information that will help you have a safe trip.

Spare Tire: You should always be sure that your car is equipped with a good spare tire and all the equipment needed to change it. Aside from potholes, the roads seem to hold many sharp metal objects like screws and nails, and getting a flat tire is unfortunately quite common. If you do change your tire, bring the flat to a tire repair shop (*reparación de llantas*) as soon as possible so you will be ready in case of another flat.

Buses and Trucks: Roads in Costa Rica are frequented by many trucks and buses, which reduce visibility, slow down traffic, and are difficult to pass. Use caution when driving near these large vehicles, as they will often stop unexpectedly. When you are driving on a multilane highway, it is usually safer to stay in the fast lane, to avoid the buses, which stop and start in the right lane.

Passing: Ticos pass often and without hesitation. Expect to be passed often, and be cautious about meeting an oncoming passer. You will also frequently find it necessary to pass others because of slow trucks and buses on the road. Use caution and patience in these situations and wait for a safe opportunity.

Beeping: When driving in Costa Rica, you are likely to hear other drivers honking quite frequently. Costa Ricans are extremely liberal with their horns, and while your initial reaction might be to think that you have done something wrong, it is far more likely they are acknowledging a friend or catcalling a woman.

Road Signs: Street signs in Costa Rica are sparse at best. Most roads do not have street names or signs, and while highway signs are more reliable, they too are often missing. Route numbers, while officially assigned to major roads, are not widely used. Instead, signs usually advertise towns and cities to which the road leads. Because the roads are so poorly marked, I have taken great care to give detailed directions to each location discussed in this guide. If you stay alert and follow these directions carefully, you should not get lost.

While most traffic signs in Costa Rica are easily understood by their symbols, two are worth clarification. Most intersections with stoplights also have a stop sign on one of the two intersecting roads. These stop signs are backup, intended to be ignored unless the lights are broken, in which case they become the primary means of directing traffic. Also, in many places, especially at small bridges, the road will be reduced to one lane. In these situations there should be a triangular sign facing one direction that says, "CEDA EL PASO." This is equivalent to a yield sign and indicates that you must yield to any oncoming traffic at this point.

Things To Bring

Binoculars: Binoculars are a must-have item, and a pair that is waterproof is highly recommended. The frequent rains and high humidity within the country can put non-waterproof binoculars out of commission very quickly. If you have been thinking about upgrading your binoculars, now is the ideal time to act.

Telescope: A telescope is a lower-priority item. While it can help to get nice looks at large, immobile birds like parrots, toucans, and raptors, it is useless when looking at smaller, active species. I find carrying a scope on hikes to be cumbersome, and bring mine along only when birding Coastline sites, marshes, or large fields.

Field Guide: This is a top-priority item, and luckily there are two excellent Costa Rican field guides. The most current is *The Birds of Costa Rica: A Field Guide*, by Richard Garrigues and Robert Dean. *A Guide to the Birds of Costa Rica,* by Gary Stiles and Alexander Skutch, is a classic that presents extensive information on natural history and behavior. Both books are discussed in greater detail later in this chapter under "Related Books of Interest."

Audio Recordings: For visitors interested in learning the vocalizations of Costa Rica's birds, the Cornell Lab of Ornithology has published two good compilations of Costa Rican birdsong. These are *Costa Rican Bird Song Sampler*, by David L. Ross Jr., and *Voices of Costa Rican Birds: Caribbean Slope*, by David L. Ross Jr. and Bret M. Whitney. Both of these recordings are discussed later in this chapter under "Related Books of Interest."

Travel Guide: Information about hotels, restaurants, and other tourist activities, which I have not covered in this guide, can easily be found in a standard travel guide. I recommend that you have at least one with you. There are many travel guides for Costa Rica, and a few of my favorites are discussed later in this chapter under "Related Books of Interest."

Maps: I highly recommend having a detailed countrywide foldout map to use in conjunction with the maps provided in this book. The *National Geographic Adventure Map: Costa Rica* is the most detailed map of the country that I have found. It uses route numbers consistent with those used in this book, so interfacing between the two should be easy. If you plan to do much driving in the country, this map is *highly* recommended, but be sure to order it online beforehand, because finding these maps in Costa Rica can be difficult. If you are in Costa Rica and looking for this map try 7th Street Books, an excellent English language bookstore located on Calle 7 and Avenida Central in San José center.

Rain Gear: Good rain gear is another vitally important item to have in Costa Rica. A few considerations are important. Your rain gear should be lightweight, breathable, and easily transportable because you will want to carry it with you on any extended outings. A small umbrella and a baseball cap or broad-rimmed hat can also be helpful.

Dry Bag: Any time you bring a camera or other electronic equipment into the field, you should be sure to have a dry bag with you. If you get caught in a downpour, you could find a lake in your pack by the time you reach shelter. You should also pack a small plastic bag to protect your field guide.

Footwear: Proper footwear is essential. You should bring a pair of hiking boots, as these are good in a variety of habitats. I also strongly recommend having a pair of high rubber boots. These are ideal to wear in wet, muddy conditions, as well as at sites where snakes are prevalent. Many people find them a bit uncomfortable, although using a sole insert can help.

Field Clothes: Costa Rica has an unfortunate combination of heat, humidity, and insects, which makes dressing properly for the field a challenge. Many new synthetics are both breathable and quick drying, and these materials work well in the tropics. I suggest using long pants in the field because they offer protection from insects and are more comfortable with tall rubber boots. The choice between long sleeves and short sleeves is less clear. I usually start in a T-shirt but carry a light, long-sleeved shirt with me in case the insects become a problem. If you are at high elevations, you will actually need to dress for the cold. If you plan to visit the mountains, be sure to bring along a sweatshirt or fleece jacket suitable for cold temperatures.

Sun Protection: Be sure to bring strong sun block and a baseball cap or other broad-brimmed hat with you.

Insect Protection: Be sure to have a good insect repellant with you, as mosquitoes and other biting insects can often be annoying. In some extreme cases you might even want to have a mosquito head net.

Flashlight: Flashlights are a must-have item in Costa Rica, and they should always be used when walking at night. Snakes are most active at night, and depending on where you are, it is not uncommon to find them in the middle of paths and walkways that might otherwise feel safe. I find a headlamp, which allows for hand-free use, to be a convenient alternative.

Water Bottles: It is always a good idea to have a bottle of water with you whenever you go out on a walk.

Small Daypack: With all of the above-mentioned items that you need to bring into the field, a small daypack is a necessity. Bring one that is lightweight and, if possible, equipped with a rain-fly.

Dangers

Costa Rica is generally very safe; however, there are certain risks of which you should be aware. The following information is not intended to scare you in any way. It is meant to draw your attention to possible dangers in the hopes of avoiding problems during your trip.

Theft: While the number of violent crimes in Costa Rica is relatively low, petty theft is a problem, especially in large cities and tourist towns. San José is notorious for pickpockets and thieves, and you should be particularly careful there. The most important thing is to stay alert and to take common-sense precautions. Keep your car locked and take your valuables with you. In the countryside the people are much more trustworthy, and you will not need to worry as much.

Water: Visitors to the tropics are always concerned about drinking the water and the possibility of "Montezuma's Revenge." Costa Rica has very high water standards, and throughout most of the country the tap water is safe for visitors. There are a few places, most of them rural, where the water is not potable, so it is always a good policy to ask. Even so, many visitors choose to play it safe and drink bottled water during their trip.

Mosquitoes: Many people envision the rainforests as being infested with swarms of vicious mosquitoes. In reality, this is far from the truth—a June walk through a temperate forest would expose you to far more mosquitoes than you will likely find in Costa Rica. This is not to say that they do not exist. Depending on how sensitive you are, mosquitoes can be a nuisance throughout the lowland regions of the country. In a few locations, for example, Palo Verde during the wet season, they can be downright awful. Tropical mosquitoes demand more respect than their northern counterparts because of the possibility of disease transmission. In Costa Rica serious mosquito-transmitted diseases, such as malaria and dengue fever, are rare, but it is prudent to take precautions against getting bitten. Wearing a long-sleeve shirt and pants and using insect repellant are the best ways to protect yourself. A head net might also be useful in extreme situations.

Ants: Ants are another potential danger. While they are not disease vectors, they often pack a painful bite. You will encounter numerous species of ants almost everywhere you go in Costa Rica, but a few types are worthy of special mention here.

The Bullet Ant is the largest ant found in Central America, reaching over 20 mm (almost an inch) long! It is found in the Caribbean Lowlands, and in some places can be quite common. The Bullet Ant gets its name from its infamous sting, which even the locals treat with respect. While the bite won't kill you, it is extremely painful and could require a trip to the hospital. This is one good reason to always be aware of where you put your hands, feet, and seat.

Army ant swarms, as discussed elsewhere, are a birder's dream, but you should use precaution when birding near one. Try to keep to edges because getting caught in the middle of the swarm could lead to many serious ant bites. If need be, you can walk quickly through a swarm, but don't stop until you get clear of the ants.

In the dry North Pacific you are likely to encounter Bullhorn Acacia Ants. These have a symbiosis with the *Acacia* tree in which they live. The tree provides nectar for the ant to feed on, as well as housing structures in the form of hollow thorns. In return for the food and shelter, the ants, with their painful bite, protect the tree from herbivores. Try to avoid brushing against these trees when in the field. Interestingly, the ants also help the *Acacia* trees by killing other plants that try to sprout up in their immediate vicinity. Fortunately, this makes *Acacia* trees easy to recognize because there is no living undergrowth below the tree.

Scorpions: These arthropods like to hide in shoes, under blankets, and in other nooks and crannies. They are most abundant in the dry North Pacific Slope, and when staying there, it is smart to check your sheets before getting into bed, and knock out your shoes each morning.

Chiggers: Chiggers are unpleasant arthropods with an irritating bite that far exceeds their tiny size. They are most often found in grassy environments, where they will climb onto your body and then embed themselves into your skin. The bites last longer and itch more than mosquito bites. Tucking your pants into your socks may look silly, but it helps prevent chiggers from getting access to your skin. Bodily areas at special risk of chigger bites include ankles, waist, and armpits. Many people apply either a DEET-based repellant or 100% sulfur powder to these areas as a second means of defense.

Snakes: Costa Rica is home to a number of poisonous snakes, including coral snakes, Fer-de-lance, and Bushmaster. While these are unlikely to be encountered, they do pose a threat that should be taken seriously. Most snakes in Costa Rica are not venomous, but if you see one, you should treat it as dangerous. The highest snake risk exists in the Caribbean and South Pacific lowlands and lower middle-elevations, although venomous snakes do occur throughout the country.

Snakes will not go out of their way to bite you, so the most likely encounter involves accidentally stepping on one. The best protection against this is to always know where you are stepping. In temperate zones birders become accustomed to walking blindly in search of birds. While in the tropics you should always know exactly where you are putting your feet (or hands, as some snakes are arboreal). Wearing a pair of high rubber boots is another good precaution. These will protect your feet and lower legs, which are most vulnerable. Such boots are often worn by Ticos who work outdoors, and are inexpensive and easily purchased in Costa Rica. If a snake does bite you, seek medical attention immediately.

Sun: The strength of the tropical sun is legendary, and for good reason. Sunlight entering tropical skies has less atmosphere to penetrate because the angle of the rays is closer to perpendicular. Thus, the UV light is stronger than in temperate zones, and it is easy to get sunburned. This is especially true at high elevations, where, even on a cloudy day, you can get seriously burned. You should always wear sun block, a hat, and ideally, long-sleeve shirts and pants.

Cars: The greatest risk you face in Costa Rica is undoubtedly automobiles. As mentioned above, the roads are dangerous and the drivers often reckless. News stories about fatal car accidents are commonplace. Keep your wits about you both when driving and when walking near roads.

Related Books of Interest

Bird Field Guides and CDs

The Birds of Costa Rica: A Field Guide, Richard Garrigues and Robert Dean (Cornell University Press / Zona Tropical Publications, 2007). This is the latest field guide for Costa Rica, and it presents the most up-to-date information on the country's avifauna. This guide follows the modern field-guide format, with five or six illustrations presented on the right page, accompanied by limited text and range maps on the left page. The illustrations are generally better than those found in the Stiles and Skutch guide, and the nomenclature is consistent with that used in this book. Highly recommended.

A Guide to the Birds of Costa Rica, Gary F. Stiles and Alexander F. Skutch (Cornell University Press, 1989). This field guide is a classic, long considered to be one of the best guides to tropical birds available. However, many users find the book's cumbersome size, awkward layout (with plates being separated from text), and lack of range maps to be bothersome. The Garrigues and Dean guide has resolved all of these issues nicely. The main advantage of Stiles and Skutch's book over Garrigues and Dean's is the text. Each species is described in detail—all plumages, behavior and habits, vocalizations, as well as nesting information, status in the country, and range throughout the world. Also, the illustrations in Stiles and Skutch better convey the relative sizes between species. Any serious birder will be armed with both guides.

A Guide to the Birds of Panama: with Costa Rica, Nicaragua, and Honduras, Robert S. Ridgely and John A. Gwynne Jr. (Princeton University Press, 1989). This guide covers the birds of all southern Central America. Similar in format to the Stiles and Skutch guide, it is an excellent third option.

Costa Rican Bird Song Sampler, David L. Ross Jr. (Cornell Laboratory of Ornithology, 1998). This CD presents the songs of 184 of the more commonly heard species from across the country. Ross divides the species into regional groups, similar to those found in this guide, which makes the songs easier to study. If you are beginning to learn the songs of Costa Rican birds, this is where you should start.

Voices of Costa Rican Birds: Caribbean Slope, David L. Ross Jr. and Bret M. Whitney (Cornell Laboratory of Ornithology, 1995). This two-CD set includes the songs of 220 species of birds found along Costa Rica's Caribbean Slope. You will find many species not included on the *Costa Rican Bird Song Sampler*, and it is a good purchase for those looking to add to their library of Costa Rican bird vocalizations.

Other Field Guides

Travellers' Wildlife Guides: Costa Rica, Les Beletsky (Interlink Publishing, 2004). This book describes many of Costa Rica's amphibians, reptiles, birds, and mammals. The text, which is separate from the plates, presents a general natural history of families rather than species. The plates, found at the end of the book, include attractive color illustrations of 350 species from across the animal kingdom.

A Field Guide to the Wildlife of Costa Rica, Carrol L. Henderson (University of Texas Press, 2002). This is another general wildlife guide that describes many of Costa Rica's insects, amphibians, reptiles, birds, and mammals. It differs from the Beletsky guide in a few important ways. The text provides fairly extensive information on the species level, rather than about families, including information about identification, distribution, and natural history. Also, a color photograph accompanies each species description. The main drawback of this guide is its hefty size.

The Mammals of Costa Rica: A Natural History and Field Guide, Mark Wainwright (Cornell University Press / Zona Tropical Publications, 2002). If you can find space for a field guide devoted only to mammals, this is an excellent choice. It includes color illustrations of all species discussed as well as information on range, vocalizations, and conservation. Each species description also includes an extensive discussion of its natural history, which sets this guide apart from its competitors.

Tropical Plants of Costa Rica: A Guide to Native and Exotic Flora, Willow Zuchowski (Cornell University Press / Zona Tropical Publications, 2007). Covering 430 of Costa Rica's most prominent plant species, this book is an excellent one to own if you have an interest in flora. Each species account includes a description of the plant, its distribution, and a beautiful color photograph. Zuchowski also presents interesting natural-history information for each species, such as pollinators, seed dispersal, and human uses.

General Natural History

Tropical Nature: Life and Death in the Rain Forests of Central and South America, Adrian Forsyth and Ken Miyata (Touchstone, 1987). Through a series of 17 short essays, the authors paint a wonderful picture of the fascinating and unique organisms and ecosystems that exist in neotropical rainforests. The authors' enthusiasm is contagious, and after reading this book you will be eager to get into the field and experience many of these things firsthand. While the topics are scientific in content, they are made easily accessible in this fun read.

Birds of Tropical America, Steven Hilty (University of Texas Press, 1994). This excellent book describes many interesting facets of breeding, behavior, and diversity exhibited in tropical birds. Much of the information is scientific in nature, but Hilty handles these topics in a light and colorful way that makes this book an easy read.

Costa Rican Natural History, edited by Daniel H. Janzen (University of Chicago Press, 1983). This classic volume presents a wealth of information on various topics important to Costa Rican biology. Top researchers from many fields helped to compile the information presented in this thick 800-page book. Be warned that this extremely academic work is not suggested for pleasure reading.

A Neotropical Companion, John Kricher (Princeton University Press, 1997). Perhaps the best general presentation of tropical ecology available, this book covers a wide range of interesting topics. While the subject matter is scientific, Kricher presents it in a way that is accessible to the layperson. The information provided in this volume will give you a deep appreciation for the plants, animals, and ecosystems that you will experience in Costa Rica.

Chasing Neotropical Birds, Vera and Bob Thornton (University of Texas Press, 2005). This fun book will inspire anyone to visit the neotropics. It colorfully describes the adventures of the authors as they search for birds at many of tropical America's most alluring birding destinations, including Costa Rica. Beautiful color photos accompany the text in this handsome hardbound volume, which would make a nice gift.

Travel Guides

Moon Handbooks: Costa Rica, Christopher P. Baker (Avalon Travel Publishing, 2004). This travel guide is my personal favorite. It is well written, the organization and maps are very clear, and Baker describes a wide range of lodging options, restaurants, and activities.

Frommer's Costa Rica, Eliot Greenspan (Frommers, 2007). This is the preferred travel guide for readers looking to frequent the more luxurious lodges and restaurants. Compared with the other two recommended travel guides, this one presents fewer budget options and uses the space for more detailed description of upscale locations.

Lonely Planet Costa Rica, Mara Vorhees (Lonely Planet Publications, 2004). Touting the well-reputed "Lonely Planet" name, this travel guide is an excellent choice that will suit the needs of all travelers.

Planning Your Trip

When to Visit

There is no bad time to visit Costa Rica, and the best time will depend on your personal priorities. The largest diversity of birds is found during the migration months, September and October, and March and April. The winter months also hold high species diversity, with the presence of many North American breeders. However, most of these migrants are low-priority birds compared with the tropical residents that are generally found within the country year-round.

Weather is another factor. The dry season runs from December through April, but this only "guarantees" dry weather in the North Pacific region. For all other parts of the country, rain is still probable, although the amount and frequency are likely to be less than during the rainy season.

The final consideration is expense. Most tourists visit during the dry season because it happens to correspond to the northern winter. Hotels, lodges, and rental companies refer to this as the "high season," and prices are usually raised during these months. Moreover, there will be larger crowds at the popular destinations, like Poás Volcano, Arenal Volcano, and the Monteverde Cloud Forest Reserve.

Modes of Transportation

Unless you are birding with a tour group, you will need to give serious thought to transportation. In my opinion, renting a car is the best way to travel on a birding trip because you have the freedom to go wherever you like, whenever you like. Without a car, you will not be able to access many of the sites described in this book. Driving yourself, however, has its challenges. You will need to navigate the country, as well as deal with Costa Rica's poor roads and aggressive drivers. If you choose to rent a car, get a high-clearance vehicle with four-wheel-drive capability. Many of the roads described in this guide are recommended only for a 4×4, and such a vehicle will provide added peace of mind while in the field. Renting from a company with multiple offices throughout the country is also a good idea because help will be closer at hand if you break down.

If you do choose to rent a car, I highly recommend looking into the option of including a GPS navigation system, now available for Costa Rica from some rental car agencies. A GPS system will give you step-by-step directions to thousands of points throughout the country. While it may not include every location described in this book, it will contain the vast majority, and is especially useful for navigating through cities like San José. Such systems are very easy to use and will give you added confidence in your ability to travel around the country.

A cheaper option than renting a car is to use a private bus service, such as Inter-bus, which runs between many of the popular tourist destinations in the country. Public buses are also an option, although not recommended. The routes and schedules of public buses are difficult to decipher, and travel on them is time-consuming and stressful. Any bus option is more limiting than renting a car, and would result in less quality time spent in the field.

At the other end of the price spectrum are in-country flights through companies such as NatureAir and Sansa. While this option is relatively expensive, it does allow for quick and painless travel over long distances. Those wishing to visit the South Pacific region, which is quite isolated from the rest of the country, may find flying (either round-trip or one-way) an especially enticing option.

Constructing Your Itinerary

A good itinerary is an essential part of a successful trip. This book, a standard travel guide, and a good map of the country are all you will need to plan your perfect birding expedition in Costa Rica. Here are a few important ideas you should consider as you develop your itinerary.

Variety is the spice of life, and you should try to visit as many different regions and subregions as possible. This will expose you to a wide variety of birds, ecosystems, and scenery.

Pacing is another important consideration when creating your itinerary. Be sure to take note of the "birding time" value supplied in each site description. Visiting a new site every day will allow you to see more places, but will be exhausting. I suggest including at least a few large sites, with "birding time" values of two or three days, on your itinerary. Not having to uproot to a new location every day will allow you time for relaxation and give you some familiarity with the sites. Also, plan your site sequence carefully in order to limit the amount of driving. Your time in Costa Rica is best spent in the field, not on the road. The map provided later in this chapter depicts driving times between major points, and will be helpful in calculating the expected travel times between sites.

Sample Itineraries

Below are eight possible itineraries. While you may choose to follow these suggestions directly, the main purpose is to inspire you to think about how to link together sites to create a well-paced trip that meets your goals.

Pacific and Caribbean

This 10-day trip encompasses four of Costa Rica's most important birding destinations—Palo Verde National Park, Carara National Park, Rancho Naturalista, and La Selva Biological Station. Following this itinerary, you are sure to rack up an impressive list of species. Unfortunately, you will spend very little time in the mountains, so you are likely to miss many of the mountain endemics.

Day 1: Fly into San José International Airport. Spend the night in San José.

Day 2: Drive to Palo Verde National Park (site 2-3) first thing in the morning (driving time: approx. 5 hours). Bird Palo Verde in the afternoon and spend the night inside the park.

Day 3: Bird Palo Verde in the morning. Drive to Carara National Park (site 3A-1) in the afternoon (driving time: approx. 3.5 hours). Spend the night in the area.

Day 4: Bird Carara and the Tárcoles River Mouth (site 5-3). Spend the night in the area.

Day 5: Bird Carara in the morning. Drive to Poás Volcano National Park (site 4B-1) in the afternoon (driving time: approx. 2.5 hours). Spend the night in the area.

Day 6: Bird Poás Volcano in the morning. Drive to Virgen del Socorro (site 1B-3), stopping for lunch at the Cinchona hummingbird feeders (driving time: approx. 30 minutes). After lunch, bird Virgen del Socorro and then drive to La Selva Biological Station (site 1A-4) in the late afternoon (driving time: approx. 1 hour). Spend the night in the area.

Day 7: Bird La Selva and spend the night in the area.

Day 8: Bird La Selva in the morning. Drive to Rancho Naturalista (site 1B-7) in the afternoon (driving time: approx. 2 hours). Spend the night on site.

Day 9: Bird Rancho Naturalista and spend the night on site.

Day 10: Return to San José in the afternoon (driving time: approx. 1.5 hours).

Northern Loop

This nine-day trip visits a number of the best birding sites in the northern part of the country and proceeds through a wide diversity of habitats. It avoids San José and the Central Valley, starting and ending at the international airport in Liberia.

Day 1: Fly into Liberia International Airport. Drive to Rincón de la Vieja National Park (site 2-2) and spend the night in the area (driving time: approx. 1 hour).

Day 2: Bird Rincón de la Vieja and spend the night in the area.

Day 3: Bird Rincón de la Vieja in the morning. Drive to Las Heliconias Lodge (site 1B-1) in the afternoon and spend the night on site (driving time: approx. 2 hours).

Day 4: Bird Las Heliconias Lodge all day and spend the night on site.

Day 5: Drive to Caño Negro National Wildlife Refuge (site 1A-1) first thing in the morning (driving time: approx. 1.5 hours) and bird it for most of the day. Drive to Arenal Volcano National Park (site 1B-2) in the afternoon and spend the night in the area (driving time: approx. 2 hours).

Day 6: Bird Arenal Volcano during the day. Drive to Bosque de Paz Biological Reserve and Lodge (site 4A-3) in the late afternoon (driving time: approx. 2 hours). Stay on site.

Day 7: Bird Bosque de Paz all day and stay on site.

Day 8: Bird Bosque de Paz in the morning. Drive to La Ensenada Wildlife Refuge (site 2-6) in the afternoon and stay on site (driving time: approx. 3 hours).

Day 9: Bird La Ensenada during the day. In the evening return to Liberia (driving time: approx. 1.5 hours).

Southern Loop

This nine-day loop focuses on the Mountains and Caribbean Slope. The pacing is relatively slow, providing plenty of time to bird Km 70 and Rancho Naturalista. If you come in October and want to see the hawk migration, Keköldi Hawk Watch (site 1A-7) could easily be added to the itinerary.

Day 1: Arrive at the San José International Airport. Drive to Km 70 (site 4B-3) and spend the night on site (driving time: approx. 2 hours).

Day 2: Bird Km 70 and Cerro de la Muerte (site 4B-5) all day. Spend another night at Km 70.

Day 3: Bird Km 70 in the morning. Drive to Tapantí National Park (site 4A-5) in the afternoon (driving time: approx 1.5 hours).

Day 4: Bird Tapantí during the day. Drive to Rancho Naturalista (site 1B-7) in the afternoon and spend the night on site (driving time: approx 2 hours).

Day 5: Bird Rancho Naturalista all day and spend the night on site.

Day 6: Bird Rancho Naturalista all day and spend the night on site.

Day 7: Bird Rancho Naturalista in the morning. Drive to Selva Bananito Lodge (site 1A-5) in the afternoon and spend the night on site (driving time: approx. 3 hours).

Day 8: Bird Selva Bananito Lodge all day and spend the night on site.

Day 9: Drive to Braulio Carrillo National Park (site 1B-4) early in the morning (driving time: approx 2.5 hours). Bird Braulio Carrillo and then return to San José (driving time: approx. 1 hour).

The South Pacific

Aside from one day in the mountains, this eight-day trip focuses on sites in the South Pacific region of the country. The pacing is comfortable because each site is relatively close to the next. This itinerary will give you a good chance to see most of the South Pacific specialties and endemics.

Day 1: Arrive at the San José International Airport. Drive to Km 70 (site 4B-3) and spend the night on site (driving time: approx. 2 hours).

Day 2: Bird Km 70 and Cerro de la Muerte (site 4B-5) during the day. Drive to Talari Mountain Lodge (site 3B-1) in the afternoon and spend the night on site (driving time: approx 1.5 hours).

Day 3: Bird Talari Mountain Lodge and nearby locations during the day. Drive to Las Cruces Biological Station (site 3B-3) in the afternoon and spend the night on site (driving time: approx. 2.5 hours).

Day 4: Leave early from Las Cruces and drive to Las Alturas (site 3B-4), and spend the day birding (driving time: approx. 1 hour). Return to Las Cruces for dinner and spend the night on site.

Day 5: Bird Las Cruces. In the afternoon, drive to Esquinas Rainforest Lodge (site 3A-7) and stay on site (driving time: approx. 1.5 hours).

Day 6: Bird Esquinas all day and spend the night on site.

Day 7: Bird Esquinas in the morning. Drive to Oro Verde Biological Reserve (site 3A-3) in the afternoon and spend the night in the area (driving time: approx 1.5 hours).

Day 8: Bird Oro Verde in the morning and then return to San José in the afternoon (driving time: approx. 3.5 hours).

Costa Rican Endemics

This exciting two-week trip is geared for the serious lister. You will visit some of the best sites across the country and are sure to spot a staggering number of birds. Also, there is a good possibility you might see 77 of Costa Rica's 89 endemic species along this route. The diversity comes at a price, however, as the trip is fast paced, necessitates a good amount of travel, and visits a number of expensive lodges.

Day 1: Arrive in the San José International Airport and spend the night in the vicinity.

Day 2: Take an early in-country flight to Puerto Jiménez, where you can pick up a rental car. Drive to Bosque del Río Tigre (site 3A-5), bird the afternoon, and spend the night on site (driving time: approx. 20 minutes).

Day 3: Bird Bosque del Río Tigre all day and spend the night on site.

Day 4: Drive to Las Cruces Biological Station (site 3B-3), birding Rincón and the Coastal Plains (refer to site 3A-5) along the way (driving time: approx. 3.5 hours). Spend the night on site.

Day 5: Bird Las Cruces all day and spend the night on site.

Day 6: Bird Las Cruces in the morning and then drive to San Gerardo de Dota (site 4B-4) in the afternoon (driving time: approx. 4 hours). Stay in the area overnight.

Day 7: Bird San Gerardo de Dota all day and spend the night in the area.

Day 8: Bird Cerro de la Muerte (site 4B-5) in the morning. In the afternoon drive to Bosque de Paz Biological Reserve and Lodge (site 4A-3) and stay on site (driving time: approx. 4 hours).

Day 9: Bird Bosque de Paz all day and stay on site.

Day 10: Bird Bosque de Paz in the morning. Drive to La Selva Biological Station (site 1A-4), making a quick detour to the Cinchona hummingbird feeders (refer to site 1B-3) along the way (driving time: approx. 2 hours).

Day 11: Bird La Selva all day and spend the night in the area.

Day 12: Drive to Horquetas (driving time: approx.15 minutes), where you will be transferred to Rara Avis Rainforest Lodge (site 1B-5). Spend the night on site.

Day 13: Bird Rara Avis all day and stay on site.

Day 14: Transfer back to Horquetas and then return to San José (driving time: approx. 1 hour).

Off the Beaten Trail

This trip is for adventurous birders who want local flavor. The facilities at two of these sites are rustic, and Spanish is useful. Note, however, that one day is spent at the Monteverde Reserve, a tourist hotspot.

Day 1: Arrive in the San José International Airport. Drive to the Monteverde/Santa Elena area (4A-1) and spend the night (driving time: approx. 3.5 hours).

Day 2: In the morning, hike to the San Gerardo Biological Station (described in 4A-2). Bird the area all day and spend the night on site.

Day 3: Bird San Gerardo all day and spend the night on site.

Day 4: Bird San Gerardo until lunch, and then hike back up to the town of Santa Elena. Spend the night in the area.

Day 5: Bird the Monteverde Reserve and surrounding area (4A-1). Then drive to San José, spending the night in Santa Anna (driving time: approx. 3.5 hours).

Day 6: Bird El Rodeo (6-1) in the morning. Then drive to El Copal Biological Reserve (1B-8), spending the night on site (driving time: approx. 2 hours).

Day 7: Bird El Copal all day, and spend the night on site.

Day 8: Bird El Copal all morning, returning to San José in the late afternoon (driving time: approx. 2 hours).

Birding with the Family

This eight-day itinerary visits a number of popular tourist destinations that also offer good birding. It combines non-birding activities with some good birding destinations and is meant for families or groups with a range of interests.

Day 1: Arrive at the San José International Airport and spend the night in the area.

Day 2: Drive to Tárcoles River Mouth (site 5-3) and take the mangrove boat tour. Continue on to Manuel Antonio National Park (site 3A-2) and spend the night in the area (driving time: approx. 3 hours total).

Day 3: Spend the day at Manuel Antonio and stay in the area.

Day 4: Drive to Monteverde Cloud Forest Reserve (site 4A-1) and stay in the area (driving time: approx. 4 hours).

Day 5: Bird the Monteverde area and enjoy the many tourist attractions with your family.

Day 6: Drive to Arenal Volcano National Park (site 1B-2) and spend the night in the area (driving time: approx. 3.5 hours).

Day 7: Bird Arenal Volcano in the morning and spend the afternoon with the family. Remain in the area for the night.

Day 8: Spend the morning at Arenal Volcano with your family or birding and return to San José in the afternoon (driving time: approx. 2.5 hours).

A Quick Trip

This itinerary is ideal for those who only have a few days to spend birding. While the formal itinerary below is only three days long, you could easily extend the trip by adding in Carara National Park (site 3A-1) or Tapantí National Park (site 4A-5).

Day 1: Drive to Poás Volcano National Park (site 4B-1) in the morning and spend a few hours birding the area (driving time: approx. 1 hour). Drive to Virgen del Socorro (site 1B-3), stopping for lunch at the Cinchona hummingbird feeders (driving time: approx. 30 minutes). After lunch, bird Virgen del Socorro and then drive to La Selva Biological Station (site 1A-4) in the late afternoon (driving time: approx. 1 hour). Spend the night in the area.

Day 2: Bird La Selva and spend the night in the area.

Day 3: Bird La Selva in the morning and then drive to Braulio Carillo National Park (site 1B-4) and bird the area thoroughly (driving time: approx. 35 minutes). Return to San José (driving time: approx. 1 hour).

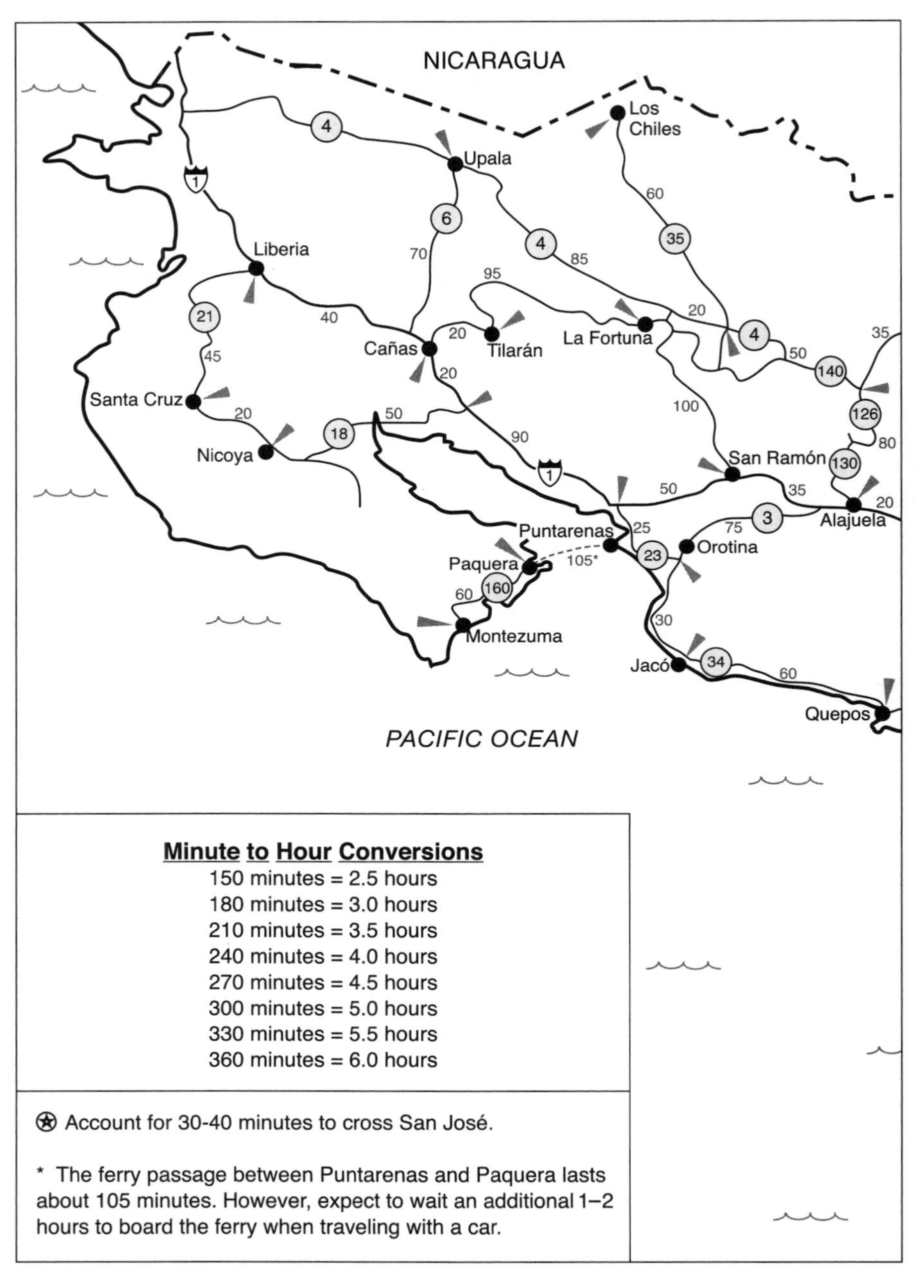

Minute to Hour Conversions

150 minutes = 2.5 hours
180 minutes = 3.0 hours
210 minutes = 3.5 hours
240 minutes = 4.0 hours
270 minutes = 4.5 hours
300 minutes = 5.0 hours
330 minutes = 5.5 hours
360 minutes = 6.0 hours

⊛ Account for 30-40 minutes to cross San José.

* The ferry passage between Puntarenas and Paquera lasts about 105 minutes. However, expect to wait an additional 1–2 hours to board the ferry when traveling with a car.

Countrywide Road Map. The numbers in between wedge markers indicate approximate driving times.

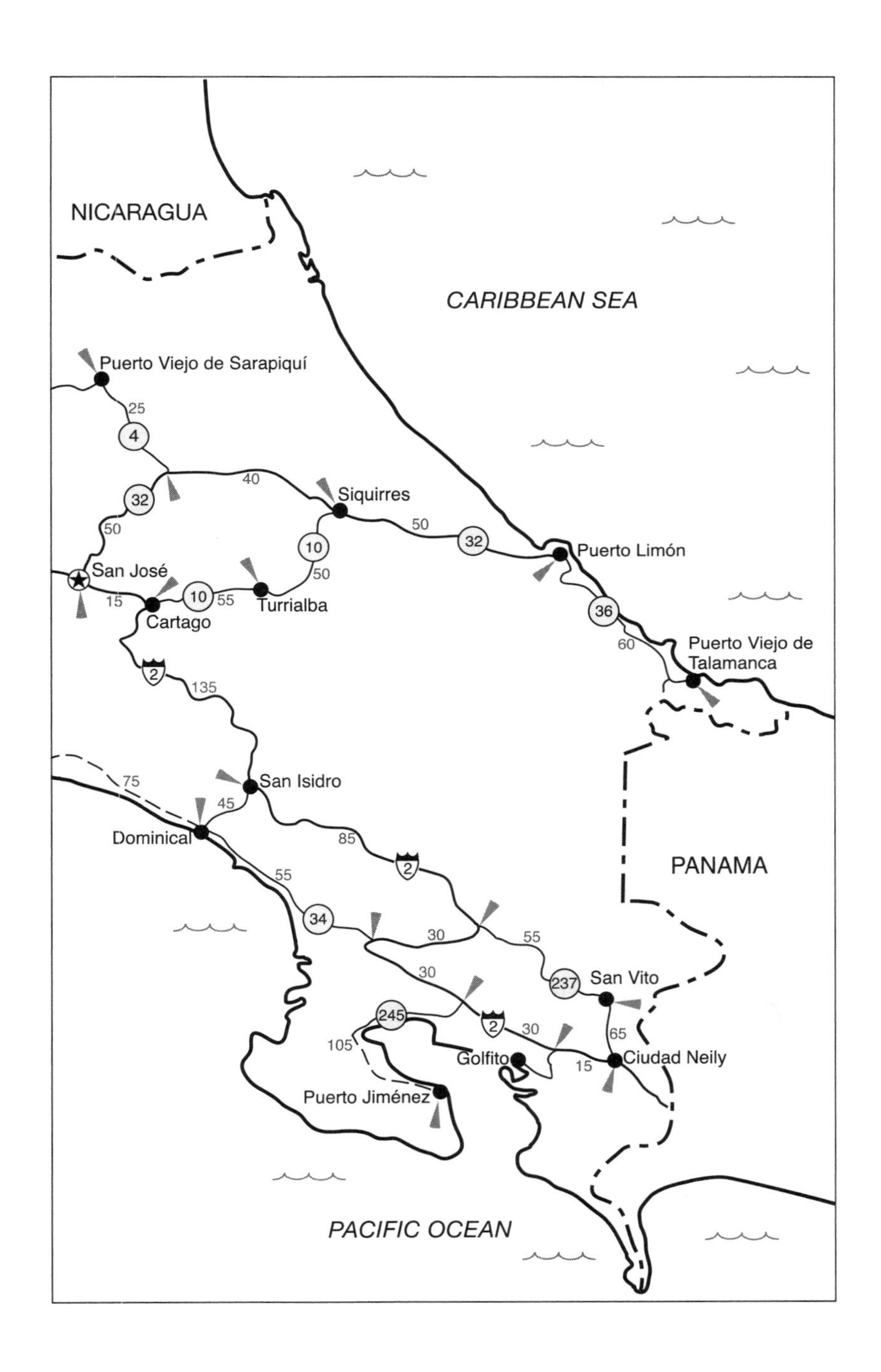
NICARAGUA
CARIBBEAN SEA
Puerto Viejo de Sarapiquí
25
4
32
40
Siquirres
50
10
50
32
Puerto Limón
San José
15
10
55
Turrialba
50
Cartago
36
60
Puerto Viejo de Talamanca
2
135
75
San Isidro
45
Dominical
85
2
PANAMA
55
34
30
55
30
237
San Vito
245
2
30
65
105
Golfito
15
Ciudad Neily
Puerto Jiménez
PACIFIC OCEAN

Region 1
The Caribbean Slope

Normally found in the Caribbean Lowlands, the **Bare-necked Umbrellabird** migrates upslope to breed between March and May. Here, in middle-elevation forests, males gather in leks to display for females by inflating their large red throat sacs. The Bare-necked Umbrellabird is most reliably found at La Selva Biological Station (1A-4) and at San Gerardo Biological Station (described in site 4A-2).

Regional Map: The Caribbean Slope

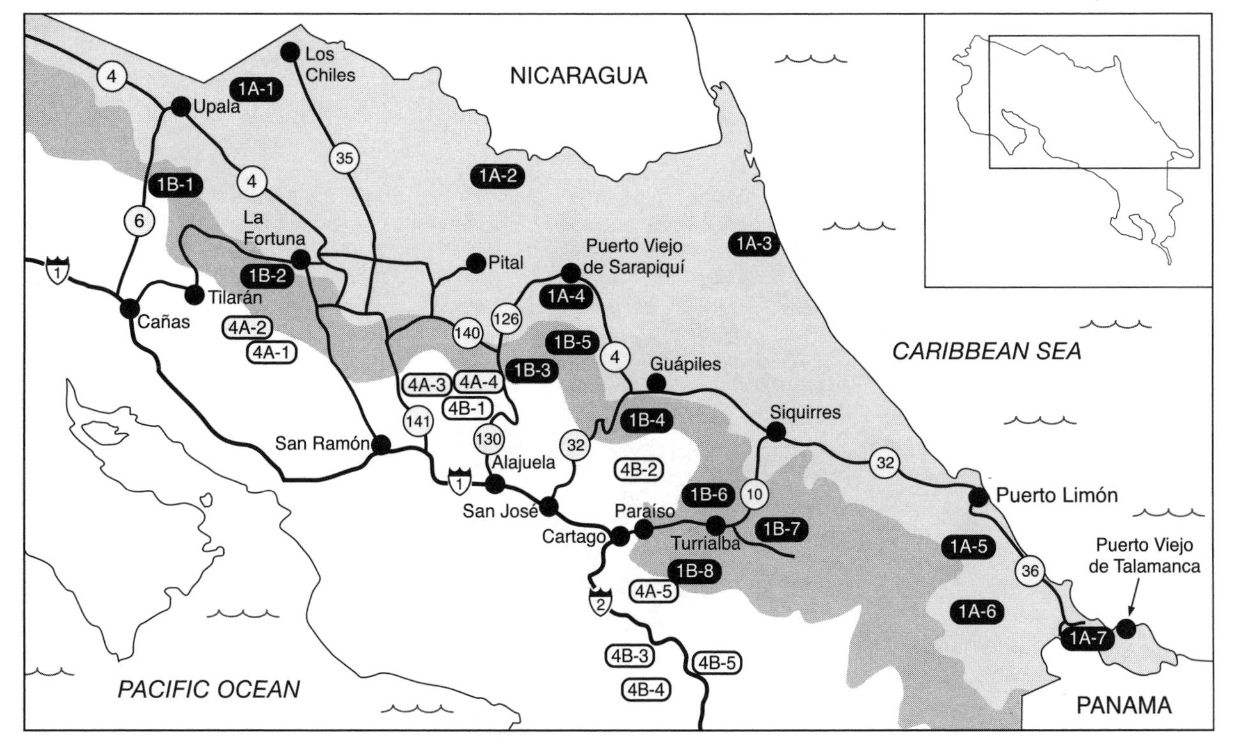

Lowlands
1A-1: Caño Negro National Refuge
1A-2: Laguna del Lagarto Lodge
1A-3: Tortuguero National Park
1A-4: La Selva Biological Station
1A-5: Selva Bananito Lodge
1A-6: Hitoy Cerere Biological Reserve
1A-7: Keköldi Hawk Watch

Middle Elevations
1B-1: Las Heliconias Lodge
1B-2: Arenal Volcano National Park
1B-3: Virgen del Socorro
1B-4: Braulio Carrillo National Park
1B-5: Rara Avis Rainforest Lodge
1B-6: Guayabo National Monument
1B-7: Rancho Naturalista
1B-8: El Copal Biological Reserve

The Caribbean Slope includes everything east of the Continental Divide and lower than 1,200 meters in elevation. It is the largest region in this guide, encompassing nearly half of the country and covering 15 sites. Unlike the Pacific Slope of Costa Rica, the differences in avifauna between the northern and southern parts of the Caribbean Slope are relatively minor, and the entire area can be presented as one cohesive region. (The major exception is the Río Frío district in the northwestern corner, which has a strong climatic influence from the dry North Pacific region. The one site described in this area is Caño Negro National Refuge, site 1A-1.)

The Caribbean Slope has been split into two subregions, Caribbean Lowlands and Caribbean Middle-Elevations, to address the distinct differences in avifauna caused by elevation. Lowlands are generally considered to run from sea level to about 500 meters. At this point, some of the lowland specialists, such as Blue-chested Hummingbird, Chestnut-colored Woodpecker, and Long-tailed Tyrant, begin to disappear, while middle-elevation species, such as Black-and-yellow Tanager, Blue-and-gold Tanager, and Dull-mantled Antbird, start to show up. The Caribbean Middle-Elevations extend from 500 meters up to the start of the Mountains region, at 1,200 meters.

The vast majority of the Caribbean Slope is home to the quintessential rainforest that visitors imagine in the neotropics. These forests are some of the most diverse ecosystems in the world, home to an incredible wealth of life. For example, the well-studied forest at La Selva Biological Station (1A-4) has recorded 2,016 species of plants, 116 species of mammals, 87 species of reptiles, and 414 species of birds! The Caribbean Slope, as a whole, is home to some 25 species of hummingbirds and 50 species of flycatchers alone. For birders new to the neotropics, this diversity can be overwhelming, although exciting at the same time.

A wealth of water helps to support this diversity, and the majority of the region receives 3 to 6 meters (10 to 20 feet) of rainfall annually. While there is generally less rainfall during the dry season (December through April), there are no guarantees with the weather at any time of year, so be sure to come prepared. Unfortunately, deforestation has played a major role in creating the landscape that visitors now find when they arrive at the Caribbean Slope. By the 1970s, pioneers had cut down most of the forest, especially in the lowlands where the flat land was easier to cultivate. Today, while many areas are being allowed to regenerate, it is difficult to find extensive primary forest in the Caribbean Lowlands.

As the forest was cut, the land was taken over by agriculture, including the raising of cattle and the production of banana, pineapple, yuca, papaya, and heart of palm. Bananas are an especially important product in this region. Enormous banana plantations cover much of the Caribbean Lowlands, and large international corporations like Chiquita and Dole have major operations here in Costa Rica. Bananas are one of Costa Rica's most important exports, and there is a good chance that the bananas you eat at home were grown in the Caribbean Lowlands.

The city of Puerto Limón is worthy of note, as it is the largest in the region and home to Costa Rica's principal port. Aside from the large cruise ships that often dock here, most of the country's imports and exports pass through these docks. The huge freighters, fleets of trucks, and large container depots found in and around Limón confirm the important role that this city plays in Costa Rica's international trade.

Puerto Limón is also where many immigrants have entered the country, and the surrounding area has a strong Afro-Caribbean influence. Many Jamaicans disembarked here in search of work on the Braulio Carrillo Highway and in the banana plantations. This combination of Costa Rican and Afro-Caribbean influences creates a unique culture that is well represented by one of its most distinctive recipes—"rice and beans." This popular dish is very similar to the typical Costa Rican breakfast *gallo pinto*, but is seasoned with coconut milk and other Caribbean flavors. Although Spanish is the principal language, many of the locals speak a unique Creole language called Patuá, which is a combination of English and French. If you visit beach towns like Cahuita and Puerto Viejo, you will surely notice a distinct Afro-Caribbean flavor to the culture.

In the Caribbean Slope, the two most popular tourist destinations are Tortuguero National Park (1A-3) and Arenal Volcano National Park (1B-2). Both areas offer good birding as well as a wide variety of accommodations and tourist attractions. The real standout locations for birding, however, are La Selva Biological Station (1A-4), Rancho Naturalista (1B-7), and El Copal Biological Reserve (1B-8). All three of these sites offer beautiful habitat, extensive trail systems, and a huge diversity of species. Keköldi Hawk Watch (1A-7) is also worthy of special mention as it is one of the three most important hawk-monitoring stations in the world.

Regional Specialties

Slaty-breasted Tinamou (3) 12:8
*Black-eared Wood-Quail (13) 12:12
Rufescent Tiger-Heron (21) 5:17
Agami Heron (21) 5:1
Green Ibis (29) 4:10
Lesser Yellow-headed Vulture (31) 13:2
Semiplumbeous Hawk (41) 16:10
Slaty-backed Forest-Falcon (55) 16:6
Sungrebe (63) 7:9
Olive-backed Quail-Dove (93) 18:22
Purplish-backed Quail-Dove (93) 18:21
Olive-throated Parakeet (95) 19:11
Great Green Macaw (99) 19:2
*Red-fronted Parrotlet (97) 19:15
Rufous-vented Ground-Cuckoo (103) 21:10
Central American Pygmy-Owl (109) 20:16
Great Potoo (115) 20:5
Chimney Swift (119) 22:8
Gray-rumped Swift (119) 22:10
Black-crested Coquette (137) 25:6
Blue-chested Hummingbird (125) 24:14
*Snowcap (137) 25:8
Bronze-tailed Plumeleteer (129) 23:12
*Lattice-tailed Trogon (145) 26:3
Rufous Motmot (147) 27:7
Keel-billed Motmot (147) 27:10
Broad-billed Motmot (147) 27:9
Green-and-rufous Kingfisher (149) 27:3
Pied Puffbird (151) 28:7
Lanceolated Monklet (151) 28:5

White-fronted Nunbird (151) 28:4
Great Jacamar (153) 26:11
*Yellow-eared Toucanet (155) 27:20
Black-cheeked Woodpecker (157) 28:14
Cinnamon Woodpecker (161) 28:9
Chestnut-colored Woodpecker (161) 28:8
Plain-brown Woodcreeper (171) 29:11
Fasciated Antshrike (175) 31:1
Western Slaty-Antshrike (177) 31:5
*Streak-crowned Antvireo (181) 32:7
Spot-crowned Antvireo (181) 32:8
Checker-throated Antwren (183) 32:5
White-flanked Antwren (183) 32:2
Dull-mantled Antbird (181) 31:14
Spotted Antbird (181) 31:12
Ocellated Antbird (179) 31:7
Black-headed Antthrush (187) 30:17
*Black-crowned Antpitta (185) 30:19
*Thicket Antpitta (185) 30:14
Brown-capped Tyrannulet (189) 37:13
Rufous-browed Tyrannulet (193) 37:12
Black-capped Pygmy-Tyrant (197) 37:5
Black-headed Tody-Flycatcher (197) 37:8
Yellow-margined Flycatcher (195) 37:17
*Tawny-chested Flycatcher (199) 36:10
Long-tailed Tyrant (201) 36:6
White-ringed Flycatcher (211) 35:16
*Gray-headed Piprites (215) 33:3
Lovely Cotinga (219) 34:6
*Snowy Cotinga (219) 34:4
Purple-throated Fruitcrow (219) 34:11
*Bare-necked Umbrellabird (221) 34:13
White-collared Manakin (221) 33:1
White-crowned Manakin (223) 33:8
Sharpbill (225) 34:7
Black-chested Jay (231) 39:20
Band-backed Wren (239) 38:3
*Black-throated Wren (243) 38:13
Bay Wren (241) 38:12
*Stripe-breasted Wren (241) 38:5
Spot-breasted Wren (239) 38:7
Nightingale Wren (245) 38:21
Song Wren (245) 38:22
Tawny-faced Gnatwren (237) 32:14
Black-headed Nightingale-Thrush (247) 38:25
Pale-vented Thrush (251) 39:5
Olive-crowned Yellowthroat (273) 42:11
Ashy-throated Bush-Tanager (279) 45:12
*Black-and-yellow Tanager (283) 46:14
Dusky-faced Tanager (281) 45:18
Carmiol's Tanager (279) 45:16
*Sulphur-rumped Tanager (283) 45:15
Tawny-crested Tanager (281) 46:17
Red-throated Ant-Tanager (277) 47:12
Crimson-collared Tanager (287) 47:3
Passerini's Tanager (287) 47:4
*Blue-and-gold Tanager (283) 45:11
*Plain-colored Tanager (291) 46:11
Emerald Tanager (289) 46:5
Rufous-winged Tanager (289) 46:10
*Nicaraguan Seed-Finch (295) 49:13
Black-headed Saltator (309) 48:1
Slate-colored Grosbeak (309) 48:6
Black-faced Grosbeak (311) 48:5
*Nicaraguan Grackle (317) 44:17
Black-cowled Oriole (319) 44:5
Yellow-tailed Oriole (319) 44:3
Chestnut-headed Oropendola (323) 44:9
Olive-backed Euphonia (325) 45:3

Subregion 1A

The Caribbean Lowlands

The handsome **Great Green Macaw** feeds primarily in wild almond trees (*Dipteryx panamensis*) and nomadically wanders across the Caribbean Lowlands in search of ripening fruit. Laguna del Lagarto Lodge (1A-2) is an excellent location to look for this species.

Common Birds to Know

Black Vulture (31) 13:4
Turkey Vulture (31) 13:3
Orange-chinned Parakeet (97) 19:14
Rufous-tailed Hummingbird (129) 24:10
Keel-billed Toucan (155) 27:18
Chestnut-mandibled Toucan (155) 27:19
Black-cheeked Woodpecker (157) 28:14
Great Kiskadee (211) 35:13
Boat-billed Flycatcher (211) 35:12
Social Flycatcher (211) 35:14
Gray-capped Flycatcher (211) 35:15
Tropical Kingbird (213) 35:1
Cinnamon Becard (215) 33:11
House Wren (243) 38:18
Clay-colored Thrush (251) 39:8
Chestnut-sided Warbler (259) 43:4
Bananaquit (277) 40:24
Passerini's Tanager (287) 47:4
Blue-gray Tanager (291) 46:15
Palm Tanager (291) 45:19
Golden-hooded Tanager (289) 46:13
Variable Seedeater (295) 49:3
Black-striped Sparrow (303) 50:14
Buff-throated Saltator (309) 48:2
Montezuma Oropendola (323) 44:8

Site 1A-1: Caño Negro National Refuge

Birding Time: 5 hours
Elevation: 50 meters
Trail Difficulty: 1
Reserve Hours: N/A
Entrance Fee: $10 (US 2008) plus boat fee
4×4 Recommended

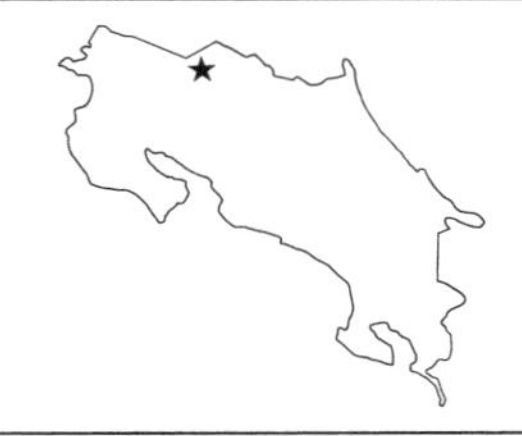

The Río Frío region of Costa Rica, located in the northwestern corner of the Caribbean Slope, is quite unique. While the area is in the Caribbean, its weather pattern is more closely tied to that of the North Pacific region, with a distinct dry season running from December through April. The Río Frío, which runs through this area, experiences dramatic shifts in water level due to the seasonal rains. During the dry months of the year, the river is reduced to a trickle, barely navigable by boat. During the wet months, rains fill the riverbed, causing the river to overflow its banks. Caño Negro National Refuge, created in 1984, protects 24,620 acres of land along the Río Frío where this flooding creates a large seasonal lake and series of lagoons. These marshes are best visited during the dry months, when much of the water dries up and the birds congregate in huge numbers around the few remaining waterholes.

The wetlands in Caño Negro National Refuge provide an ideal habitat for many water birds, including herons, kingfishers, and ibises, and, in Costa Rica, are only rivaled by those in the Tempisque-Bebedero Watershed of Guanacaste. There is even a small breeding population of the rare Jabiru living in the area. The Caño Negro region also holds a few species that cannot be reliably found anywhere else in the country, such as Lesser Yellow-headed Vulture, Spot-breasted Wren, and the endemic Nicaraguan Grackle.

Unfortunately the area has experienced heavy deforestation in recent years, and most of the surrounding countryside has been converted to cattle pasture and farmland, principally for growing beans. The refuge itself provides the best birding opportunities but is only accessible through hired boat, making the unique area frustratingly difficult to explore. However, you should be able to find most of the region's specialty species from the boat.

Target Birds

Pinnated Bittern (21) 5:11
Boat-billed Heron (27) 5:2
Glossy Ibis (29) 4:9
Green Ibis (29) 4:10
Jabiru (19) 4:5
Lesser Yellow-headed Vulture (31) 13:2
Snail Kite (35) 14:6
Plumbeous Kite (37) 14:8
Black-collared Hawk (33) 17:5
Northern Harrier (41) 15:7
Sungrebe (63) 7:9
Limpkin (63) 5:5
Plain-breasted Ground-Dove (89) 18:9
Gray-headed Dove (91) 18:15
Mangrove Cuckoo (101) 21:4
Striped Owl (105) 20:2
Great Potoo (115) 20:5
American Pygmy Kingfisher (149) 27:6
Olivaceous Piculet (159) 29:1
Bare-crowned Antbird (179) 31:13
Spot-breasted Wren (239) 38:7
Ruddy-breasted Seedeater (297) 49:1
*Nicaraguan Seed-Finch (295) 49:13
*Nicaraguan Grackle (317) 44:17
Yellow-tailed Oriole (319) 44:3

Access

From Los Chiles: Take Rt. 35 south for 7.2 km and then turn right (west) onto a dirt road signed for Caño Negro. Follow this road straight for 19.4 km to a three-way intersection. Continue straight through this intersection and you will immediately enter the town of Caño Negro. (Driving time from Los Chiles: 40 min)

From Upala: Head south on Rt. 4 for 11.0 km to the small town of Colonia Puntarenas. Turn left (east) onto a dirt road and proceed 26.0 km to a T-intersection. Turn right here and you will immediately enter the town of Caño Negro. (Driving time from Upala: 60 min)

Logistics: Most visitors to Caño Negro access the area as a day trip from La Fortuna and Arenal Volcano National Park (1B-2), which is a 2-hour drive. You may also choose to stay in Upala, Los Chiles, or Caño Negro itself, where there are a few intermediate hotels as well as some economical options. This will allow you to get an earlier start to the day.

Birding Sites

Aside from the refuge itself, which is only accessible by boat, most of the land around Caño Negro has been converted to farmland, which could yield

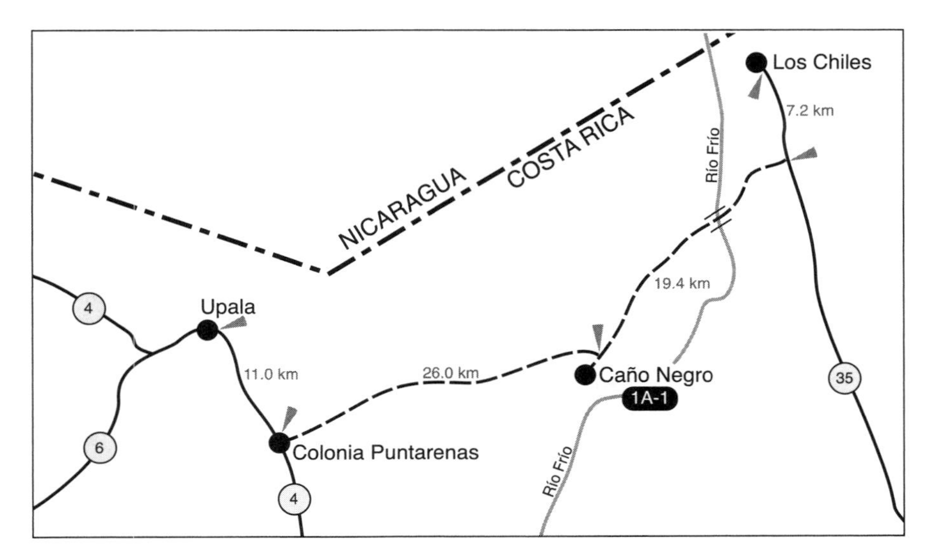

Caño Negro National Refuge Driving Map

interesting species such as Ruddy-breasted Seedeater, Nicaraguan Seed-Finch, and Plain-breasted Ground-Dove. Looking for birds around the town of Caño Negro can also be productive. The trees and shrubs along any of the small streets may yield Green-breasted Mango, Olive-throated Parakeet, Olivaceous Piculet, Greenish Elaenia, and Black-headed Trogon. The road connecting the park and the soccer field is one of the better areas. Here you may find the Spot-breasted Wren, Northern Beardless-Tyrannulet, and Yellow Tyrannulet.

Depending on water level, you may find water birds—Green Ibis, White Ibis, Gray-necked Wood-Rail, Least Sandpiper, and Roseate Spoonbill, among others—in the three seasonal lagoons around town (refer to map). From the dock a trail leads out to the river and along the bank where the locals often moor their boats. This is a great place to find Nicaraguan Grackle. You can also scan the lagoons on either side of you for Lesser Yellow-headed Vulture. This species is almost identical to the more abundant Turkey Vulture, except the color of its head is orangey yellow instead of red. It is helpful to note that this species often flies low over the lagoons, like a Northern Harrier, whereas the Turkey Vulture is usually higher.

To enter the Caño Negro National Refuge you must first stop at the park headquarters in town and purchase your tickets (open 8 a.m. to 4 p.m.). Next, you must hire a boat and guide to take you down the river. River guides wait for clients in the Caño Negro town park, around a set of benches at the corner closest to the dock. If a guide is not readily obvious, ask around. It should not be too hard to find one who will be able to take you into the park immediately.

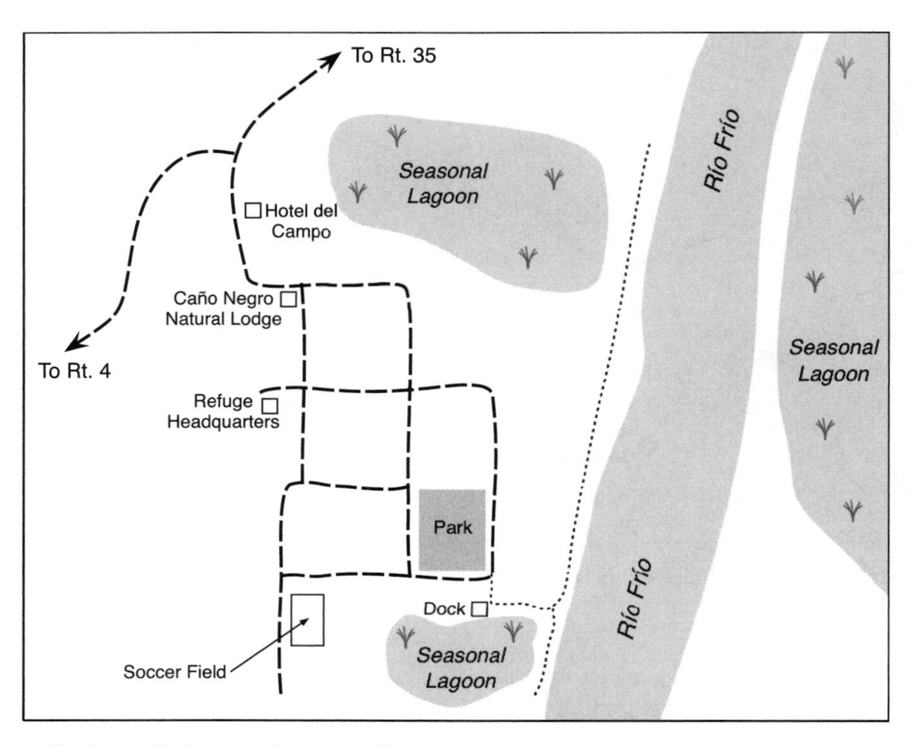

Caño Negro National Refuge Area Map

The location and density of the water birds at Caño Negro change depending on river level, and the local guides will have up-to-date knowledge of where birds are being seen. Heading upriver, however, is generally more productive than heading downriver. You will come to a number of lagoons and marshes that can be teaming with birds, including Bare-throated Tiger-Heron, White-Ibis, Purple Gallinule, Limpkin, and Wood Stork. These lagoons are also good places to find Jabiru, Black-collared Hawk, Osprey, and Lesser Yellow-headed Vulture. The guides will often park the boats along the riverbank and have you get off to view the lagoons from land. For this reason, it is a good idea to bring a telescope if you have one.

As you cruise along the river you should see a number of kingfishers, including Amazon Kingfisher and American Pygmy Kingfisher. Boat-billed Heron, Yellow-crowned Night-Heron, and Black-crowned Night-Heron can often be seen roosting in the tree branches overhanging the river. Gray-breasted Martin, Mangrove Swallow, and Southern Rough-winged Swallow swoop over the river. You also have a good chance of finding a Sungrebe, which likes to swim around roots, sticks, and debris along the riverbank, and your guide may be able to take you to the roost of a potoo.

Species to Expect

Little Tinamou (3) 12:17
Black-bellied Whistling-Duck (5) 8:7
Blue-winged Teal (7) 8:5
Neotropic Cormorant (17) 4:4
Anhinga (17) 4:3
Bare-throated Tiger-Heron (21) 5:16
Great Blue Heron (23) 5:6
Great Egret (23) 5:14
Snowy Egret (23) 5:10
Little Blue Heron (23) 5:9
Tricolored Heron (25) 5:8
Cattle Egret (23) 5:13
Green Heron (25) 6:2
Black-crowned Night-Heron (27) 5:4
Yellow-crowned Night-Heron (27) 5:3
Boat-billed Heron (27) 5:2
White Ibis (29) 4:8
Glossy Ibis (29) 4:9
Green Ibis (29) 4:10
Roseate Spoonbill (27) 4:7
Jabiru (19) 4:5
Wood Stork (19) 4:6
Black Vulture (31) 13:4
Turkey Vulture (31) 13:3
Lesser Yellow-headed Vulture (31) 13:2
Osprey (33) 17:14
Black-collared Hawk (33) 17:5
Roadside Hawk (43) 16:12
Crested Caracara (53) 14:12
Laughing Falcon (53) 15:8
Peregrine Falcon (51) 15:5
White-throated Crake (57) 6:9
Gray-necked Wood-Rail (55) 6:13
Purple Gallinule (59) 6:15
Sungrebe (63) 7:9
Limpkin (63) 5:5
Spotted Sandpiper (73) 11:8
Solitary Sandpiper (71) 11:4
Least Sandpiper (75) 11:14
Rock Pigeon (85)
Pale-vented Pigeon (87) 18:3
Ruddy Ground-Dove (89) 18:7
Blue Ground-Dove (91) 18:6
White-tipped Dove (91) 18:14
Olive-throated Parakeet (95) 19:11
Orange-chinned Parakeet (97) 19:14
Squirrel Cuckoo (103) 21:7
Mangrove Cuckoo (101) 21:4
Striped Cuckoo (101) 21:5
Groove-billed Ani (103) 21:9
Striped Owl (105) 20:2
Common Pauraque (111) 21:18
Great Potoo (115) 20:5
Green-breasted Mango (129) 23:13
Rufous-tailed Hummingbird (129) 24:10
Black-headed Trogon (141) 26:10
Ringed Kingfisher (149) 27:1
Amazon Kingfisher (149) 27:4
Green Kingfisher (149) 27:5
American Pygmy Kingfisher (149) 27:6
Olivaceous Piculet (159) 29:1
Black-cheeked Woodpecker (157) 28:14
***Hoffmann's Woodpecker (157) 28:16**
Cinnamon Woodpecker (161) 28:9
Lineated Woodpecker (161) 27:13
Pale-billed Woodpecker (161) 27:14
Cocoa Woodcreeper (173) 29:17
Streak-headed Woodcreeper (173) 29:8
Barred Antshrike (175) 31:3
Northern Beardless-Tyrannulet (191) 37:19
Yellow Tyrannulet (189) 37:11
Greenish Elaenia (195) 37:21
Yellow-bellied Elaenia (193) 37:26
Common Tody-Flycatcher (197) 37:7
Yellow-olive Flycatcher (195) 37:16
Tropical Pewee (203) 36:9
Bright-rumped Attila (201) 35:6
Great Crested Flycatcher (209) 35:17
Great Kiskadee (211) 35:13
Boat-billed Flycatcher (211) 35:12
Social Flycatcher (211) 35:14
Gray-capped Flycatcher (211) 35:15
Tropical Kingbird (213) 35:1
Cinnamon Becard (215) 33:11
Masked Tityra (217) 34:1
Yellow-throated Vireo (225) 40:6
Lesser Greenlet (229) 40:7
Gray-breasted Martin (235) 22:15
Mangrove Swallow (233) 22:23
Southern Rough-winged Swallow (235) 22:19
Barn Swallow (237) 22:12
Bay Wren (241) 38:12

Spot-breasted Wren (239) 38:7
House Wren (243) 38:18
Clay-colored Thrush (251) 39:8
Tennessee Warbler (255) 40:22
Yellow Warbler (259) 42:2
Chestnut-sided Warbler (259) 43:4
Prothonotary Warbler (267) 42:1
Northern Waterthrush (269) 43:14
Olive-crowned Yellowthroat (273) 42:11
Gray-crowned Yellowthroat (273) 42:13
Bananaquit (277) 40:24
Summer Tanager (285) 47:5
Passerini's Tanager (287) 47:4
Blue-gray Tanager (291) 46:15
Golden-hooded Tanager (289) 46:13
Red-legged Honeycreeper (293) 46:2
Blue-black Grassquit (297) 49:7
Variable Seedeater (295) 49:3
White-collared Seedeater (295) 49:2
Black-striped Sparrow (303) 50:14
Grayish Saltator (309) 48:3
Indigo Bunting (313) 48:12
Red-winged Blackbird (315) 44:14
Eastern Meadowlark (315) 50:16
Great-tailed Grackle (317) 44:16
*Nicaraguan Grackle (317) 44:17
Bronzed Cowbird (317) 44:15
Black-cowled Oriole (319) 44:5
Orchard Oriole (319) 44:6
Baltimore Oriole (321) 44:7
Montezuma Oropendola (323) 44:8
***Yellow-crowned Euphonia (327) 45:1**

Site 1A-2: Laguna del Lagarto Lodge

Birding Time: 2 days
Elevation: 50 meters
Trail Difficulty: 2–3
Reserve Hours: N/A
Entrance Fee: Included with overnight
4×4 Recommended
Trail Map Available on Site

Laguna del Lagarto Lodge is a small lodge set on 250 acres of primary forest in northern Costa Rica, just 15 kilometers from the Nicaraguan border. The area is extremely remote, requiring nearly two hours of travel along a dirt road. Once you get there, though, you will find the accommodations reasonably comfortable. The lodge has 20 simply furnished rooms, decks to relax on, and an open-air dining area, from which you can watch parrots and toucans feeding at the fruit feeders. Fifteen kilometers of trails wind through the forest behind the lodge. While these trails are relatively flat, they are often muddy, so be sure to have good boots. The lagoons, which are home to varied wildlife including turtles, caiman, and many species of birds, are the most unusual aspect of Laguna del Lagarto. Some of these lagoons are accessible by canoe, and it is quite easy to lose all sense of the civilized world as you float silently through the forest.

The Great Green Macaw, which can often be found around the hotel grounds, is the principal attraction for birders. While the general volume and diversity of birds may not be as high as at other sites, Laguna del Lagarto is one of the best places to find a number of unusual species, including Agami Heron, Gray-headed

Piprites, and American Pygmy Kingfisher. It is also a nice place to get easy looks at many of the larger, more colorful tropical birds such as kingfishers, toucans, and parrots. For this reason, Laguna del Lagarto can make an excellent destination for less experienced tropical birders, or for birders traveling with non-birding companions.

Target Birds

Tawny-faced Quail (13) 12:15
Agami Heron (21) 5:1
Green Ibis (29) 4:10
King Vulture (31) 13:5
Semiplumbeous Hawk (41) 16:10
Sungrebe (63) 7:9
Olive-backed Quail-Dove (93) 18:22
Great Green Macaw (99) 19:2
Scarlet Macaw (99) 19:1
Vermiculated Screech-Owl (107) 20:12
Central American Pygmy-Owl (109) 20:16
Common Potoo (115) 20:4
American Pygmy Kingfisher (149) 27:6
Pied Puffbird (151) 28:7
Scaly-throated Leaftosser (167) 30:9
Plain-brown Woodcreeper (171) 29:11
Long-tailed Woodcreeper (171) 29:10
Black-striped Woodcreeper (173) 29:16
*Streak-crowned Antvireo (181) 32:7
Checker-throated Antwren (183) 32:5
White-flanked Antwren (183) 32:2
Bare-crowned Antbird (179) 31:13
Black-capped Pygmy-Tyrant (197) 37:5
Royal Flycatcher (201) 35:23
White-ringed Flycatcher (211) 35:16
*Gray-headed Piprites (215) 33:3
Speckled Mourner (215) 34:8
*Snowy Cotinga (219) 34:4
Tawny-faced Gnatwren (237) 32:14
*Nicaraguan Seed-Finch (295) 49:13

Access

From Pital: Take the main road through Pital until it turns to dirt, by a gas station. (It is probably a good idea to fill up here.) From the gas station proceed down the dirt road for 2.0 km to a fork and bear right. Continue for 8.4 km and bear right again at another intersection. In 1.6 km you will come to an intersection with another gas station. Stay straight, and in 6.8 km, bear right at another fork. Proceed 17.0 km, passing through the small town of Boca Tapada, and turn right on a dirt road signed for Laguna del Lagarto. You will arrive at the lodge in 1.4 km. (Driving time from Pital: 105 min)

Logistics: To bird this area you must stay at the lodge. Here they serve three meals a day, but it is a good idea to bring in bottled water because drinking water is not provided free of charge. If you don't have a car, Laguna del Lagarto offers a pick-up and drop-off service for an added fee.

Laguna del Lagarto Lodge
Tel: (506) 2289 8163
Website: www.lagarto-lodge-costa-rica.com
E-mail: lagarto@racsa.co.cr

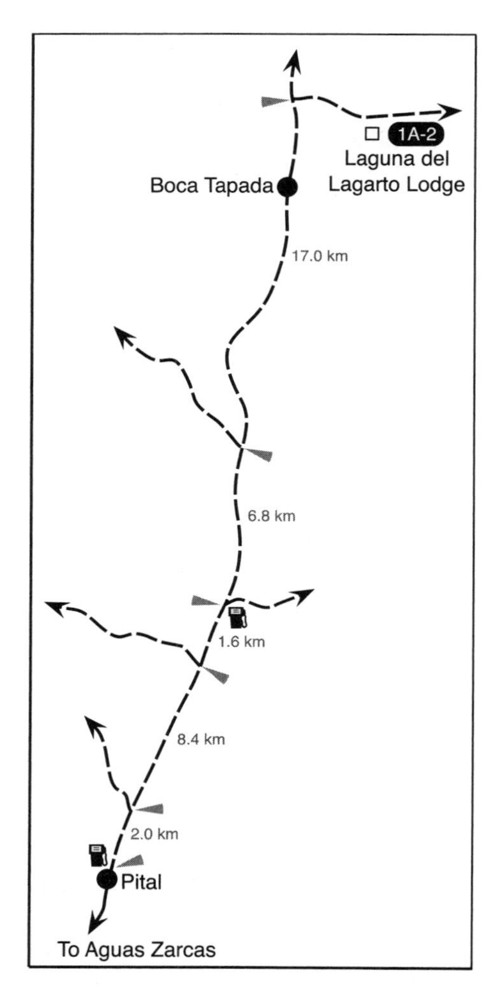

Laguna del Lagarto Lodge Driving Map

Birding Sites

Birding the grounds around the lodge is a good place to start. Many trees and shrubs have been planted in the area, including a small garden in the back, which always seems to be teaming with parrots, parakeets, and macaws. Keep an eye open for Brown-hooded Parrot, Olive-throated Parakeet, Red-lored Parrot, and Mealy Parrot. Both Scarlet Macaw and Great Green Macaw are often seen perching in the large wild almond trees along the forest edge. The fruit feeders visible from the restaurant are a good place to observe toucans. In the small garden, look for Slaty-tailed Trogon, Golden-hooded Tanager, Tropical Pewee, and Streak-headed Woodcreeper. At dawn this is a good place to hear Central American Pygmy-Owl. Common Pauraque and Mottled Owl are regulars at night.

At the edge of the forest behind the lodge is a small compost pile. This is worth checking for Gray-chested Dove, Olive-backed Quail-Dove, and Gray-necked Wood-Rail.

There are two lagoons near the lodge, one on either side. One of these can be accessed by canoe, while the other has a trail leading around its edge. These are good places to look for Ringed Kingfisher, American Pygmy Kingfisher, Agami Heron, and Sungrebe. Another lagoon located just up the road can also be canoed and offers many of the species just mentioned, as well as Mangrove Swallow, Green Ibis, and Least Grebe.

The trails that lead through the forest behind Laguna del Lagarto Lodge are a good place to find a number of interesting birds. Regulars include Scarlet-rumped Cacique, Black-throated Trogon, Black-striped Woodcreeper, Plain-brown Woodcreeper, and Western Slaty-Antshrike. Others, such as Gray-headed Piprites and Royal Flycatcher, are possibilities. Sendero La Danta is an especially good trail from which to encounter small mixed flocks of Tawny-crowned Greenlet, Checker-throated Antwren, White-flanked Antwren, and Streak-crowned Antvireo. Keep an eye on the trail and you may see the large, three-toed, footprint of the *danta* (tapir).

Finally, it is often productive to spend some time birding the entrance road to the lodge. This accesses some new-growth and forest-edge habitat where one can often find Cinnamon Becard, Red-capped Manakin, Bay Wren, Shining Honeycreeper, and White-shouldered Tanager.

Species to Expect

Great Tinamou (3) 12:6
Slaty-breasted Tinamou (3) 12:8
Least Grebe (9) 7:4
Anhinga (17) 4:3
Great Blue Heron (23) 5:6
Great Egret (23) 5:14
Snowy Egret (23) 5:10
Little Blue Heron (23) 5:9
Cattle Egret (23) 5:13
Green Heron (25) 6:2
Agami Heron (21) 5:1
Green Ibis (29) 4:10
Wood Stork (19) 4:6
Black Vulture (31) 13:4
Turkey Vulture (31) 13:3
King Vulture (31) 13:5
Double-toothed Kite (35) 16:1
White Hawk (41) 17:2
Gray Hawk (43) 16:14
Short-tailed Hawk (43) 16:11
Collared Forest-Falcon (55) 16:7
Crested Caracara (53) 14:12
Laughing Falcon (53) 15:8
Bat Falcon (51) 15:14
White-throated Crake (57) 6:9
Gray-necked Wood-Rail (55) 6:13
Pale-vented Pigeon (87) 18:3
Short-billed Pigeon (87) 18:5
Ruddy Ground-Dove (89) 18:7
White-tipped Dove (91) 18:14
Gray-chested Dove (91) 18:16
Olive-backed Quail-Dove (93) 18:22
*Crimson-fronted Parakeet (95) 19:10
Olive-throated Parakeet (95) 19:11
Great Green Macaw (99) 19:2
Scarlet Macaw (99) 19:1
Orange-chinned Parakeet (97) 19:14
Brown-hooded Parrot (97) 19:9
White-crowned Parrot (97) 19:7
Red-lored Parrot (99) 19:4
Mealy Parrot (99) 19:3
Squirrel Cuckoo (103) 21:7

Groove-billed Ani (103) 21:9
Vermiculated Screech-Owl (107) 20:12
Central American Pygmy-Owl (109) 20:16
Mottled Owl (107) 20:6
Common Pauraque (111) 21:18
Gray-rumped Swift (119) 22:10
Long-billed Hermit (121) 23:2
Stripe-throated Hermit (121) 23:1
White-necked Jacobin (125) 23:17
Violet-crowned Woodnymph (127) 24:4
Blue-chested Hummingbird (125) 24:14
Rufous-tailed Hummingbird (129) 24:10
Black-throated Trogon (141) 26:9
Slaty-tailed Trogon (145) 26:2
Broad-billed Motmot (147) 27:9
Ringed Kingfisher (149) 27:1
Amazon Kingfisher (149) 27:4
Green Kingfisher (149) 27:5
American Pygmy Kingfisher (149) 27:6
White-whiskered Puffbird (151) 28:6
Collared Aracari (155) 27:15
Keel-billed Toucan (155) 27:18
Chestnut-mandibled Toucan (155) 27:19
Black-cheeked Woodpecker (157) 28:14
Plain Xenops (169) 29:3
Plain-brown Woodcreeper (171) 29:11
Wedge-billed Woodcreeper (171) 29:6
Northern Barred-Woodcreeper (169) 29:19
Cocoa Woodcreeper (173) 29:17
Black-striped Woodcreeper (173) 29:16
Streak-headed Woodcreeper (173) 29:8
Western Slaty-Antshrike (177) 31:5
*Streak-crowned Antvireo (181) 32:7
Checker-throated Antwren (183) 32:5
White-flanked Antwren (183) 32:2
Chestnut-backed Antbird (179) 31:11
Ocellated Antbird (179) 31:7
Black-faced Antthrush (187) 30:15
Yellow-bellied Elaenia (193) 37:26
Ochre-bellied Flycatcher (199) 36:25
Paltry Tyrannulet (191) 37:10
Black-capped Pygmy-Tyrant (197) 37:5
Yellow-olive Flycatcher (195) 37:16
Royal Flycatcher (201) 35:23
Tropical Pewee (203) 36:9
Yellow-bellied Flycatcher (205) 36:20
Long-tailed Tyrant (201) 36:6
Bright-rumped Attila (201) 35:6
Rufous Mourner (201) 34:9
Great Crested Flycatcher (209) 35:17
Great Kiskadee (211) 35:13
Boat-billed Flycatcher (211) 35:12
Social Flycatcher (211) 35:14
Gray-capped Flycatcher (211) 35:15
Tropical Kingbird (213) 35:1
*Gray-headed Piprites (215) 33:3
Cinnamon Becard (215) 33:11
Masked Tityra (217) 34:1
White-collared Manakin (221) 33:1
Red-capped Manakin (223) 33:6
Yellow-throated Vireo (225) 40:6
Philadelphia Vireo (227) 40:15
Tawny-crowned Greenlet (229) 32:4
Lesser Greenlet (229) 40:7
Green Shrike-Vireo (229) 40:1
Gray-breasted Martin (235) 22:15
Mangrove Swallow (233) 22:23
Northern Rough-winged Swallow (235) 22:18
Southern Rough-winged Swallow (235) 22:19
Bay Wren (241) 38:12
*Stripe-breasted Wren (241) 38:5
House Wren (243) 38:18
White-breasted Wood-Wren (245) 38:15
Long-billed Gnatwren (237) 32:15
Tropical Gnatcatcher (237) 41:2
Wood Thrush (249) 39:4
Clay-colored Thrush (251) 39:8
Golden-winged Warbler (257) 41:5
Tennessee Warbler (255) 40:22
Yellow Warbler (259) 42:2
Chestnut-sided Warbler (259) 43:4
Black-and-white Warbler (267) 41:13
Gray-crowned Yellowthroat (273) 42:13
Bananaquit (277) 40:24
White-shouldered Tanager (281) 46:18
Summer Tanager (285) 47:5
Passerini's Tanager (287) 47:4
Blue-gray Tanager (291) 46:15
Palm Tanager (291) 45:19
Golden-hooded Tanager (289) 46:13
Green Honeycreeper (293) 46:7
Shining Honeycreeper (293) 46:1
Blue-black Grassquit (297) 49:7
Variable Seedeater (295) 49:3
Black-striped Sparrow (303) 50:14

Buff-throated Saltator (309) 48:2
Blue-black Grosbeak (311) 48:10
Red-winged Blackbird (315) 44:14
Great-tailed Grackle (317) 44:16
Bronzed Cowbird (317) 44:15
Baltimore Oriole (321) 44:7
Scarlet-rumped Cacique (321) 44:11
Chestnut-headed Oropendola (323) 44:9
Montezuma Oropendola (323) 44:8
Olive-backed Euphonia (325) 45:3

Site 1A-3: Tortuguero National Park

Birding Time: 3 days
Elevation: Sea level
Trail Difficulty: Most birding by boat
Reserve Hours: 6 a.m. to 4 p.m.
Entrance Fee: $10 (US 2008)

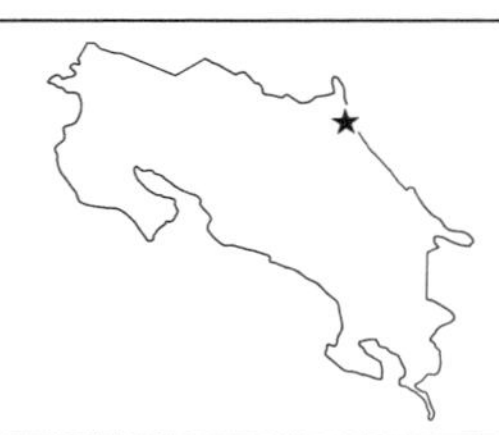

Tortuguero National Park is one of Costa Rica's most popular and unique ecotourist destinations, seeing roughly 100,000 visitors per year. The small town of Tortuguero, home to only about 700 permanent residents, sits on a narrow strip of land, bordered on one side by the Tortuguero Lagoon and on the other by the Caribbean Sea. There are no roads in the area, and other than a small airstrip, the only means of transportation available to both tourists and locals is boat. An extensive series of natural canals winds through the forest of the national park and offers convenient and comfortable nature viewing. Many upscale rainforest lodges have sprung up along the banks of the river, offering tourists the enticing combination of comfort and adventure.

While the national park was not created until 1970, conservation efforts at Tortuguero date back to the 1950s, when a researcher named Archie Carr identified the beach as an extremely important nesting colony for sea turtles. Archie Carr and others after him have worked tirelessly to protect the turtles, which had been hunted by the locals for their meat, eggs, and shells. Today, Tortuguero National Park is known to be the largest nesting colony of Green Sea Turtles in the Western Hemisphere, and Leatherback, Hawksbill, and Loggerhead turtles also lay their eggs in her sands. The nesting season runs from mid-July through mid-October and is the best time to visit if you want to see these massive creatures of the sea.

Tortuguero National Park protects over 185,000 acres, although only a third of this is terrestrial. Most of the forest inside the national park is second growth, having been selectively logged before the park was created. Nonetheless, it is one of the best areas in the country for observing animals. Monkeys, sloth, and caiman are regularly encountered, and rarities, such as tapir and manatee, are also possibilities. Even jaguars are occasionally seen swimming across the canals or prowling along the beach in search of turtle meat.

Serious birders may find Tortuguero restricting. Birding by boat puts you at the mercy of the driver and makes many of the small interior-forest birds difficult to access. Furthermore, Tortuguero is remote, and a lot of time is lost traveling into and out of the area. However, it is a great place to find a few difficult species, including Green-and-rufous Kingfisher and Rufescent Tiger-Heron. And for birders who are less mobile or simply looking for a more relaxing vacation, the boat rides through Tortuguero offer excellent looks at many water-based species and other wildlife.

Target Birds

Great Curassow (11) 12:3
Rufescent Tiger-Heron (21) 5:17
Agami Heron (21) 5:1
Green Ibis (29) 4:10
Hook-billed Kite (35) 16:9
Black-collared Hawk (33) 17:5
Crane Hawk (41) 14:4
Common Black-Hawk (45) 13:6
Crested Eagle (47) 17:10
Slaty-backed Forest-Falcon (55) 16:6
Sungrebe (63) 7:9
White-crowned Pigeon (86) 51:1
Great Green Macaw (99) 19:2
Yellow-billed Cuckoo (101) 21:2
Short-tailed Nighthawk (113) 21:14
Great Potoo (115) 20:5
Green-and-rufous Kingfisher (149) 27:3
American Pygmy Kingfisher (149) 27:6
White-fronted Nunbird (151) 28:4
Western Slaty-Antshrike (177) 31:5
Brown-capped Tyrannulet (189) 37:13
Gray Kingbird (213) 35:7
Purple-throated Fruitcrow (219) 34:11
Hooded Warbler (269) 42:5

Access

While I generally encourage renting a car during a birding trip to Costa Rica, this site may be easier and cheaper to visit without one. Most of the upscale lodges will arrange transportation into and out of Tortuguero (from as far away as San José), and you should ask about this when you make your reservation. If you are staying at one of the less expensive hotels, you will most likely enter and leave via a public river taxi from the Pavona Boat Landing. The landing is accessible by public bus from the nearby town of Cariari, and it also offers a guarded parking lot where you can leave a car for a small fee. The directions below describe how to get to the Pavona Landing.

From the intersection of Rt. 32 and Rt. 4: Take Rt. 32 south toward Puerto Limón for 13.2 km and then turn left at a gas station. (**From Siquirres:** Take Rt. 32 north toward Guápiles for 33.8 km and then turn right at a gas station.) Go straight for 8.2 km and then turn left just before some old train tracks. Follow parallel to the train tracks for 0.6 km and then turn right. Proceed 17.8 km, passing through the town of Cariari, to a T-intersection. Turn right, proceed 1.2 km, and then turn left at a soccer field. Take this road for 15.0 km to a fork, although it will turn to dirt midway. Take the left fork, and in 5.4 km you will arrive at the Pavona Boat Landing on your left. (Driving time from Rt. 4: 80 min; from Siquirres: 110 min)

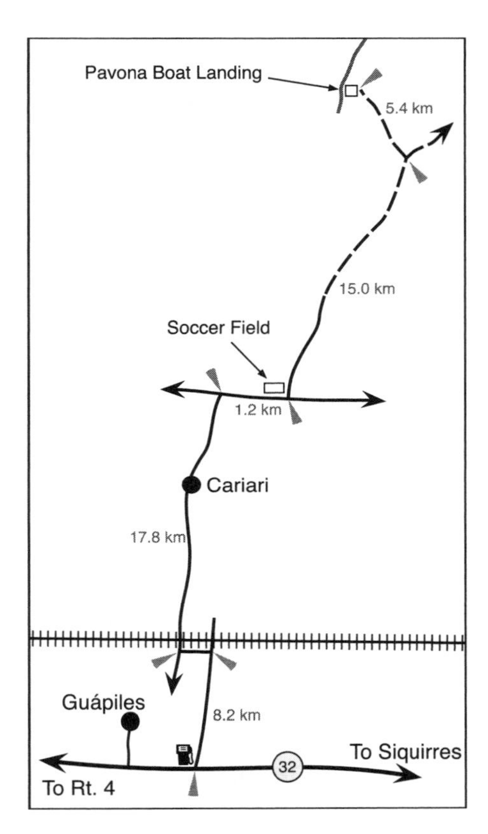

Tortuguero National Park Driving Map. The route to the Pavona Boat Landing is depicted.

Logistics: I suggest looking into the higher-quality hotels set along the banks of the river outside of town. Most of these offer packages that include meals, transportation, and guided boat tours. Downtown Tortuguero offers cheaper lodging options, although it is not as picturesque.

Birding Sites

Most of the birding at Tortuguero National Park is by boat, exploring the maze of waterways that runs through the forest. Depending on where you stay, different boat options are available. Many lodges offer 15-person boats, equipped with a quiet four-stroke engine, driver, and natural history guide. It is also possible to rent a canoe or kayak and explore the area under your own power.

The larger waterways, like the Tortuguero Lagoon and the Tortuguero River, hold a wealth of wildlife along their banks. These are especially good places to spot monkeys and sloth. Keel-billed Toucan, Chestnut-mandibled Toucan, Collared Aracari, and Mealy Parrot often perch high in the treetops. You may also find Green Ibis and Slaty-tailed Trogon. Floating grasses along the banks of the river hold Northern Jacana and White-throated Crake. Hawks such as Osprey and Common Black-Hawk are also good possibilities here.

The smaller tributary canals offer good birding and usually produce forest-based species along with herons and kingfishers. Caño Herold, Caño Chiquero, and Caño Las Palmas are some of the best areas for birds. Here you might be able to find Sungrebe, Green-and-rufous Kingfisher, American Pygmy Kingfisher, and if you are lucky, Rufescent Tiger-Heron or Agami Heron feeding along the still waters. Red-capped Manakin, Western Slaty-Antshrike, and Bay Wren often call from within the forest, while Scarlet-rumped Cacique and Cinnamon Woodpecker can be seen in the trees overhead.

If you tire of the confines of birding by boat, there are a few trails worth exploring. The most productive area is called Cerro Tortuguero (Tortuguero Hill). Since it is accessible only by boat, you will need to arrange for drop-off and pick-up, planning on about three hours to bird it thoroughly. At the base of the hill is a small community, and you should spend some time checking the fields and new growth around the settlement, where you might find Passerini's Tanager, White-collared Manakin, Plain Wren, and Black-cowled Oriole. From this area you can also scan the river mouth of the Tortuguero Lagoon, which often yields Royal Tern, Laughing Gull, and a few shorebirds. The hill itself is forested and has some flat muddy trails around its base and one trail that leads steeply to its summit. The latter offers great views of the surrounding countryside, although the hike is challenging. All of these trails offer good opportunities to find forest species, including Black-throated Trogon, Brown-capped Tyrannulet, White-fronted Nunbird, and Purple-throated Fruitcrow.

Another walking option is located at the national park headquarters at the south side of the town of Tortuguero. This trail, called Sendero El Gavilán, is a loop that takes about an hour to bird. It proceeds mostly through forest, although part of the trail is along the beach.

Species to Expect

Great Curassow (11) 12:3
Neotropic Cormorant (17) 4:4
Anhinga (17) 4:3
Magnificent Frigatebird (19) 1:6
Bare-throated Tiger-Heron (21) 5:16
Great Blue Heron (23) 5:6
Great Egret (23) 5:14
Snowy Egret (23) 5:10
Little Blue Heron (23) 5:9
Tricolored Heron (25) 5:8
Green Heron (25) 6:2
Yellow-crowned Night-Heron (27) 5:3
Green Ibis (29) 4:10
Roseate Spoonbill (27) 4:7
Black Vulture (31) 13:4
Turkey Vulture (31) 13:3
Osprey (33) 17:14
Gray-headed Kite (33) 17:3
Swallow-tailed Kite (37) 15:2
Common Black-Hawk (45) 13:6
Broad-winged Hawk (43) 16:13
Laughing Falcon (53) 15:8
White-throated Crake (57) 6:9
Black-bellied Plover (67) 9:1
Collared Plover (65) 10:3
Killdeer (65) 10:1

Northern Jacana (61) 6:18
Spotted Sandpiper (73) 11:8
Whimbrel (71) 9:14
Sanderling (75) 11:11
Laughing Gull (81) 3:8
Royal Tern (81) 3:2
Pale-vented Pigeon (87) 18:3
Ruddy Ground-Dove (89) 18:7
Gray-chested Dove (91) 18:16
Olive-throated Parakeet (95) 19:11
Great Green Macaw (99) 19:2
Brown-hooded Parrot (97) 19:9
White-crowned Parrot (97) 19:7
Red-lored Parrot (99) 19:4
Mealy Parrot (99) 19:3
Squirrel Cuckoo (103) 21:7
Groove-billed Ani (103) 21:9
Gray-rumped Swift (119) 22:10
Long-billed Hermit (121) 23:2
Stripe-throated Hermit (121) 23:1
Rufous-tailed Hummingbird (129) 24:10
Black-throated Trogon (141) 26:9
Slaty-tailed Trogon (145) 26:2
Ringed Kingfisher (149) 27:1
Belted Kingfisher (149) 27:2
Amazon Kingfisher (149) 27:4
Green Kingfisher (149) 27:5
White-necked Puffbird (151) 28:3
White-fronted Nunbird (151) 28:4
Collared Aracari (155) 27:15
Keel-billed Toucan (155) 27:18
Chestnut-mandibled Toucan (155) 27:19
Black-cheeked Woodpecker (157) 28:14
Cinnamon Woodpecker (161) 28:9
Pale-billed Woodpecker (161) 27:14
Slaty Spinetail (163) 32:9
Northern Barred-Woodcreeper (169) 29:19
Streak-headed Woodcreeper (173) 29:8
Western Slaty-Antshrike (177) 31:5
Chestnut-backed Antbird (179) 31:11
Brown-capped Tyrannulet (189) 37:13
Yellow-bellied Elaenia (193) 37:26
Paltry Tyrannulet (191) 37:10
Common Tody-Flycatcher (197) 37:7
Western Wood-Pewee (203) 36:7
Bright-rumped Attila (201) 35:6
Great Crested Flycatcher (209) 35:17
Great Kiskadee (211) 35:13
Boat-billed Flycatcher (211) 35:12
Social Flycatcher (211) 35:14
Tropical Kingbird (213) 35:1
Cinnamon Becard (215) 33:11
Purple-throated Fruitcrow (219) 34:11
White-collared Manakin (221) 33:1
Red-capped Manakin (223) 33:6
Red-eyed Vireo (227) 40:3
Lesser Greenlet (229) 40:7
Gray-breasted Martin (235) 22:15
Mangrove Swallow (233) 22:23
Northern Rough-winged Swallow (235) 22:18
Bay Wren (241) 38:12
***Stripe-breasted Wren (241) 38:5**
Plain Wren (241) 38:17
House Wren (243) 38:18
White-breasted Wood-Wren (245) 38:15
Tropical Gnatcatcher (237) 41:2
Gray-cheeked Thrush (249) 39:1
Swainson's Thrush (249) 39:2
Wood Thrush (249) 39:4
Clay-colored Thrush (251) 39:8
Gray Catbird (253) 39:12
Yellow Warbler (259) 42:2
Chestnut-sided Warbler (259) 43:4
Bay-breasted Warbler (263) 41:9
Common Yellowthroat (273) 42:10
Olive-crowned Yellowthroat (273) 42:11
Bananaquit (277) 40:24
Summer Tanager (285) 47:5
Passerini's Tanager (287) 47:4
Blue-gray Tanager (291) 46:15
Palm Tanager (291) 45:19
Green Honeycreeper (293) 46:7
Shining Honeycreeper (293) 46:1
Blue-black Grassquit (297) 49:7
Variable Seedeater (295) 49:3
Black-striped Sparrow (303) 50:14
Buff-throated Saltator (309) 48:2
Great-tailed Grackle (317) 44:16
Black-cowled Oriole (319) 44:5
Baltimore Oriole (321) 44:7
Scarlet-rumped Cacique (321) 44:11
Montezuma Oropendola (323) 44:8
*Yellow-crowned Euphonia (327) 45:1
Olive-backed Euphonia (325) 45:3
House Sparrow (307) 50:17

Birding Time: 2–3 days
Elevation: 40 meters
Trail Difficulty: 1
Reserve Hours: 7 a.m. to 6 p.m.
Entrance Fee: $28 (US 2008)
Bird Guides Available
Trail Map Available on Site

La Selva Biological Station is one of tropical America's most important ecological research facilities. It is also one of Costa Rica's most renowned birding destinations. The volume and diversity of birds are impressive, access is easy, and many rarities are regularly sighted. It is truly an exciting place to bird. The station sits on 4,050 acres of lowland forest bordering Braulio Carrillo National Park. These two protected areas create an important biological corridor that runs from lowlands to high mountain peaks. The trail system at La Selva is extensive, and while most visitors only see a small portion of the property, many diverse habitats are available. Forests of all ages, from new growth to primary, can be observed from well-maintained trails. It is not difficult to list over 100 species in a few hours of birding. An introductory three-hour guided walk is included with the entrance fee.

Since its creation in 1968, more than 3,100 scientific papers have been published based on research conducted within La Selva's forest. Each year the station hosts around 300 researchers and 100 university-level courses, and you will undoubtedly notice evidence of ongoing research as you walk the grounds. Flagging tape of various colors marks many plants, researchers carrying samples of leaves and insects often pass you on the trails, and small groups of students can be seen listening to lectures in the field. The station is owned by the Organization for Tropical Studies (OTS), a consortium of over 60 universities from across the world, although it is principally managed by Duke University. OTS runs two other research stations in Costa Rica, Palo Verde (2-3) and Las Cruces (3B-3), both of which also offer excellent birding. (If you are interested in tropical biology, OTS offers a range of excellent courses.)

Target Birds

Great Tinamou (3) 12:6
Slaty-breasted Tinamou (3) 12:8
Great Curassow (11) 12:3
Agami Heron (21) 5:1
Green Ibis (29) 4:10
Gray-headed Kite (33) 17:3
Tiny Hawk (39) 16:2
Semiplumbeous Hawk (41) 16:10
Slaty-backed Forest-Falcon (55) 16:6
Uniform Crake (57) 6:8
Sungrebe (63) 7:9
Sunbittern (29) 6:16
Olive-backed Quail-Dove (93) 18:22
Great Green Macaw (99) 19:2
Vermiculated Screech-Owl (107) 20:12
Crested Owl (105) 20:3

Black-and-white Owl (107) 20:7
Great Potoo (115) 20:5
Lesser Swallow-tailed Swift (119) 22:7
Blue-chested Hummingbird (125) 24:14
Bronze-tailed Plumeleteer (129) 23:12
Purple-crowned Fairy (125) 23:14
Rufous Motmot (147) 27:7
Broad-billed Motmot (147) 27:9
Pied Puffbird (151) 28:7
*Rufous-winged Woodpecker (159) 28:10
Chestnut-colored Woodpecker (161) 28:8
Fasciated Antshrike (175) 31:1
Great Antshrike (177) 31:2
Western Slaty-Antshrike (177) 31:5
Bare-crowned Antbird (179) 31:13
Ocellated Antbird (179) 31:7
Black-capped Pygmy-Tyrant (197) 37:5
*Tawny-chested Flycatcher (199) 36:10
White-ringed Flycatcher (211) 35:16
Speckled Mourner (215) 34:8
*Snowy Cotinga (219) 34:4
Purple-throated Fruitcrow (219) 34:11
*Bare-necked Umbrellabird (221) 34:13
Red-capped Manakin (223) 33:6
*Black-throated Wren (243) 38:13
Crimson-collared Tanager (287) 47:3
*Plain-colored Tanager (291) 46:11
Yellow-tailed Oriole (319) 44:3

Access

From Puerto Viejo de Sarapiquí: Take Rt. 4 heading toward Guápiles. After 2.6 km, turn right onto a dirt road and very quickly turn left at the outer guardhouse. You will arrive at La Selva Biological Station in another 0.6 km. (Driving time from Puerto Viejo: 5 min)

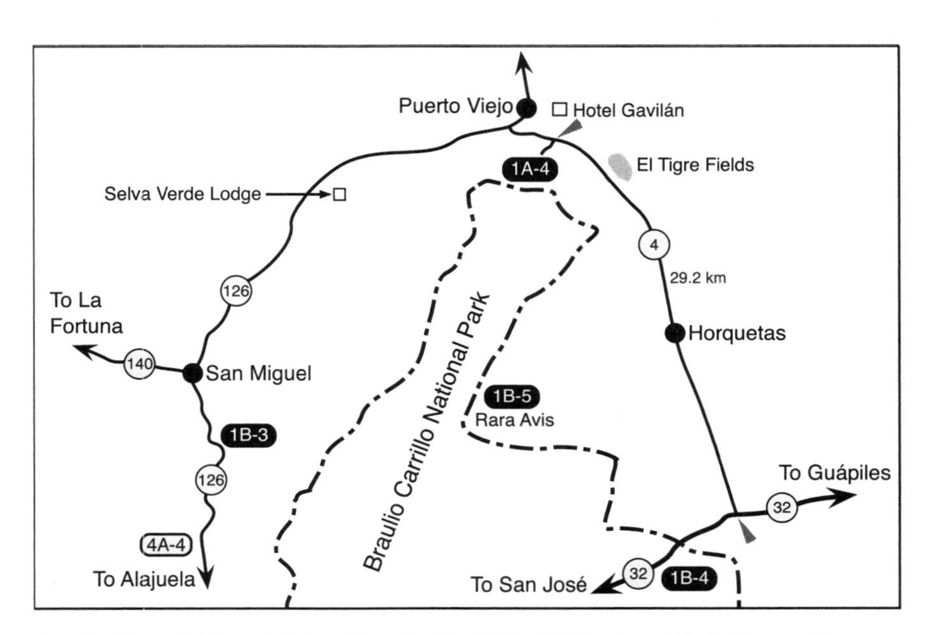

Greater Sarapiqí Area Driving Map. La Paz Waterfall Gardens (4A-4), Virgen del Socorro (1B-3), La Selva Biological Station (1A-4), Rara Avis Rainforest Lodge (1B-5), and Braulio Carrillo National Park (1B-4) are included.

From the intersection of Rt. 32 and Rt. 4: Take Rt. 4 toward Puerto Viejo de Sarapiquí for 29.2 km. Turn left onto a dirt road and very quickly turn left again at the outer guardhouse. You will arrive at La Selva Biological Station in another 0.6 km. (Driving time from Rt. 32: 25 min)

Logistics: There are a few noteworthy lodging options when birding La Selva. For serious birders the best choice is to stay at the station itself. Only overnight guests are allowed unguided access to the extensive trail system on the far side of the river. It also allows you to do some owling at night and to get an early start in the morning. The downside is a relatively high cost for spartan accommodations and average food.

Alternatively, Selva Verde Lodge, about 20 minutes from La Selva, is the most comfortable hotel in the area. Here you will find first-rate accommodations, food, and service, set within a beautiful tropical forest on the banks of the Río Sarapiquí. The lodge also protects a large tract of forest that you can walk with a guide, although the birding is not as good as at La Selva.

Finally, many birders stay at the more reasonably priced Hotel Gavilán. It is set on a small piece of land next to the Río Sarapiquí, with open forest and new-growth habitat, where you will find many disturbed-habitat species.

La Selva Biological Station
Tel: (506) 2524-0628
Website: www.ots.ac.cr
E-mail: nat-hist@ots.ac.cr

Hotel Gavilán
Tel: (506) 2766-6743
Website: www.gavilanlodge.com
E-mail: gavilan@racsa.co.cr

Selva Verde Lodge
Tel: (506) 2766-6800
Website: www.selvaverde.com
E-mail: selvaver@racsa.co.cr

Birding Sites

Upon arriving at La Selva, many people find it hard to move beyond the grounds around the *comedor* (dining room) and reception buildings, where flowering and fruiting plants attract a wide range of birds. Band-backed Wren, Paltry Tyrannulet, Ochre-bellied Flycatcher, and Passerini's Tanager can be easily found around the low shrubs. A small patch of porterweed, planted around the comedor patio, attracts Bananaquit, Violet-headed Hummingbird, and Blue-chested Hummingbird. You may also find Collared Aracari, Keel-billed Toucan, Crested Guan, White-ringed Flycatcher, and Shining Honeycreeper perched in the tops of the surrounding trees. Also, keep an eye open for Snowy Cotinga and Plain-colored Tanager, which are often seen in the area.

My favorite loop to walk at La Selva starts at the comedor and follows the entrance road back out for about 500 meters. The birding along this stretch of road is excellent, and you should look for Blue Ground-Dove, Olive-crowned Yellowthroat, Fasciated Antshrike, Gray-headed Chachalaca, Rufous-winged Woodpecker, Black-cowled Oriole, and Yellow-tailed Oriole. You will come to a gate on your left with "OET" inscribed on the top. Pass through the gate and proceed straight down a short grassy path that connects to the SAZ trail. White-collared Manakin, Rufous-tailed Jacamar, Black-headed Saltator, Yellow Tyrannulet, and the uncommon Bare-crowned Antbird can all be found here. Turning left on SAZ trail will complete the loop and bring you back to the comedor and reception. Birds to look for along this final stretch include Black-throated Wren, Great Antshrike, and White-flanked Antwren.

From the large bridge over the Río Puerto Viejo you may be able to spot Green Kingfisher, Buff-rumped Warbler, Sungrebe, and Anhinga near the water. White-necked Puffbird and Rufous Motmot are frequently seen here, and it is a good place to encounter mixed flocks moving along the riverbank. The planted grounds on the far side of the bridge can also be productive. Look for Olive-backed Euphonia, Cinnamon Becard, Slaty-tailed Trogon, and White-ruffed Manakin. Bronze-tailed Plumeleteer has been reliably found in a heliconia patch next to the researchers' lounge, which is the second building on your left.

On this side of the bridge is the beginning of an extensive trail system, which provides researchers with access to the entire La Selva property. Your entrance fee into La Selva includes a three-hour guided tour, which will give you a chance to experience these trails. However, only overnight guests are allowed individual access to the trails after your tour ends. It would be almost impossible to explore all of these trails on a single visit to La Selva, so I have described the most notable and accessible areas below.

The Arboretum, an open area with tagged trees located a short distance into the forest, is a nice place to bird. On your way there you might encounter Long-billed Hermit, Band-tailed Barbthroat, or Black-capped Pygmy-Tyrant. In the Arboretum itself, look for Green Ibis, Great Curassow, Broad-billed Motmot, Slaty-breasted Tinamou, and Eye-ringed Flatbill. At night you can hear Great Potoo, as well as Central American Pygmy-Owl, which normally calls at first light.

Another nearby area that can be quite interesting is the short Cantarrana trail, where you will find a boardwalk through a small swampy pond. Depending on the water level, you may find Buff-rumped Warbler, Northern Waterthrush, and Green Kingfisher. The rare Agami Heron and Uniform Crake are also possibilities here, and at night it is a good place to try for owls.

If you continue down the SOR trail, you will enter an area of experimental new-growth plots. A different range of species can be found in this habitat. Bronzy Hermit, Yellow-billed Cacique, Northern Bentbill, and Red-throated Ant-Tanager are all possible. Great Tinamou is often seen walking along the trail. From this point you can continue deeper into the forest or return back to the station.

A loop that is quite productive starts on the CES trail, connects to the CEN trail, and then loops back to the station via the STR trail. The CES trail proceeds through a relatively old forest and often produces Western Slaty-Antshrike, Ochre-bellied Flycatcher, and Black-throated Trogon. It is also well known among birders as being the most reliable place at La Selva to find Bare-necked Umbrellabird. The best spot to look for this exciting species is about 800 meters down the CES trail, where it dips into a small ravine, just before the intersection of the LOC trail.

When you arrive at the STR trail, you will notice that the habitat changes slightly. There are some very large old trees along the trail, and while the understory can be quite dense, the area is generally more open and receives more light than the interior forest. Stripe-breasted Wren and White-breasted Wood-Wren are common along the edge, and you may be able to find Slaty-breasted Tinamou, Semiplumbeous Hawk, Purple-throated Fruitcrow, and Scarlet-rumped Cacique.

While army ants can be encountered anywhere on the La Selva property, the STR trail offers an especially good chance of finding a swarm. At La Selva, some of the most regular army ant followers include Ocellated Antbird, Northern Barred-Woodcreeper, Plain-brown Woodcreeper, Bicolored Antbird, and Great Tinamou.

A number of trails lead deep into the La Selva forest, although their condition deteriorates as you get farther away from the station. Here you can find true primary forest, with enormous trees and a very sparse understory. The birdlife in the primary forest can be patchy. While it is possible to find interesting species such as Sungrebe, Red-capped Manakin, Olive-backed Quail-Dove, White-tipped Sicklebill, and Bare-necked Umbrellabird, it is not uncommon to walk for long stretches of time without hearing or seeing a single bird. If you decide to explore this area, be sure not to get lost; bring a map, good boots, and an emergency flashlight.

La Selva is a great area for night birds. Vermiculated Screech-Owl and Black-and-white Owl can often be heard calling near the large bridge. Farther into the forest you have chances at detecting Spectacled Owl, Crested Owl, and Great Potoo. At dusk or dawn the silhouette of a Short-tailed Nighthawk can sometimes be seen flying overhead.

Species to Expect

Great Tinamou (3) 12:6
Little Tinamou (3) 12:17
Slaty-breasted Tinamou (3) 12:8
Gray-headed Chachalaca (11) 12:1
Crested Guan (11) 12:4
Great Curassow (11) 12:3
Cattle Egret (23) 5:13
Green Ibis (29) 4:10
Black Vulture (31) 13:4
Turkey Vulture (31) 13:3
King Vulture (31) 13:5
Osprey (33) 17:14
Gray-headed Kite (33) 17:3
Double-toothed Kite (35) 16:1
Semiplumbeous Hawk (41) 16:10
Broad-winged Hawk (43) 16:13

Gray Hawk (43) 16:14
Short-tailed Hawk (43) 16:11
Laughing Falcon (53) 15:8
White-throated Crake (57) 6:9
Sunbittern (29) 6:16
Spotted Sandpiper (73) 11:8
Pale-vented Pigeon (87) 18:3
Short-billed Pigeon (87) 18:5
Ruddy Ground-Dove (89) 18:7
Blue Ground-Dove (91) 18:6
White-tipped Dove (91) 18:14
Gray-chested Dove (91) 18:16
*Crimson-fronted Parakeet (95) 19:10
Olive-throated Parakeet (95) 19:11
Great Green Macaw (99) 19:2
Orange-chinned Parakeet (97) 19:14
Brown-hooded Parrot (97) 19:9
White-crowned Parrot (97) 19:7
Red-lored Parrot (99) 19:4
Mealy Parrot (99) 19:3
Squirrel Cuckoo (103) 21:7
Groove-billed Ani (103) 21:9
Vermiculated Screech-Owl (107) 20:12
Crested Owl (105) 20:3
Black-and-white Owl (107) 20:7
Short-tailed Nighthawk (113) 21:14
Common Pauraque (111) 21:18
Great Potoo (115) 20:5
Chimney Swift (119) 22:8
Gray-rumped Swift (119) 22:10
Bronzy Hermit (121) 23:4
Long-billed Hermit (121) 23:2
Stripe-throated Hermit (121) 23:1
White-necked Jacobin (125) 23:17
Violet-headed Hummingbird (137) 25:11
Violet-crowned Woodnymph (127) 24:4
Blue-chested Hummingbird (125) 24:14
Rufous-tailed Hummingbird (129) 24:10
Bronze-tailed Plumeleteer (129) 23:12
Purple-crowned Fairy (125) 23:14
Violaceous Trogon (141) 26:8
Black-throated Trogon (141) 26:9
Slaty-tailed Trogon (145) 26:2
Rufous Motmot (147) 27:7
Broad-billed Motmot (147) 27:9
Green Kingfisher (149) 27:5
White-necked Puffbird (151) 28:3
Rufous-tailed Jacamar (153) 26:12
Collared Aracari (155) 27:15
Keel-billed Toucan (155) 27:18
Chestnut-mandibled Toucan (155) 27:19
Black-cheeked Woodpecker (157) 28:14
Smoky-brown Woodpecker (159) 28:13
*Rufous-winged Woodpecker (159) 28:10
Cinnamon Woodpecker (161) 28:9
Chestnut-colored Woodpecker (161) 28:8
Lineated Woodpecker (161) 27:13
Pale-billed Woodpecker (161) 27:14
Plain Xenops (169) 29:3
Plain-brown Woodcreeper (171) 29:11
Wedge-billed Woodcreeper (171) 29:6
Northern Barred-Woodcreeper (169) 29:19
Cocoa Woodcreeper (173) 29:17
Streak-headed Woodcreeper (173) 29:8
Fasciated Antshrike (175) 31:1
Great Antshrike (177) 31:2
Barred Antshrike (175) 31:3
Western Slaty-Antshrike (177) 31:5
White-flanked Antwren (183) 32:2
Dot-winged Antwren (183) 32:1
Chestnut-backed Antbird (179) 31:11
Ocellated Antbird (179) 31:7
Yellow Tyrannulet (189) 37:11
Yellow-bellied Elaenia (193) 37:26
Ochre-bellied Flycatcher (199) 36:25
Paltry Tyrannulet (191) 37:10
Black-capped Pygmy-Tyrant (197) 37:5
Common Tody-Flycatcher (197) 37:7
Eye-ringed Flatbill (201) 37:23
Yellow-olive Flycatcher (195) 37:16
Yellow-margined Flycatcher (195) 37:17
Eastern Wood-Pewee (203) 36:8
Tropical Pewee (203) 36:9
Bright-rumped Attila (201) 35:6
Rufous Mourner (201) 34:9
Dusky-capped Flycatcher (209) 35:21
Great Crested Flycatcher (209) 35:17
Great Kiskadee (211) 35:13
Boat-billed Flycatcher (211) 35:12
Social Flycatcher (211) 35:14
Gray-capped Flycatcher (211) 35:15
White-ringed Flycatcher (211) 35:16
Sulphur-bellied Flycatcher (213) 35:10
Piratic Flycatcher (193) 35:8
Tropical Kingbird (213) 35:1
Cinnamon Becard (215) 33:11
White-winged Becard (217) 33:13
Masked Tityra (217) 34:1

Black-crowned Tityra (217) 34:2
*Snowy Cotinga (219) 34:4
White-collared Manakin (221) 33:1
Red-eyed Vireo (227) 40:3
Lesser Greenlet (229) 40:7
Mangrove Swallow (233) 22:23
Southern Rough-winged Swallow (235) 22:19
Band-backed Wren (239) 38:3
***Black-throated Wren (243) 38:13**
Bay Wren (241) 38:12
***Stripe-breasted Wren (241) 38:5**
Plain Wren (241) 38:17
House Wren (243) 38:18
White-breasted Wood-Wren (245) 38:15
Long-billed Gnatwren (237) 32:15
Tropical Gnatcatcher (237) 41:2
Wood Thrush (249) 39:4
Clay-colored Thrush (251) 39:8
White-throated Thrush (251) 39:9
Golden-winged Warbler (257) 41:5
Tennessee Warbler (255) 40:22
Yellow Warbler (259) 42:2
Chestnut-sided Warbler (259) 43:4
American Redstart (267) 41:10
Olive-crowned Yellowthroat (273) 42:11
Buff-rumped Warbler (271) 40:23
Bananaquit (277) 40:24
Red-throated Ant-Tanager (277) 47:12
Summer Tanager (285) 47:5
Scarlet Tanager (287) 47:8
Crimson-collared Tanager (287) 47:3
Passerini's Tanager (287) 47:4
Blue-gray Tanager (291) 46:15
Palm Tanager (291) 45:19
*Plain-colored Tanager (291) 46:11
Silver-throated Tanager (291) 46:6
Golden-hooded Tanager (289) 46:13
Blue Dacnis (291) 46:3
Green Honeycreeper (293) 46:7
Shining Honeycreeper (293) 46:1
Red-legged Honeycreeper (293) 46:2
Blue-black Grassquit (297) 49:7
Variable Seedeater (295) 49:3
Thick-billed Seed-Finch (295) 49:10
Orange-billed Sparrow (303) 48:17
Black-striped Sparrow (303) 50:14
Buff-throated Saltator (309) 48:2
Black-headed Saltator (309) 48:1
Black-faced Grosbeak (311) 48:5
Blue-black Grosbeak (311) 48:10
Black-cowled Oriole (319) 44:5
Yellow-tailed Oriole (319) 44:3
Baltimore Oriole (321) 44:7
Scarlet-rumped Cacique (321) 44:11
Chestnut-headed Oropendola (323) 44:9
Montezuma Oropendola (323) 44:8
***Yellow-crowned Euphonia (327) 45:1**
Olive-backed Euphonia (325) 45:3

Nearby Birding Opportunities

El Tigre Fields

El Tigre Fields, only five minutes from La Selva, are wet overgrown cow pastures that harbor a number of difficult-to-find species, including Red-breasted Blackbird, Nicaraguan Seed-Finch, Pinnated Bittern, Least Bittern, Gray-breasted Crake, and Green Ibis. More common species that you should expect to see include Roadside Hawk, Red-winged Blackbird, Purple Gallinule, Gray-crowned Yellowthroat, Thick-billed Seed-Finch, and Pale-vented Pigeon.

To get to El Tigre Fields from La Selva, turn right onto Rt. 4 and proceed straight for 4.6 km toward Guápiles. At this point a small dirt road on your left leads through a strange cement arch to where you should park your car. If you see anyone around, ask permission to walk the dirt road into the fields. (There has been talk of turning these fields into a pineapple plantation, so if you arrive and find nothing but pineapples, you will know of the unfortunate demise of a very nice habitat.)

Birding Time: 1–2 days
Elevation: 50 meters
Trail Difficulty: 2
Reserve Hours: N/A
Entrance Fee: Included with overnight
4×4 Recommended

Selva Bananito Lodge is a true ecotourist destination. Created in 1994 as an alternative to logging the forest that it now protects, the lodge was constructed primarily of waste wood from nearby logging operations. The Stein family originally bought the land in 1974 for farming purposes. After finding little success with a number of crops, they turned to ecotourism as a means of supporting themselves and protecting their land. Today, two-thirds of the property (2,100 acres) is a private biological reserve, while the remainder is still farmed for wood, bananas, and cattle.

Visitors at Selva Bananito Lodge are met with a pleasant combination of comfort and seclusion. A 40-minute drive along dirt roads leaves you at the edge of the Talamanca wilderness, far from the sights and sounds of civilization. The lack of electricity makes for peaceful evenings by candlelight. The facilities, however, are well maintained and very comfortable. The cabins are built on stilts, which helps to keep out insects and allows for an open-air environment. You will also find hammocks, solar-heated showers, and comfortable beds in each of the 11 cabins. Three meals are served daily in the dining room, and aside from hiking and birding, the lodge offers horseback rides as well as a short canopy tour.

For birding, Selva Bananito Lodge is an excellent destination. It is famous among locals for the Great Jacamar, an extremely rare bird in Costa Rica that is regularly reported from the forest behind the lodge. Many other interesting species can also be found in the area.

Target Birds

*Black-eared Wood-Quail (13) 12:12
Sunbittern (29) 6:16
Olive-backed Quail-Dove (93) 18:22
Blue-headed Parrot (97) 19:8
Spectacled Owl (105) 20:8
Central American Pygmy-Owl (109) 20:16
Black-and-white Owl (107) 20:7
Blue-chested Hummingbird (125) 24:14
Rufous Motmot (147) 27:7
Broad-billed Motmot (147) 27:9
Pied Puffbird (151) 28:7
White-fronted Nunbird (151) 28:4
Great Jacamar (153) 26:11
Fasciated Antshrike (175) 31:1
Western Slaty-Antshrike (177) 31:5
Brown-capped Tyrannulet (189) 37:13
*Snowy Cotinga (219) 34:4
Purple-throated Fruitcrow (219) 34:11
Red-capped Manakin (223) 33:6
Tawny-faced Gnatwren (237) 32:14
*Plain-colored Tanager (291) 46:11
Red-breasted Blackbird (315) 44:13

Access

From Puerto Limón: Go south on Rt. 36 for 16.0 km and then turn right onto a paved road. (**From Puerto Viejo de Talamanca:** Head north toward Puerto Limón for 42.2 km and then turn left onto a paved road.) After 4.2 km, in the town of Bananito, you will cross train tracks as the road turns to dirt. From the tracks, proceed 5.4 km and bear left at a fork in the road. Just 0.6 km from the fork you will cross a bridge over the Río Bananito. In another 1.2 km, make a right turn, and you will come to a T-intersection in 1.0 km (after crossing the Río Bananito again, and this time there is no bridge). Turn left here and go 2.0 km to another T-intersection. Turn right, proceeding past some farm buildings, and in 0.2 km make a left turn. You will cross the Río Bananito one final time, and come to Selva Bananito Lodge in 0.4 km. (Driving time from Limón: 50 min; from Puerto Viejo de Talamanca: 85 min)

Logistics: Birding Selva Bananito Lodge is only recommended if you plan to spend the night on site. Three meals a day are served in the dining room, but there are no stores of any kind around. Be sure to bring in whatever personal items you may need.

Selva Bananito Lodge
Tel: (506) 2253-8118
Website: http://www.selvabananito.com
E-mail: conserva@racsa.co.cr

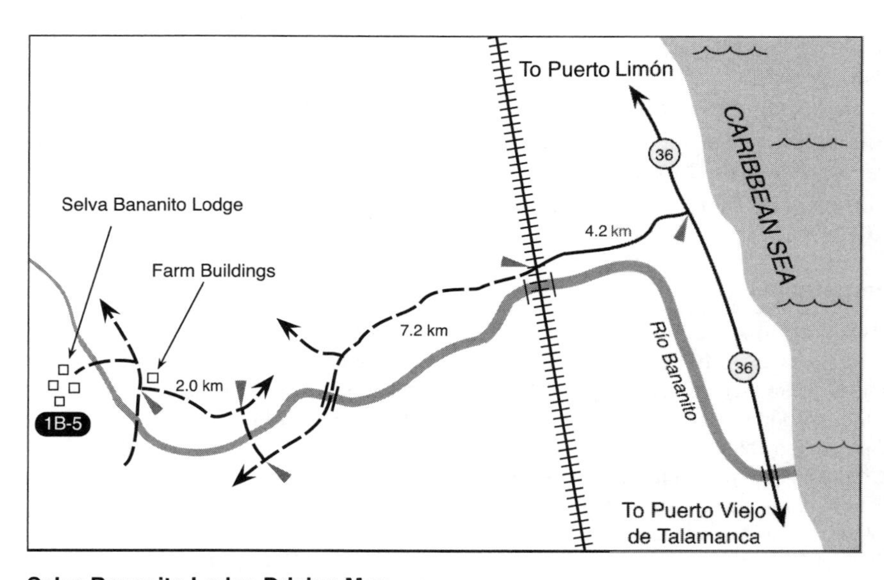

Selva Bananito Lodge Driving Map

Birding Sites

Unlike many other eco-lodges in Costa Rica, Bananito was not built in primary forest. Instead, in an effort to minimize the impact on the remaining old-growth forest, the owners constructed the lodge near the forest, on pastureland and new growth. These pastures and scrub habitats offer excellent birding and often produce rare species. Unfortunately, there are not many good paths through the area, although a few discernable footpaths do exist.

To get to the best pastures, walk the trail leading out of the back of the lodge and past a small pond, where you can often find Purple Gallinule, Green Heron, Green Kingfisher, and Ringed Kingfisher. The trail continues up through the pastures, crosses a small stream, where Sunbittern is sometimes seen, and into more pastureland. Here you should get good looks at Collared Aracari, Black-crowned Tityra, Chestnut-headed Oropendola, Blue-headed Parrot, and Brown-hooded Parrot, all of which like to perch in the tops of the tall trees scattered throughout the pastures. Check dead trees for Long-tailed Tyrant, which often nests in old woodpecker holes. Two birds that occur regularly around Bananito Lodge, but are challenging to find elsewhere, are Snowy Cotinga and Pied Puffbird. Both species often sit unobtrusively high in the trees, so look carefully. Other birds to watch for in the pastureland are Yellow Tyrannulet, Plain-colored Tanager, Fasciated Antshrike, White-winged Becard, Blue-chested Hummingbird, and Giant Cowbird.

The Cristalina Fields, located just south of the lodge, offer access to uninterrupted open-field habitat and often hold some interesting birds. This is a great place to see Red-breasted Blackbird, which can be difficult to find on the Caribbean Slope. You also may see Eastern Meadowlark, Gray-breasted Martin, and Green-breasted Mango. These fields offer a large sky and are a nice place to scan for raptors, especially during September and October when hawks are migrating in large numbers.

Around the lodge itself, look for White-necked Jacobin, Violaceous Trogon, Cinnamon Becard, Shining Honeycreeper, and Black-cheeked Woodpecker. Dawn and dusk are good times to listen for Collared Forest-Falcon, Great Tinamou, and Little Tinamou. A number of owls are possible in the area, including Central American Pygmy-Owl, Mottled Owl, and Spectacled Owl.

A beautiful section of the old-growth forest can be accessed by Sendero Historia Natural, a three-hour loop. The trailhead is hard to find, so I suggest having one of the natural history guides at the lodge show you the entrance point, or even accompany you on the walk. This trail is one of the only places in Costa Rica where you have a fighting chance to find the very rare Great Jacamar. This sit-and-wait predator may be quite unobtrusive, so you will need to be very observant. There is also a lek of Red-capped Manakins about halfway around the loop. Olive-backed Quail-Dove, Scaly-breasted Wren, and Black-faced Antthrush inhabit the understory of this forest, although they are difficult to see.

The canopy holds Black-striped Woodcreeper and Purple-throated Fruitcrow. Other birds to watch for include Black-throated Trogon, Broad-billed Motmot, Rufous Motmot, and Western Slaty-Antshrike.

Species to Expect

Great Tinamou (3) 12:6
Little Tinamou (3) 12:17
Gray-headed Chachalaca (11) 12:1
Crested Guan (11) 12:4
Green Heron (25) 6:2
Black Vulture (31) 13:4
Turkey Vulture (31) 13:3
King Vulture (31) 13:5
White-tailed Kite (37) 15:1
Double-toothed Kite (35) 16:1
Common Black-Hawk (45) 13:6
Broad-winged Hawk (43) 16:13
Short-tailed Hawk (43) 16:11
Collared Forest-Falcon (55) 16:7
White-throated Crake (57) 6:9
Gray-necked Wood-Rail (55) 6:13
Purple Gallinule (59) 6:15
Northern Jacana (61) 6:18
Spotted Sandpiper (73) 11:8
Pale-vented Pigeon (87) 18:3
Short-billed Pigeon (87) 18:5
Ruddy Ground-Dove (89) 18:7
Blue Ground-Dove (91) 18:6
Gray-chested Dove (91) 18:16
Olive-backed Quail-Dove (93) 18:22
*Crimson-fronted Parakeet (95) 19:10
Brown-hooded Parrot (97) 19:9
Blue-headed Parrot (97) 19:8
White-crowned Parrot (97) 19:7
Squirrel Cuckoo (103) 21:7
Groove-billed Ani (103) 21:9
Spectacled Owl (105) 20:8
Central American Pygmy-Owl (109) 20:16
Mottled Owl (107) 20:6
Common Pauraque (111) 21:18
White-collared Swift (117) 22:1
Gray-rumped Swift (119) 22:10
Long-billed Hermit (121) 23:2
Stripe-throated Hermit (121) 23:1
White-necked Jacobin (125) 23:17
Green-breasted Mango (129) 23:13
Blue-chested Hummingbird (125) 24:14
Rufous-tailed Hummingbird (129) 24:10
Violaceous Trogon (141) 26:8
Black-throated Trogon (141) 26:9
Slaty-tailed Trogon (145) 26:2
Rufous Motmot (147) 27:7
Broad-billed Motmot (147) 27:9
Ringed Kingfisher (149) 27:1
Green Kingfisher (149) 27:5
Pied Puffbird (151) 28:7
Collared Aracari (155) 27:15
Keel-billed Toucan (155) 27:18
Chestnut-mandibled Toucan (155) 27:19
Black-cheeked Woodpecker (157) 28:14
Cinnamon Woodpecker (161) 28:9
Lineated Woodpecker (161) 27:13
Plain Xenops (169) 29:3
Plain-brown Woodcreeper (171) 29:11
Wedge-billed Woodcreeper (171) 29:6
Northern Barred-Woodcreeper (169) 29:19
Cocoa Woodcreeper (173) 29:17
Black-striped Woodcreeper (173) 29:16
Streak-headed Woodcreeper (173) 29:8
Fasciated Antshrike (175) 31:1
Western Slaty-Antshrike (177) 31:5
Dusky Antbird (177) 31:8
Chestnut-backed Antbird (179) 31:11
Black-faced Antthrush (187) 30:15
Yellow Tyrannulet (189) 37:11
Yellow-bellied Elaenia (193) 37:26
Ochre-bellied Flycatcher (199) 36:25
Paltry Tyrannulet (191) 37:10
Northern Bentbill (197) 37:6
Common Tody-Flycatcher (197) 37:7
Yellow-olive Flycatcher (195) 37:16
Olive-sided Flycatcher (203) 36:4
Western Wood-Pewee (203) 36:7
Eastern Wood-Pewee (203) 36:8
Tropical Pewee (203) 36:9
Yellow-bellied Flycatcher (205) 36:20
Long-tailed Tyrant (201) 36:6
Bright-rumped Attila (201) 35:6
Rufous Mourner (201) 34:9
Dusky-capped Flycatcher (209) 35:21

Great Crested Flycatcher (209) 35:17
Great Kiskadee (211) 35:13
Boat-billed Flycatcher (211) 35:12
Social Flycatcher (211) 35:14
Gray-capped Flycatcher (211) 35:15
Piratic Flycatcher (193) 35:8
Tropical Kingbird (213) 35:1
Eastern Kingbird (213) 35:3
Cinnamon Becard (215) 33:11
White-winged Becard (217) 33:13
Masked Tityra (217) 34:1
Black-crowned Tityra (217) 34:2
*Snowy Cotinga (219) 34:4
Purple-throated Fruitcrow (219) 34:11
White-collared Manakin (221) 33:1
Red-capped Manakin (223) 33:6
Red-eyed Vireo (227) 40:3
Lesser Greenlet (229) 40:7
Brown Jay (231) 39:19
Gray-breasted Martin (235) 22:15
Southern Rough-winged Swallow (235) 22:19
Barn Swallow (237) 22:12
Band-backed Wren (239) 38:3
***Black-throated Wren (243) 38:13**
Bay Wren (241) 38:12
*Stripe-breasted Wren (241) 38:5
House Wren (243) 38:18
White-breasted Wood-Wren (245) 38:15
Scaly-breasted Wren (245) 38:20
Long-billed Gnatwren (237) 32:15
Tropical Gnatcatcher (237) 41:2
Swainson's Thrush (249) 39:2
Clay-colored Thrush (251) 39:8
Tennessee Warbler (255) 40:22
Chestnut-sided Warbler (259) 43:4
Bay-breasted Warbler (263) 41:9
Olive-crowned Yellowthroat (273) 42:11
Bananaquit (277) 40:24
Dusky-faced Tanager (281) 45:18
Tawny-crested Tanager (281) 46:17
White-lined Tanager (281) 46:16
Red-throated Ant-Tanager (277) 47:12
Crimson-collared Tanager (287) 47:3
Passerini's Tanager (287) 47:4
Blue-gray Tanager (291) 46:15
Palm Tanager (291) 45:19
***Plain-colored Tanager (291) 46:11**
Golden-hooded Tanager (289) 46:13
Shining Honeycreeper (293) 46:1
Blue-black Grassquit (297) 49:7
Variable Seedeater (295) 49:3
Thick-billed Seed-Finch (295) 49:10
Black-striped Sparrow (303) 50:14
Buff-throated Saltator (309) 48:2
Black-headed Saltator (309) 48:1
Red-breasted Blackbird (315) 44:13
Great-tailed Grackle (317) 44:16
Bronzed Cowbird (317) 44:15
Giant Cowbird (317) 44:10
Black-cowled Oriole (319) 44:5
Scarlet-rumped Cacique (321) 44:11
Chestnut-headed Oropendola (323) 44:9
Montezuma Oropendola (323) 44:8
***Yellow-crowned Euphonia (327) 45:1**
Olive-backed Euphonia (325) 45:3

Site 1A-6: Hitoy Cerere Biological Reserve

Birding Time: 5 hours
Elevation: 125 meters
Trail Difficulty: 2
Reserve Hours: 8 a.m. to 4 p.m.
Entrance Fee: $10 (US 2008)
4×4 Recommended

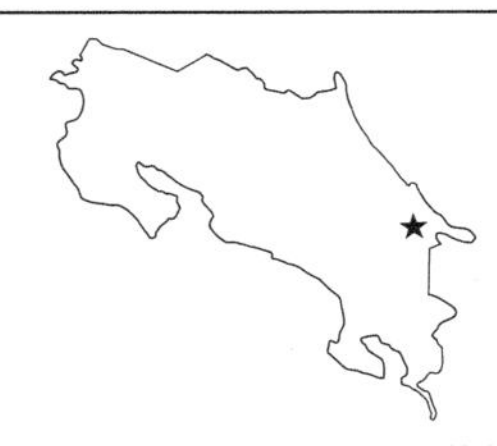

Hitoy Cerere Biological Reserve sits in one of the most diverse and least explored areas in all of Costa Rica. It protects 24,575 acres of forest along the lower Caribbean slopes of the Talamanca Mountains. While the station primarily accommodates the needs of researchers and students, ecotourists and birders are also welcome.

The ranger station at Hitoy Cerere comprises five small buildings cut into the edge of a vast forest that extends all the way through the enormous Talamanca wilderness into Panama. The facilities include forest trails, clean restrooms, and a picnic ground. Even though some surrounding areas have been cleared for cattle grazing, one gets a sense of real isolation out here, and you can find some excellent birding. Be on the lookout for species such as Purple-throated Fruitcrow, Sunbittern, and Lattice-tailed Trogon.

The drive to Hitoy Cerere will give you a firsthand look at a banana plantation owned by Dole Fruit Company. Two processing plants along the road may be of special interest. You can watch the workers screen, clean, and package the bananas to be shipped across the world.

Access

From Puerto Limón: Head south on Rt. 36 for 33.8 km to a gas station on your right. Turn right here, following signs for Pandora. (**From Puerto Viejo de Talamanca:** Head north on Rt. 36 for 24.6 km to a gas station. Turn left here, following signs for Pandora.) Proceed straight for 14.2 km (the road turns to dirt after 9.8 km). Turn left and go 1.0 km to a T-intersection. Turn right and go 4.8 km to a fork. At the fork bear left and follow the road, which quickly takes a sharp left turn. After 1.6 km, bear left at another fork on this two-track dirt road and you will come to the reserve in 3.6 km. (Driving time from Limón: 80 min; from Puerto Viejo de Talamanca: 80 min)

Logistics: This reserve is quite isolated so be sure to bring in food and whatever else you may need. The easiest place to stay nearby is the beach town of Cahuita, located on Rt. 36 (heading toward Puerto Viejo de Talamanca) about 55 minutes away.

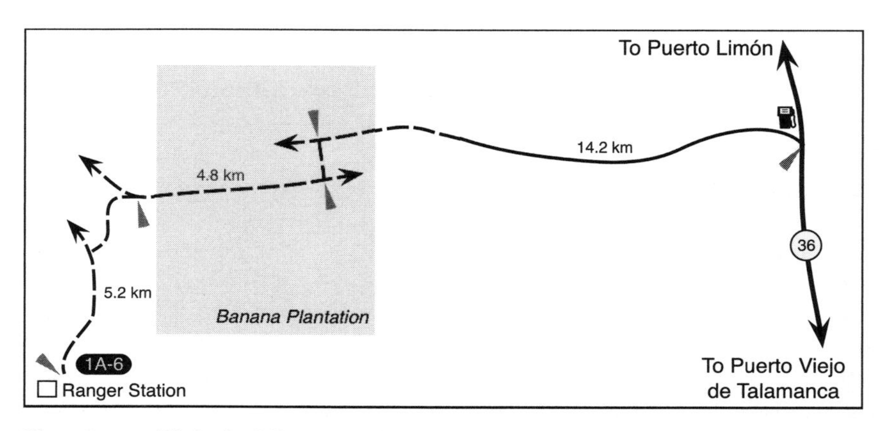

Hitoy Cerere Biological Reserve Driving Map

Birding Sites

When you arrive at Hitoy Cerere I suggest you start birding the second-growth and edge habitat around the ranger station. This is a great place to find Black-cowled Oriole, Violaceous Trogon, Collared Aracari, Cinnamon Becard, Blue Dacnis, and Plain-colored Tanager. The small stream that runs immediately in front of the ranger station is also worth checking. Look for Buff-rumped Warbler, Sunbittern, Royal Flycatcher, and Red-throated Ant-Tanager.

A short trail leading down to the Hitoy Cerere River from the station often yields many of the aforementioned species, as well as Brown-capped Tyrannulet, Stripe-throated Hermit, White-necked Jacobin, Dot-winged Antwren, and Northern Bentbill. At the river, be sure to scan the large vista of sky for any raptors such as Barred Hawk, Double-toothed Kite, Short-tailed Hawk, and King Vulture.

Behind the ranger station there are more productive birding trails leading through the forest. The trail system starts unobtrusively behind one of the small buildings at the right side of the station. The first 150 meters is a great place to find White-collared Manakins, which often display at leks in the area. As the trail leads into older forest you may find Rufous Motmot, Lattice-tailed Trogon, Western Slaty-Antshrike, Checker-throated Antwren, and Scaly-breasted Wren. Also, keep an eye on the canopy, as Purple-throated Fruitcrow, Black-striped Woodcreeper, and Scarlet-rumped Cacique all prefer the upper elevations of the forest.

Along the road just before the station you can find some dense new-growth habitat as well as pastureland. This is a good place to hear the songs of Little Tinamou, Dusky Antbird, and Thicket Antpitta, although seeing them is a challenge. You should also keep an eye out for Bay Wren, Long-tailed Tyrant, Blue Ground-Dove, Giant Cowbird, and White-lined Tanager.

Site 1A-7: Keköldi Hawk Watch

Birding Time: 1 day
Elevation: 0–200 meters
Trail Difficulty: 3
Reserve Hours: N/A
Entrance Fee: $15 (US 2008)

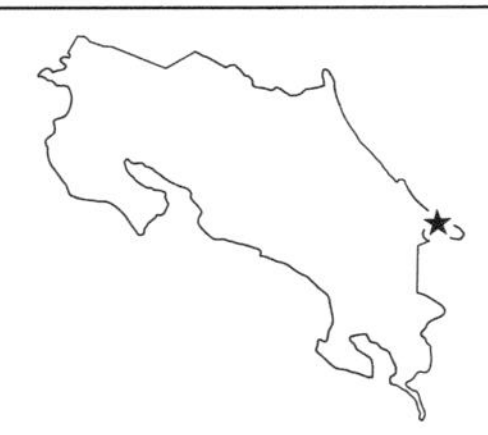

In 2000, scientists established an official raptor-monitoring station atop a small hill within the Keköldi Indigenous Reserve of southeastern Costa Rica. Costa Rica is an ideal country for observing bird migration because most transcontinental migrants follow its narrow landmass as they travel between North and South America. Keköldi is especially suited to the hawk watch because its location between the Talamanca Mountains and the Caribbean Sea amplifies this funneling effect. During the few years since its creation, the numbers counted at this modest facility have been astounding. Every year the station records well over two million raptors, placing it among the top three hawk-watch stations in the world! While the well-known Veracruz Station in Mexico produces higher numbers, it encompasses multiple monitoring sites, none of which, individually, see the numbers that Keköldi's single tower does.

At Keköldi the hawks are observed from a 15-meter-tall wooden tower constructed atop a hill, affording panoramic views of the surrounding countryside. To the east, the Caribbean Sea lies less than two kilometers away, while a ridge of mountains creates the western view. The hawks are funneled through this small gap where the mountains approach the sea, and on large-migration days the entire sky around Keköldi is swirling with birds.

The Keköldi Indigenous Reserve was created in the late 1970s as a home for the native Bri Bri people. The reserve encompasses 12,800 acres where about 70 families live in modest houses dispersed throughout the forest. These houses are built on stilts, and each is surrounded by the family's animals, including chickens, turkeys, and pigs. While most of the Bri Bri speak Spanish and make their living outside the reserve, within the borders of the reserve their native tongue is still used. Even without the attraction of the hawk watch, it is a wonderful experience to see the unique lifestyle of these indigenous people.

Target Birds

King Vulture (31) 13:5
Hook-billed Kite (35) 16:9
Mississippi Kite (37) 15:3
Plumbeous Kite (37) 14:8
Sharp-shinned Hawk (39) 16:3
Cooper's Hawk (39) 16:4
Semiplumbeous Hawk (41) 16:10
Common Black-Hawk (45) 13:6
Swainson's Hawk (47) 17:7
Black Hawk-Eagle (49) 13:9
Slaty-backed Forest-Falcon (55) 16:6
Merlin (51) 15:15
Peregrine Falcon (51) 15:5
Uniform Crake (57) 6:8
Blue-headed Parrot (97) 19:8
Central American Pygmy-Owl (109) 20:16
Striped Owl (105) 20:2
Common Nighthawk (113) 21:13
Great Potoo (115) 20:5
Pied Puffbird (151) 28:7
Spot-crowned Antvireo (181) 32:8
White-flanked Antwren (183) 32:2
Ocellated Antbird (179) 31:7
Brown-capped Tyrannulet (189) 37:13
Black-headed Tody-Flycatcher (197) 37:8
*Snowy Cotinga (219) 34:4
Purple-throated Fruitcrow (219) 34:11
Red-capped Manakin (223) 33:6
Black-chested Jay (231) 39:20
Purple Martin (235) 22:16
Veery (249) 39:3
*Sulphur-rumped Tanager (283) 45:15
*Plain-colored Tanager (291) 46:11
Rufous-winged Tanager (289) 46:10
Shiny Cowbird (329)

Access

From Puerto Viejo de Talamanca: Head north toward Puerto Limón for 4.0 km and then turn left onto a small dirt road next to a small store. (If you are heading south, this turn is 0.8 km beyond the turn that would take you to Sixaola.) In only a few hundred meters you will come to a small entrance to the Keköldi reserve on your left known as Finca Iguanas. There are a few indigenous houses here, and it is a good place to park your car for the day, as well as meet the guide who will lead you up to the tower.

Logistics: Prior reservations are required to visit the Keköldi Indigenous Reserve. You can also arrange to take meals near the hawk station, which is convenient, saving you the hassle of bringing in a bag lunch. The nearby beach town of Puerto Viejo de Talamanca has many lodging options. There is also a small station with dormitory rooms on site, although the facilities are very rustic. You can make reservations to visit through the Keköldi reserve itself or ACTUAR, a local travel agency.

Keköldi Indigenous Reserve
Tel: (506) 2756-8036 (Spanish)
Website: www.kekoldi.org
E-mail: centrocientifico@kekoldi.org

ACTUAR Travel Agency
Tel: (506) 2248-9470
Website: www.actuarcostarica.com
E-mail: info@actuarcostarica.com

Birding Sites

The hawk migration is monitored in the spring and fall, although the fall season sees far more hawks moving through the area. Official counts take place daily, 7 a.m. to 5 p.m., from August 15 to December 15 and from February 15 to May 15. While raptors are seen throughout the count seasons, mid to late October is the time when the highest concentrations are moving through. Although daily numbers vary, during this time it is not uncommon for the station to record between 100,000 and 200,000 birds in a single day! The information I provide below focuses on the fall migration season.

The bulk of the numbers migrating over Keköldi comprise four species: Broad-winged Hawk, Swainson's Hawk, Turkey Vulture, and Mississippi Kite. The migration starts with Mississippi Kites, which usually reach peak numbers around the middle of September. During the first half of October, Broad-winged Hawks move through in huge numbers, and large movements of Swainson's Hawks are seen mid to late October. The final major migrant to pass by is the Turkey Vulture, whose numbers peak from late October to early November. During the fall season the station records on average about 250,000 Mississippi Kites, 1,100,000 Broad-winged Hawks, 450,000 Swainson's Hawks, and 1,000,000 Turkey Vultures.

Other raptors migrate through the area in much smaller numbers. The Peregrine Falcon, which is recorded mid-September to mid-November, is worthy of special note because Keköldi annually records the largest Peregrine migration of any station in the world—about 2,500 birds. Merlins move through during the same time period, although in lower numbers. Plumbeous Kite and Swallow-tailed Kite are seen in August and September. Cooper's Hawk, Zone-tailed Hawk, and Hook-billed Kite start to show up in late September and are seen in very low numbers throughout the rest of the fall. Osprey is recorded throughout the season, although they reach their peak in early to mid October. Other possible migrants include Northern Harrier, Sharp-shinned Hawk, Red-tailed Hawk, and American Kestrel.

Raptors are not the only noteworthy migrants to pass over the Keköldi reserve. Swallows move through in staggering numbers. During the first half of the fall season these are principally Cliff Swallows, although by October Bank Swallows are the most abundant. Barn Swallows can also be seen throughout the season, but in lower numbers. Large flocks of Purple Martins and Chimney Swifts travel over the station as well. Passerine migration is notable, and a bird-banding program has been started to monitor their movements. Look for Swainson's Thrush, American Redstart, Cerulean Warbler, Black-and-white Warbler, Canada Warbler, and Blackburnian Warbler. Also, keep an eye open for Olive-sided Flycatcher, Eastern Kingbird, and Common Nighthawk.

The resident species that can be found at Keköldi do not deserve to be completely overshadowed by the migrants. Birding along the short dirt road that passes in front of Finca Iguanas can produce a number of interesting species. Look for Purple-throated Fruitcrow, Blue-chested Hummingbird, Long-tailed Tyrant, Snowy Cotinga, as well as the South Caribbean specialty bird, Black-chested Jay. The forest in the area is mostly second growth, and old cacao plantation. The maze of trails that lead through the indigenous reserve are home to many Caribbean Slope forest species such as Slaty-backed Forest-Falcon, Checker-throated Antwren, White-flanked Antwren, Western Slaty-Antshrike, and Red-capped Manakin. The hawk observation tower offers a rare eye-level view of the forest canopy and allows for good looks at species such as Blue Dacnis, Shining Honeycreeper, Plain-colored Tanager, and Black-headed Tody-Flycatcher.

Species to Expect

Great Tinamou (3) 12:6
Little Tinamou (3) 12:17
Gray-headed Chachalaca (11) 12:1
Crested Guan (11) 12:4
Brown Pelican (17) 4:1
Magnificent Frigatebird (19) 1:6
Black Vulture (31) 13:4
Turkey Vulture (31) 13:3
King Vulture (31) 13:5
Osprey (33) 17:14
Double-toothed Kite (35) 16:1
Mississippi Kite (37) 15:3
Sharp-shinned Hawk (39) 16:3
Common Black-Hawk (45) 13:6
Broad-winged Hawk (43) 16:13
Short-tailed Hawk (43) 16:11
Swainson's Hawk (47) 17:7
Zone-tailed Hawk (45) 14:2
Black Hawk-Eagle (49) 13:9
Slaty-backed Forest-Falcon (55) 16:6
Collared Forest-Falcon (55) 16:7
Merlin (51) 15:15
Bat Falcon (51) 15:14
Peregrine Falcon (51) 15:5
Gray-necked Wood-Rail (55) 6:13
Pale-vented Pigeon (87) 18:3
Short-billed Pigeon (87) 18:5
Ruddy Ground-Dove (89) 18:7
Gray-chested Dove (91) 18:16
Olive-throated Parakeet (95) 19:11

Orange-chinned Parakeet (97) 19:14
White-crowned Parrot (97) 19:7
Red-lored Parrot (99) 19:4
Mealy Parrot (99) 19:3
Squirrel Cuckoo (103) 21:7
Groove-billed Ani (103) 21:9
Great Potoo (115) 20:5
White-collared Swift (117) 22:1
Lesser Swallow-tailed Swift (119) 22:7
Bronzy Hermit (121) 23:4
Band-tailed Barbthroat (121) 23:5
Long-billed Hermit (121) 23:2
Stripe-throated Hermit (121) 23:1
White-necked Jacobin (125) 23:17
Violet-crowned Woodnymph (127) 24:4
Blue-chested Hummingbird (125) 24:14
Rufous-tailed Hummingbird (129) 24:10
Violaceous Trogon (141) 26:8
Black-throated Trogon (141) 26:9
Slaty-tailed Trogon (145) 26:2
Rufous Motmot (147) 27:7
Broad-billed Motmot (147) 27:9
White-necked Puffbird (151) 28:3
White-whiskered Puffbird (151) 28:6
Collared Aracari (155) 27:15
Keel-billed Toucan (155) 27:18
Chestnut-mandibled Toucan (155) 27:19
Black-cheeked Woodpecker (157) 28:14
Pale-billed Woodpecker (161) 27:14
Plain Xenops (169) 29:3
Wedge-billed Woodcreeper (171) 29:6
Northern Barred-Woodcreeper (169) 29:19
Cocoa Woodcreeper (173) 29:17
Black-striped Woodcreeper (173) 29:16
Streak-headed Woodcreeper (173) 29:8
Western Slaty-Antshrike (177) 31:5
Checker-throated Antwren (183) 32:5
White-flanked Antwren (183) 32:2
Dot-winged Antwren (183) 32:1
Dusky Antbird (177) 31:8
Chestnut-backed Antbird (179) 31:11
Black-faced Antthrush (187) 30:15
Brown-capped Tyrannulet (189) 37:13
Yellow-bellied Elaenia (193) 37:26
Ochre-bellied Flycatcher (199) 36:25
Paltry Tyrannulet (191) 37:10
Common Tody-Flycatcher (197) 37:7
Black-headed Tody-Flycatcher (197) 37:8
Ruddy-tailed Flycatcher (199) 36:23
Olive-sided Flycatcher (203) 36:4
Western Wood-Pewee (203) 36:7
Eastern Wood-Pewee (203) 36:8
Long-tailed Tyrant (201) 36:6
Bright-rumped Attila (201) 35:6
Dusky-capped Flycatcher (209) 35:21
Great Crested Flycatcher (209) 35:17
Great Kiskadee (211) 35:13
Boat-billed Flycatcher (211) 35:12
Social Flycatcher (211) 35:14
Gray-capped Flycatcher (211) 35:15
Tropical Kingbird (213) 35:1
Cinnamon Becard (215) 33:11
White-winged Becard (217) 33:13
Masked Tityra (217) 34:1
Purple-throated Fruitcrow (219) 34:11
White-collared Manakin (221) 33:1
Red-capped Manakin (223) 33:6
Yellow-throated Vireo (225) 40:6
Red-eyed Vireo (227) 40:3
Lesser Greenlet (229) 40:7
Black-chested Jay (231) 39:20
Gray-breasted Martin (235) 22:15
Bank Swallow (235) 22:17
Cliff Swallow (233) 22:13
Barn Swallow (237) 22:12
Band-backed Wren (239) 38:3
Bay Wren (241) 38:12
***Stripe-breasted Wren (241) 38:5**
House Wren (243) 38:18
White-breasted Wood-Wren (245) 38:15
Scaly-breasted Wren (245) 38:20
Long-billed Gnatwren (237) 32:15
Tropical Gnatcatcher (237) 41:2
Swainson's Thrush (249) 39:2
Wood Thrush (249) 39:4
Clay-colored Thrush (251) 39:8
Tennessee Warbler (255) 40:22
Yellow Warbler (259) 42:2
Chestnut-sided Warbler (259) 43:4
Blackburnian Warbler (261) 41:8
Bay-breasted Warbler (263) 41:9
Black-and-white Warbler (267) 41:13
American Redstart (267) 41:10
Canada Warbler (269) 42:8
Buff-rumped Warbler (271) 40:23
Bananaquit (277) 40:24
Carmiol's Tanager (279) 45:16

Red-throated Ant-Tanager (277) 47:12
Summer Tanager (285) 47:5
Scarlet Tanager (287) 47:8
Passerini's Tanager (287) 47:4
Blue-gray Tanager (291) 46:15
Palm Tanager (291) 45:19
Golden-hooded Tanager (289) 46:13
Blue Dacnis (291) 46:3
Green Honeycreeper (293) 46:7
Shining Honeycreeper (293) 46:1
Variable Seedeater (295) 49:3
Black-striped Sparrow (303) 50:14
Buff-throated Saltator (309) 48:2
Blue-black Grosbeak (311) 48:10
Great-tailed Grackle (317) 44:16
Black-cowled Oriole (319) 44:5
Baltimore Oriole (321) 44:7
Scarlet-rumped Cacique (321) 44:11
Chestnut-headed Oropendola (323) 44:9
Montezuma Oropendola (323) 44:8
***Yellow-crowned Euphonia (327) 45:1**
Olive-backed Euphonia (325) 45:3
White-vented Euphonia (327) 45:7

Subregion 1B

The Caribbean Middle-Elevations

With its pure white cap and dark purple body, the **Snowcap** is a striking endemic hummingbird of middle-elevation forests on the Caribbean Slope. This tiny bird is most easily found at Rancho Naturalista (1B-7) and El Copal Biological Reserve (1B-8), where it often visits porterweed flowers (*Stachytarpheta frantzii*-as illustrated).

Common Birds to Know

Black Vulture (31) 13:4
Turkey Vulture (31) 13:3
White-tipped Dove (91) 18:14
White-crowned Parrot (97) 19:7
White-collared Swift (117) 22:1
Rufous-tailed Hummingbird (129) 24:10
Keel-billed Toucan (155) 27:18
Black-cheeked Woodpecker (157) 28:14
Great Kiskadee (211) 35:13
Social Flycatcher (211) 35:14
Tropical Kingbird (213) 35:1
Lesser Greenlet (229) 40:7
Blue-and-white Swallow (233) 22:20
*Stripe-breasted Wren (241) 38:5
House Wren (243) 38:18
White-breasted Wood-Wren (245) 38:15
Clay-colored Thrush (251) 39:8
Chestnut-sided Warbler (259) 43:4
Summer Tanager (285) 47:5
Passerini's Tanager (287) 47:4
Blue-gray Tanager (291) 46:15
Palm Tanager (291) 45:19
Green Honeycreeper (293) 46:7
Variable Seedeater (295) 49:3
Baltimore Oriole (321) 44:7

Site 1B-1: Las Heliconias Lodge

Birding Time: 1 day
Elevation: 800 meters
Trail Difficulty: 1–3
Reserve Hours: N/A
Entrance Fee: $13 (US 2008)
Trail Map Available on Site

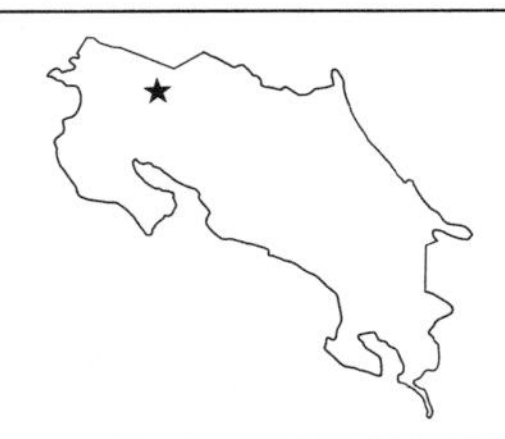

The Guanacaste Mountain Range, which divides the wet slopes of the Caribbean from the dry forest of the North Pacific, is much smaller than the other ranges in the country, and has a number of low passes between the major peaks. The low passes are ecologically interesting locations where flora and fauna, from either slope, are able to mix. Las Heliconias Lodge is located in just such a low pass along the lower flanks of Tenorio Volcano, one of two active volcanoes in the Guanacaste range. Here you can find Pacific Slope species, including Long-tailed Manakin, Hoffmann's Woodpecker, White-winged Dove, and Inca Dove, alongside Caribbean Slope specialties, such as Black-crested Coquette, Yellow-eared Toucanet, Black-cowled Oriole, and Black-cheeked Woodpecker. This is also the best location in the country to find the rare Tody Motmot, and many people come here solely in search of this bird.

The lodge itself, run by a private association of local owners, sits adjacent to Tenorio Volcano National Park and affords stunning views of the countryside below. Facilities are basic, although clean, and include lodging for over 30 people, a restaurant, and a small gift shop. Las Heliconias maintains about five kilometers of forest trails within its 180-acre property. One of the trails proceeds over a series of three giant hanging bridges, the largest of which is 105 meters long and 37 meters high.

Target Birds

Black-and-white Hawk-Eagle (49) 17:13
Ornate Hawk-Eagle (49) 17:11
Rufous-vented Ground-Cuckoo (103) 21:10
Black-crested Coquette (137) 25:6
*Orange-bellied Trogon (143) 26:4
Tody Motmot (147) 27:12
White-fronted Nunbird (151) 28:4
*Yellow-eared Toucanet (155) 27:20
*Rufous-winged Woodpecker (159) 28:10
*Streak-crowned Antvireo (181) 32:7
Checker-throated Antwren (183) 32:5
Spotted Antbird (181) 31:12
Bicolored Antbird (179) 31:9
Black-headed Antthrush (187) 30:17
Thrush-like Schiffornis (215) 33:10
Lovely Cotinga (219) 34:6
Sharpbill (225) 34:7
Nightingale Wren (245) 38:21
Song Wren (245) 38:22
Hepatic Tanager (285) 47:6
Rufous-winged Tanager (289) 46:10

Access

From Upala: Take Rt. 4 toward Cañas. Only 1.2 km outside of Upala, Rt. 4 turns right and heads to La Cruz. You should continue straight on what becomes Rt. 6 toward Cañas. In 26.6 km turn left at the Banco Nacional in the center of the town of Bijagua. Go 0.4 km to a fork and keep left. Go 0.6 km to a second fork, where you should keep left again. From here Las Heliconias Lodge will be on your left in 2.0 km. (Driving time from Upala: 40 min)

From the intersection of Rt. 1 and Rt. 6: Follow Rt. 6 straight for 33.6 km to the town of Bijagua and turn right at the small Banco Nacional. Go 0.4 km to a fork and keep left. Go 0.6 km to a second fork, where you should keep left again. From here Las Heliconias Lodge will be on your left in 2.0 km. (Driving time from Rt. 1: 40 min)

Logistics: While there are both more economical and upscale accommodations in the area, I find it easiest to stay at Las Heliconias Lodge itself. You can make reservations either through the lodge directly or through COOPRENA Travel Agency.

Las Heliconias Lodge
Tel: (506) 2466-8483 (Spanish)
E-mail: lasheliconiaslodge@yahoo.com

COOPRENA Travel Agency
Tel: (506) 2290-8646
Website: www.turismoruralcr.com
E-mail: cooprena@racsa.co.cr

Birding Sites

Two well-maintained trails, Sendero Las Heliconias and Sendero Puentes Colgantes, lead through the forest around the lodge. These trails take about 45 minutes and 1.5 hours to bird, respectively. The birding here is very good, and you should find a nice mix of species. Along these trails is the best place to find Tody Motmot, which likes to perch quietly in the understory of the forest. Other birds you may find foraging in the understory include Spotted Antbird, Song Wren, Nightingale Wren, and Gray-chested Dove. Noisy flocks of Carmiol's Tanagers are common and are often associated with Golden-crowned Warbler, Streak-crowned Antvireo, and Wedge-billed Woodcreeper. Other interesting species to keep an eye out for include Long-tailed Manakin, Thrushlike Schiffornis, White-fronted Nunbird, and Yellow-eared Toucanet.

A third trail, Sendero Lago Danta, leads steeply upslope for two and a half kilometers to a small lake in an old volcano crater. The top of this trail reaches over 1,300 meters, and the forest is very different from what is found below. Some of the species you may encounter here include Orange-bellied Trogon, Gray-throated Leaftosser, Orange-billed Nightingale-Thrush, and Black-thighed Grosbeak. While the birding can be interesting, the hike is demanding, and I generally do not recommend birding this trail unless you are looking for a strenuous hike.

You should also spend some time birding the edge, second-growth, and pasture habitats found along the road near the hotel. The first 300 meters of road leading back toward Bijagua is often very productive. Here you could find Hepatic Tanager, Scarlet-thighed Dacnis, Red-legged Honeycreeper, and Rufous-winged Tanager. It is also a good place to look for Black-cowled Oriole, Rufous-tailed Jacamar, White-necked Jacobin, Piratic Flycatcher, and Double-toothed Kite. Black-crested Coquette is consistently found in this stretch of road, and you may also find Black-hooded Antshrike, a species normally associated with the South Pacific Slope. Near the cow pastures farther down the road you may find Gray Hawk, Grayish Saltator, White-winged Dove, and Northern Rough-winged Swallow.

Species to Expect

Crested Guan (11) 12:4
Cattle Egret (23) 5:13
Black Vulture (31) 13:4
Turkey Vulture (31) 13:3
Swallow-tailed Kite (37) 15:2
Double-toothed Kite (35) 16:1
Roadside Hawk (43) 16:12
Gray Hawk (43) 16:14
Crested Caracara (53) 14:12
Red-billed Pigeon (87) 18:2
Short-billed Pigeon (87) 18:5
White-winged Dove (89) 18:12
Inca Dove (89) 18:11
Ruddy Ground-Dove (89) 18:7
White-tipped Dove (91) 18:14
Gray-chested Dove (91) 18:16
Orange-chinned Parakeet (97) 19:14
Brown-hooded Parrot (97) 19:9
Red-lored Parrot (99) 19:4
Mealy Parrot (99) 19:3
Squirrel Cuckoo (103) 21:7
Groove-billed Ani (103) 21:9
Crested Owl (105) 20:3
Mottled Owl (107) 20:6
Common Pauraque (111) 21:18
White-collared Swift (117) 22:1
Vaux's Swift (119) 22:9
Green Hermit (121) 23:3
Stripe-throated Hermit (121) 23:1
Violet Sabrewing (123) 23:9
White-necked Jacobin (125) 23:17
Black-crested Coquette (137) 25:6
Rufous-tailed Hummingbird (129) 24:10
*Purple-throated Mountain-gem (135) 24:7
Long-billed Starthroat (135) 23:18
Violaceous Trogon (141) 26:8

*Orange-bellied Trogon (143) 26:4
Black-throated Trogon (141) 26:9
Slaty-tailed Trogon (145) 26:2
Tody Motmot (147) 27:12
Blue-crowned Motmot (147) 27:8
Broad-billed Motmot (147) 27:9
Rufous-tailed Jacamar (153) 26:12
*Prong-billed Barbet (153) 28:1
Collared Aracari (155) 27:15
*Yellow-eared Toucanet (155) 27:20
Keel-billed Toucan (155) 27:18
Chestnut-mandibled Toucan (155) 27:19
Black-cheeked Woodpecker (157) 28:14
***Hoffmann's Woodpecker (157) 28:16**
Buff-throated Foliage-gleaner (165) 30:3
Gray-throated Leaftosser (167) 30:10
Wedge-billed Woodcreeper (171) 29:6
Spotted Woodcreeper (173) 29:20
Streak-headed Woodcreeper (173) 29:8
*Black-hooded Antshrike (177) 31:6
***Streak-crowned Antvireo (181) 32:7**
Chestnut-backed Antbird (179) 31:11
Spotted Antbird (181) 31:12
Bicolored Antbird (179) 31:9
Yellow-bellied Elaenia (193) 37:26
Ochre-bellied Flycatcher (199) 36:25
Paltry Tyrannulet (191) 37:10
Scale-crested Pygmy-Tyrant (197) 37:4
Bright-rumped Attila (201) 35:6
Rufous Mourner (201) 34:9
Dusky-capped Flycatcher (209) 35:21
Great Kiskadee (211) 35:13
Boat-billed Flycatcher (211) 35:12
Social Flycatcher (211) 35:14
Gray-capped Flycatcher (211) 35:15
Piratic Flycatcher (193) 35:8
Tropical Kingbird (213) 35:1
Thrush-like Schiffornis (215) 33:10
White-winged Becard (217) 33:13
Masked Tityra (217) 34:1
White-ruffed Manakin (223) 33:9
Long-tailed Manakin (223) 33:4
Tawny-crowned Greenlet (229) 32:4
Lesser Greenlet (229) 40:7
Gray-breasted Martin (235) 22:15
Blue-and-white Swallow (233) 22:20
Northern Rough-winged Swallow (235) 22:18
Southern Rough-winged Swallow (235) 22:19
***Stripe-breasted Wren (241) 38:5**
House Wren (243) 38:18
White-breasted Wood-Wren (245) 38:15
Gray-breasted Wood-Wren (245) 38:14
Nightingale Wren (245) 38:21
Song Wren (245) 38:22
Orange-billed Nightingale-Thrush (247) 38:26
Swainson's Thrush (249) 39:2
Mountain Thrush (251) 39:6
Clay-colored Thrush (251) 39:8
Tropical Parula (257) 41:3
Yellow Warbler (259) 42:2
Chestnut-sided Warbler (259) 43:4
Blackburnian Warbler (261) 41:8
Black-and-white Warbler (267) 41:13
Kentucky Warbler (271) 42:15
Canada Warbler (269) 42:8
Slate-throated Redstart (273) 42:7
Golden-crowned Warbler (275) 40:18
Common Bush-Tanager (279) 45:13
Carmiol's Tanager (279) 45:16
Hepatic Tanager (285) 47:6
Summer Tanager (285) 47:5
Scarlet Tanager (287) 47:8
Passerini's Tanager (287) 47:4
Blue-gray Tanager (291) 46:15
Palm Tanager (291) 45:19
Golden-hooded Tanager (289) 46:13
Scarlet-thighed Dacnis (291) 46:4
Green Honeycreeper (293) 46:7
Red-legged Honeycreeper (293) 46:2
Variable Seedeater (295) 49:3
White-collared Seedeater (295) 49:2
Yellow-faced Grassquit (297) 49:6
Black-striped Sparrow (303) 50:14
Grayish Saltator (309) 48:3
*Black-thighed Grosbeak (311) 48:7
Blue-black Grosbeak (311) 48:10
Eastern Meadowlark (315) 50:16
Melodious Blackbird (315) 52:6
Great-tailed Grackle (317) 44:16
Black-cowled Oriole (319) 44:5
Baltimore Oriole (321) 44:7
Yellow-throated Euphonia (327) 45:5
Olive-backed Euphonia (325) 45:3

Site 1B-2: Arenal Volcano National Park

Birding Time: 1–2 days
Elevation: 600 meters
Trail Difficulty: 1–3
Reserve Hours: 8 a.m. to 4 p.m.
Entrance Fee: $10 (US 2008)
Trail Map Available on Site

Arenal Volcano, one of the most active volcanoes in the world, is an especially exciting destination in Costa Rica. The large, perfectly shaped cone reaches 1,630 meters (5,350 feet) in the air and erupts about every two hours on average! Huge pyroclastic boulders are shot out the top of the volcano and come tumbling down its steep slopes. At night the show is even more spectacular, as these rocks glow a fiery orange. But don't worry, the eruptions, although frequent, are small and controlled.

Because the rich volcanic soil is some of the best in the country, much of the land surrounding Arenal has been deforested for agriculture. However, there is still sizable forest around the skirts of the volcano, and it continues all the way into the Monteverde Cloud Forest Reserve. Most of the forest accessible at this site is second growth. It harbors a large variety of birds, including many of the more colorful species, such as toucans, parrots, and tanagers.

The town of La Fortuna is the economic center of the Arenal area and holds just about everything a tourist might desire. If you have non-birding members in your group, the La Fortuna area offers zip lines, hot springs, spas, ATV tours, and many more fun activities. Consult any travel guide for more details on tourist activities.

Two safety concerns are worth special note. First, although the volcano is generally safe, you should remember that it is a *volcano* and should be treated with respect. The last major eruption, which took place in 1968, killed 87 people. Today the volcano is monitored with state-of-the-art equipment, and safety instructions are issued by the park service, depending on current conditions. On a different note, many thieves have been attracted to La Fortuna because of all the tourist activity. Be sure to keep your car locked and take special care of your belongings.

Target Birds

Great Curassow (11) 12:3
Fasciated Tiger-Heron (21) 5:15
Violaceous Quail-Dove (93) 18:20
Green-fronted Lancebill (135) 23:11
Brown Violetear (131) 23:6
Black-crested Coquette (137) 25:6
Green Thorntail (137) 25:1
*Snowcap (137) 25:8
Bronze-tailed Plumeleteer (129) 23:12
*Orange-bellied Trogon (143) 26:4
*Lattice-tailed Trogon (145) 26:3
Keel-billed Motmot (147) 27:10
White-fronted Nunbird (151) 28:4
*Yellow-eared Toucanet (155) 27:20

*Streak-crowned Antvireo (181) 32:7
Bare-crowned Antbird (179) 31:13
Dull-mantled Antbird (181) 31:14
Spotted Antbird (181) 31:12
Ocellated Antbird (179) 31:7
*Thicket Antpitta (185) 30:14
Lovely Cotinga (219) 34:6
*Bare-necked Umbrellabird (221) 34:13
*Three-wattled Bellbird (221) 34:12
White-throated Magpie-Jay (231) 39:18
Tree Swallow (233) 22:21
*Black-throated Wren (243) 38:13
Nightingale Wren (245) 38:21
Song Wren (245) 38:22
Black-headed Nightingale-Thrush (247) 38:25
Tropical Mockingbird (253)
Hepatic Tanager (285) 47:6
Crimson-collared Tanager (287) 47:3
Emerald Tanager (289) 46:5
Rufous-winged Tanager (289) 46:10

Access

From La Fortuna: Take the road that heads west out of town toward Arenal Volcano and Tilarán for 14.2 km. Turn left onto a dirt road opposite a small police building, and in 2.0 km the national park entrance will be on your left. (Driving time from La Fortuna: 20 min)

Logistics: Because the Arenal area is one of the most popular tourist destinations in the country, there is no lack of facilities. Hotels and restaurants in all price ranges are easily available if you make reservations early enough. The luxurious Arenal Observatory Lodge is the standout hotel, and the grounds, described below, offer some of the best birding in the area.

Arenal Observatory Lodge
Tel: (506) 2290-7011
Website: www.arenalobservatorylodge.com
E-mail: info@arenalobservatorylodge.com

Birding Sites

Road to the Arenal Observatory Lodge

The dirt road that starts at a small police building along the main road and runs to the Observatory Lodge passes through a range of habitats and offers an impressive array of species. The first section of dirt road, which runs through pastures and new-growth habitat, is a good place to find open-country species such as Tropical Pewee, White-collared Seedeater, Thick-billed Seed-Finch, Gray-crowned Yellowthroat, and Olive-crowned Yellowthroat.

After 2.0 km you will come to the main entrance to Arenal National Park on your left. This area is described in detail below. Just beyond the park entrance, another dirt road forks back to the right and leads to the Arenal Dam. Some interesting birds can be seen along this road, including Bare-crowned Antbird, an uncommon inhabitant of thick undergrowth. Also look for Dusky Antbird, Great Curassow, Bay Wren, and Long-tailed Tyrant.

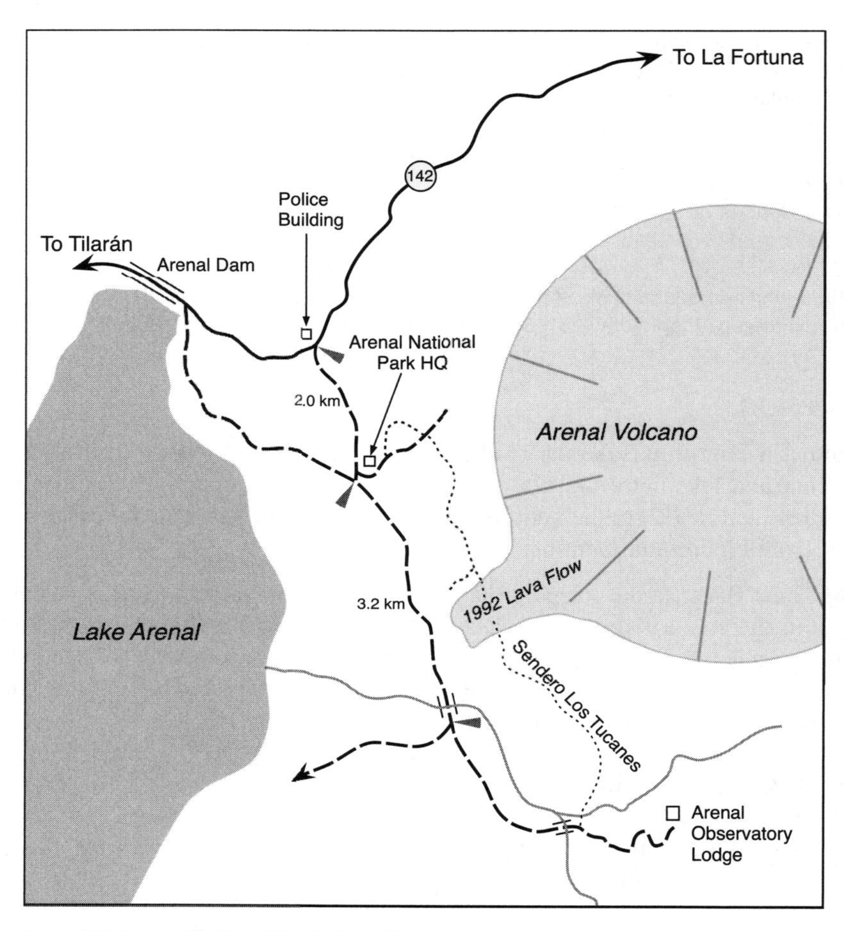

Arenal Volcano National Park Area Map

Continuing toward the Observatory Lodge you will enter a second-growth forest with dense thickets lining the road about 1.5 km beyond the park entrance. Early in the morning you are likely to encounter a diverse array of species feeding along the edge. Look for Yellow-billed Cacique, Grayish Saltator, Black-headed Saltator, Crimson-collared Tanager, and Keel-billed Toucan. Also, the hollow whistles of Thicket Antpitta are frequently heard coming from the forest edge, although seeing this species is a challenge.

As the road proceeds up toward the Arenal Observatory Lodge, the habitat becomes older and wetter. Here species such as Bright-rumped Attila, Long-tailed Tyrant, White-collared Manakin, Nightingale Wren, and Violaceous Trogon are more likely to be found. You will arrive at the gate to the Arenal Observatory Lodge 5.2 km beyond the main national park entrance.

Arenal National Park

At the main entrance to Arenal National Park you will find a volcano lookout as well as two trails. The short Sendero Heliconias offers decent birding. Look for Gray-headed Chachalaca, White-throated Magpie-Jay, Scarlet-thighed Dacnis, and Pale-vented Pigeon. The longer Sendero Coladas leads to the impressive 1992 lava flow. While the first section is usually quiet, you will find more activity once you reach the secondary forest near the lava flow. Just before the lava flow there is a small trail branching off to the right that is worth exploring.

Sendero Los Tucanes continues beyond the 1992 lava flow and connects back to the road 100 meters before the entrance to the Arenal Observatory Lodge. (I like to enter the trail from this end.) While most of the trail is flat and in good condition, there is a small river that must be crossed very close to the dirt road. Sendero Los Tucanes offers the best birding within the national park and is a good place to find Spotted Antbird, Streak-crowned Antvireo, Slaty Antwren, Sulphur-rumped Flycatcher, Rufous-tailed Jacamar, and Buff-throated Foliage-gleaner. The simple repeated whistles of the Song Wren are commonly heard, and if you can whistle back, it will often approach quite close.

Arenal Observatory Lodge

The Arenal Observatory Lodge, 5.2 km beyond the main national park entrance, sits alone atop a small ridge next to the volcano. The beautiful grounds and trails of the hotel offer excellent birding, including your best chance of seeing many hummingbirds and tanagers. If you are not staying at the Observatory Lodge, you can still bird the area by paying a $4 (US 2008) entrance fee.

From the entrance gate proceed 1.2 km up to reception and the restaurant. Check the fruit feeders behind the restaurant first. Here you can get fabulous views of many colorful species such as Passerini's Tanager, Crimson-collared Tanager, Yellow-throated Euphonia, Red-legged Honeycreeper, and Montezuma Oropendola.

Many flowering trees and bushes have been planted around the buildings to attract hummingbirds. Porterweed flowers are especially popular and almost always have Rufous-tailed Hummingbirds hovering nearby. Other species move through the area seasonally, and you should look for Brown Violetear, Violet-headed Hummingbird, Steely-vented Hummingbird, White-necked Jacobin, Blue-throated Goldentail, and Black-crested Coquette.

The lodge also maintains a number of trails that can be birded, and you should pick up a complimentary map when you enter the hotel grounds. One of the best trails is the short Saino Trail, which starts by taking you through some semi-cleared forest. Here you may see Slate-colored Grosbeak, Lineated Woodpecker,

Tawny-capped Euphonia, and Spotted Woodcreeper, as well as an assortment of tanager species, including Emerald Tanager, Hepatic Tanager, and White-shouldered Tanager. The concrete trail then enters into secondary forest, where you may encounter Olive-striped Flycatcher, White-throated Spadebill, Golden-crowned Warbler, and Carmiol's Tanager.

Two other trails are worth exploring. The Waterfall Trail, although a bit muddy, accesses nice second-growth forest and offers many of the same species found on the Saino Trail, as well as Dull-mantled Antbird and Green-fronted Lancebill, two species that are often found near the river. The Old Lava Trail produces birds similar to those found on Sendero Los Tucanes, in the national park, except that it is steep and muddy at its start.

Species To Expect

Gray-headed Chachalaca (11) 12:1
Crested Guan (11) 12:4
Great Curassow (11) 12:3
Neotropic Cormorant (17) 4:4
Little Blue Heron (23) 5:9
Cattle Egret (23) 5:13
Black Vulture (31) 13:4
Turkey Vulture (31) 13:3
Swallow-tailed Kite (37) 15:2
White Hawk (41) 17:2
Broad-winged Hawk (43) 16:13
Gray Hawk (43) 16:14
Short-tailed Hawk (43) 16:11
Laughing Falcon (53) 15:8
White-throated Crake (57) 6:9
Gray-necked Wood-Rail (55) 6:13
Northern Jacana (61) 6:18
Spotted Sandpiper (73) 11:8
Pale-vented Pigeon (87) 18:3
White-winged Dove (89) 18:12
Ruddy Ground-Dove (89) 18:7
White-tipped Dove (91) 18:14
Orange-chinned Parakeet (97) 19:14
White-crowned Parrot (97) 19:7
Red-lored Parrot (99) 19:4
Squirrel Cuckoo (103) 21:7
Groove-billed Ani (103) 21:9
Short-tailed Nighthawk (113) 21:14
White-collared Swift (117) 22:1
Green Hermit (121) 23:3
Long-billed Hermit (121) 23:2
Stripe-throated Hermit (121) 23:1
White-necked Jacobin (125) 23:17
Brown Violetear (131) 23:6
Violet-headed Hummingbird (137) 25:11
Black-crested Coquette (137) 25:6
Violet-crowned Woodnymph (127) 24:4
Steely-vented Hummingbird (127) 24:15
Rufous-tailed Hummingbird (129) 24:10
Bronze-tailed Plumeleteer (129) 23:12
Purple-crowned Fairy (125) 23:14
Violaceous Trogon (141) 26:8
Slaty-tailed Trogon (145) 26:2
Broad-billed Motmot (147) 27:9
Green Kingfisher (149) 27:5
Rufous-tailed Jacamar (153) 26:12
Keel-billed Toucan (155) 27:18
Chestnut-mandibled Toucan (155) 27:19
Black-cheeked Woodpecker (157) 28:14
Smoky-brown Woodpecker (159) 28:13
Golden-olive Woodpecker (159) 28:11
Lineated Woodpecker (161) 27:13
Pale-billed Woodpecker (161) 27:14
Buff-throated Foliage-gleaner (165) 30:3
Plain Xenops (169) 29:3
Wedge-billed Woodcreeper (171) 29:6
Spotted Woodcreeper (173) 29:20
Streak-headed Woodcreeper (173) 29:8
Russet Antshrike (177) 31:4
*Streak-crowned Antvireo (181) 32:7
Slaty Antwren (183) 32:3
Dusky Antbird (177) 31:8
Bare-crowned Antbird (179) 31:13

Spotted Antbird (181) 31:12
Bicolored Antbird (179) 31:9
*Thicket Antpitta (185) 30:14
Yellow Tyrannulet (189) 37:11
Yellow-bellied Elaenia (193) 37:26
Olive-striped Flycatcher (195) 36:24
Ochre-bellied Flycatcher (199) 36:25
Paltry Tyrannulet (191) 37:10
Scale-crested Pygmy-Tyrant (197) 37:4
Northern Bentbill (197) 37:6
Common Tody-Flycatcher (197) 37:7
Yellow-olive Flycatcher (195) 37:16
Sulphur-rumped Flycatcher (199) 36:21
Western Wood-Pewee (203) 36:7
Eastern Wood-Pewee (203) 36:8
Tropical Pewee (203) 36:9
Yellow-bellied Flycatcher (205) 36:20
Black Phoebe (207) 36:5
Long-tailed Tyrant (201) 36:6
Bright-rumped Attila (201) 35:6
Rufous Mourner (201) 34:9
Dusky-capped Flycatcher (209) 35:21
Great Kiskadee (211) 35:13
Boat-billed Flycatcher (211) 35:12
Social Flycatcher (211) 35:14
Gray-capped Flycatcher (211) 35:15
Piratic Flycatcher (193) 35:8
Tropical Kingbird (213) 35:1
Cinnamon Becard (215) 33:11
*Three-wattled Bellbird (221) 34:12
White-collared Manakin (221) 33:1
White-ruffed Manakin (223) 33:9
Yellow-throated Vireo (225) 40:6
Philadelphia Vireo (227) 40:15
Red-eyed Vireo (227) 40:3
Tawny-crowned Greenlet (229) 32:4
Lesser Greenlet (229) 40:7
White-throated Magpie-Jay (231) 39:18
Brown Jay (231) 39:19
Gray-breasted Martin (235) 22:15
Blue-and-white Swallow (233) 22:20
Southern Rough-winged Swallow (235) 22:19
Band-backed Wren (239) 38:3
***Black-throated Wren (243) 38:13**
Bay Wren (241) 38:12
***Stripe-breasted Wren (241) 38:5**
House Wren (243) 38:18
White-breasted Wood-Wren (245) 38:15
Nightingale Wren (245) 38:21
Song Wren (245) 38:22
Long-billed Gnatwren (237) 32:15
Tropical Gnatcatcher (237) 41:2
Swainson's Thrush (249) 39:2
Wood Thrush (249) 39:4
Clay-colored Thrush (251) 39:8
White-throated Thrush (251) 39:9
Golden-winged Warbler (257) 41:5
Tennessee Warbler (255) 40:22
Tropical Parula (257) 41:3
Yellow Warbler (259) 42:2
Chestnut-sided Warbler (259) 43:4
American Redstart (267) 41:10
Olive-crowned Yellowthroat (273) 42:11
Gray-crowned Yellowthroat (273) 42:13
Canada Warbler (269) 42:8
Golden-crowned Warbler (275) 40:18
Buff-rumped Warbler (271) 40:23
Bananaquit (277) 40:24
*Black-and-yellow Tanager (283) 46:14
Carmiol's Tanager (279) 45:16
White-shouldered Tanager (281) 46:18
Hepatic Tanager (285) 47:6
Summer Tanager (285) 47:5
Scarlet Tanager (287) 47:8
Crimson-collared Tanager (287) 47:3
Passerini's Tanager (287) 47:4
Blue-gray Tanager (291) 46:15
Palm Tanager (291) 45:19
Emerald Tanager (289) 46:5
Silver-throated Tanager (291) 46:6
Bay-headed Tanager (289) 46:9
Scarlet-thighed Dacnis (291) 46:4
Green Honeycreeper (293) 46:7
Red-legged Honeycreeper (293) 46:2
Blue-black Grassquit (297) 49:7
Variable Seedeater (295) 49:3
White-collared Seedeater (295) 49:2
Thick-billed Seed-Finch (295) 49:10
Yellow-faced Grassquit (297) 49:6
Orange-billed Sparrow (303) 48:17
Black-striped Sparrow (303) 50:14
Buff-throated Saltator (309) 48:2
Black-headed Saltator (309) 48:1

Slate-colored Grosbeak (309) 48:6
Blue-black Grosbeak (311) 48:10
Red-winged Blackbird (315) 44:14
Great-tailed Grackle (317) 44:16
Black-cowled Oriole (319) 44:5
Baltimore Oriole (321) 44:7
Yellow-billed Cacique (321) 44:12
Chestnut-headed Oropendola (323) 44:9
Montezuma Oropendola (323) 44:8
Yellow-throated Euphonia (327) 45:5

Nearby Birding Opportunities

Coter Lake Eco-Lodge

Forty-five minutes northwest of Arenal Volcano you will find Coter Lake Eco-Lodge, a comfortable hotel offering a number of attractions, including a canopy tour, horseback riding, and kayaking, as well as day trips to the volcano and hot springs. The lodge affords beautiful views of the volcano, Lake Arenal, and the smaller Coter Lake. For birders, Eco-Lodge provides some wonderful trails amid middle-elevation forest that lies at about 800 meters. The two most notable species for the area are Keel-billed Motmot and Rufous-winged Tanager.

Productive birding can be found on the hotel grounds. If you walk along the main road that connects reception to the cabins, you may find Smoky-brown Woodpecker, White-lined Tanager, Yellow-throated Euphonia, and Scarlet-thighed Dacnis. The rare Rufous-winged Tanager is also regularly found here, and Short-tailed Hawk often soars overhead. Alternatively, there is a longer foot-trail, Sendero Las Cabinas, which connects reception with the cabins. This leads through a small patch of forest and may yield Blue-black Grosbeak, Great Antshrike, Spotted Antbird, Bay Wren, and White-collared Manakin.

The best birding in the area, however, is off Sendero Ilusión and the smaller Sendero Heliconias, but be prepared to pay for it. Unfortunately, the management feels that they can charge $15 (US 2008) per person to enter these trails—even if you are staying at the lodge. Both trails lead off the main dirt road that continues past reception, and will provide a good half-day of birding. Here you can encounter Orange-bellied Trogon, Streak-crowned Antvireo, White-flanked Antwren, Tawny-faced Gnatwren, Song Wren, Black-headed Nightingale-Thrush, and Black-and-yellow Tanager. This is where the Keel-billed Motmot should be looked for as well.

To get to Coter Lake Eco-Lodge, begin at the Lake Arenal Dam and proceed 32.6 km toward Tilarán. Here (a few kilometers beyond the small town of Arenal), turn right onto a dirt road. In 2.8 km you will arrive at the lodge.

Coter Lake Eco-Lodge
Tel: (506) 2694-4480
Website: www.ecolodgecostarica.com
E-mail: recepcion@ecolodgecostarica.com

Birding Time: 4 hours
Elevation: 800 meters
Trail Difficulty: 2
Reserve Hours: N/A
Entrance Fee: N/A
An earthquake ravaged this area in 2009.
Check www.birdingcr.com for updated conditions.

Unlike many of the other popular birding sites in Costa Rica, Virgen del Socorro is not a large park or preserve. The site consists of a small dirt road that winds down into a beautiful forested gorge, crosses the rushing Río Sarapiquí, and climbs up the other side. The birding can be fantastic. With an elevation of 800 meters, Virgen del Socorro offers a wide variety of tanagers, flycatchers, and hummingbirds. The little-used road provides excellent viewing, and because of the steep embankment, you are at eye level with the canopy on one side. Here it is especially true that productivity is highest early in the morning or on cool, overcast days. Even a light rain can help the bird activity. When the weather gets hot and sunny, everything becomes quiet, although it is often easier to spot soaring raptors under these conditions.

Target Birds

Barred Hawk (41) 17:1
White Hawk (41) 17:2
Ornate Hawk-Eagle (49) 17:11
*Red-fronted Parrotlet (97) 19:15
Black Swift (117) 22:2
Brown Violetear (131) 23:6
Black-crested Coquette (137) 25:6
*Coppery-headed Emerald (133) 24:20
*White-bellied Mountain-gem (135) 24:5
Purple-crowned Fairy (125) 23:14
Lanceolated Monklet (151) 28:5
Scaly-throated Foliage-gleaner (165) 30:2
Rufous-rumped Antwren (183) 32:10
Immaculate Antbird (179) 31:10
Torrent Tyrannulet (201) 36:1
Rufous-browed Tyrannulet (193) 37:12
Black-headed Tody-Flycatcher (197) 37:8
Black-and-white Becard (217) 33:15
American Dipper (251) 39:11
Pale-vented Thrush (251) 39:5
*Black-and-yellow Tanager (283) 46:14
Crimson-collared Tanager (287) 47:3
*Blue-and-gold Tanager (283) 45:11
Emerald Tanager (289) 46:5
*Sooty-faced Finch (301) 48:21
Slate-colored Grosbeak (309) 48:6

Access

From the intersection of Rt. 126 and Rt. 140: (Refer to site 4B-1 for a map.) Take Rt. 126 toward Vara Blanca and Heredia for 10.2 km, at which point the dirt road leading into Virgen del Socorro branches off to the left. (Driving time from Rt. 140: 15 min)

From Alajuela: (Refer to site 4B-1 for a map.) Start at the central park and go 4 blocks north to a large four-way intersection with a traffic light. Proceed straight through the intersection onto Rt. 130 and drive for 17.8 km, at which point you will come to a T-intersection. Turn right, and in 2.8 km turn right again in the town of Poasito. Proceed 5.8 km to Vara Blanca, where you should turn left just before a gas station. In 15.6 km you will come to the turn for Virgen del Socorro on your right, which is a hard-to-see cutback onto a dirt road. (Driving time from Alajuela: 65 min)

Logistics: There are no easy lodging options in the immediate vicinity of Virgen del Socorro. Staying in Vara Blanca (refer to site 4B-1) is your best bet, although the area can also be accessed from Puerto Viejo (refer to site 1A-4).

Birding Sites

After turning off Rt. 126, proceed down the dirt road until you can find a safe place to park. The road dives down into the gorge for about a kilometer, and since this is the principal birding area you should take the time to walk it. Look for Black-crested Coquette, Brown Violetear, Spotted Woodcreeper, Bay Wren, and Immaculate Antbird. A variety of tanagers are also possibilities, including Scarlet-thighed Dacnis, Crimson-collared Tanager, Emerald Tanager, and Black-and-yellow Tanager. If the day gets hot, activity level drops precipitously, but you can often see raptors soaring over the ridges. Barred Hawk, White Hawk, Short-tailed Hawk, and Double-toothed Kite are all likely.

About 50 meters before the bridge over the Río Sarapiquí a trail leads off to the right, parallel with the river. The trailhead is not obvious, but once on the trail you should find it in good condition. It leads to a water-monitoring station a few hundred meters upriver. This area often yields Green-crowned Brilliant, Russet Antshrike, White-ruffed Manakin, Pale-vented Thrush, and Sooty-faced Finch among others.

The bridge over the river is a consistent place to find American Dipper, Torrent Tyrannulet, and Black Phoebe. Beyond this the road starts to climb and the habitat is similar to that on the other side.

Species to Expect

Black Vulture (31) 13:4
Turkey Vulture (31) 13:3
Swallow-tailed Kite (37) 15:2
Double-toothed Kite (35) 16:1
Barred Hawk (41) 17:1
White Hawk (41) 17:2
Broad-winged Hawk (43) 16:13
Short-tailed Hawk (43) 16:11
Short-billed Pigeon (87) 18:5
***Crimson-fronted Parakeet (95) 19:10**
Brown-hooded Parrot (97) 19:9
White-crowned Parrot (97) 19:7
Squirrel Cuckoo (103) 21:7
White-collared Swift (117) 22:1
Vaux's Swift (119) 22:9
Green Hermit (121) 23:3
Stripe-throated Hermit (121) 23:1
Violet Sabrewing (123) 23:9
White-necked Jacobin (125) 23:17
Brown Violetear (131) 23:6
Violet-headed Hummingbird (137) 25:11
Black-crested Coquette (137) 25:6
Green Thorntail (137) 25:1
Violet-crowned Woodnymph (127) 24:4

Rufous-tailed Hummingbird (129) 24:10
*Coppery-headed Emerald (133) 24:20
*White-bellied Mountain-gem (135) 24:5
Green-crowned Brilliant (123) 23:15
Purple-crowned Fairy (125) 23:14
Collared Trogon (143) 26:5
Red-headed Barbet (153) 28:2
Emerald Toucanet (153) 27:17
Collared Aracari (155) 27:15
Keel-billed Toucan (155) 27:18
Chestnut-mandibled Toucan (155) 27:19
Smoky-brown Woodpecker (159) 28:13
Red-faced Spinetail (163) 30:12
Spotted Barbtail (163) 29:5
Plain Xenops (169) 29:3
Olivaceous Woodcreeper (171) 29:7
Spotted Woodcreeper (173) 29:20
Russet Antshrike (177) 31:4
Immaculate Antbird (179) 31:10
Torrent Tyrannulet (201) 36:1
Olive-striped Flycatcher (195) 36:24
Slaty-capped Flycatcher (195) 36:26
Paltry Tyrannulet (191) 37:10
Scale-crested Pygmy-Tyrant (197) 37:4
Black-headed Tody-Flycatcher (197) 37:8
Eye-ringed Flatbill (201) 37:23
Yellow-olive Flycatcher (195) 37:16
Tufted Flycatcher (207) 36:11
Black Phoebe (207) 36:5
Bright-rumped Attila (201) 35:6
Dusky-capped Flycatcher (209) 35:21
Great Kiskadee (211) 35:13
Social Flycatcher (211) 35:14
*Golden-bellied Flycatcher (211) 35:9
Tropical Kingbird (213) 35:1
White-ruffed Manakin (223) 33:9
Yellow-throated Vireo (225) 40:6
Lesser Greenlet (229) 40:7
Blue-and-white Swallow (233) 22:20
Bay Wren (241) 38:12
House Wren (243) 38:18
Gray-breasted Wood-Wren (245) 38:14
Nightingale Wren (245) 38:21
American Dipper (251) 39:11
Black-headed Nightingale-Thrush (247) 38:25
Pale-vented Thrush (251) 39:5
Clay-colored Thrush (251) 39:8
Tropical Parula (257) 41:3
Chestnut-sided Warbler (259) 43:4
Black-throated Green Warbler (261) 43:9
Black-and-white Warbler (267) 41:13
Slate-throated Redstart (273) 42:7
Golden-crowned Warbler (275) 40:18
Buff-rumped Warbler (271) 40:23
Bananaquit (277) 40:24
Common Bush-Tanager (279) 45:13
***Black-and-yellow Tanager (283) 46:14**
Crimson-collared Tanager (287) 47:3
Passerini's Tanager (287) 47:4
Blue-gray Tanager (291) 46:15
Palm Tanager (291) 45:19
Emerald Tanager (289) 46:5
Silver-throated Tanager (291) 46:6
Speckled Tanager (289) 46:8
Bay-headed Tanager (289) 46:9
Golden-hooded Tanager (289) 46:13
Scarlet-thighed Dacnis (291) 46:4
Green Honeycreeper (293) 46:7
Shining Honeycreeper (293) 46:1
Variable Seedeater (295) 49:3
Yellow-faced Grassquit (297) 49:6
*Sooty-faced Finch (301) 48:21
Orange-billed Sparrow (303) 48:17
Slate-colored Grosbeak (309) 48:6
Baltimore Oriole (321) 44:7
Chestnut-headed Oropendola (323) 44:9
Montezuma Oropendola (323) 44:8
***Tawny-capped Euphonia (325) 45:6**

Nearby Birding Opportunities

Cinchona Hummingbird Feeders

Whether birding Virgen del Socorro or just traveling down Rt. 126, you should always make a stop at the hummingbird feeders in the town of

Cinchona. From Virgen del Socorro, the feeders are located 4.8 km up Rt. 126 heading toward Vara Blanca. There are actually two adjacent facilities that offer roughly the same services, hummingbirds and food. One has a large sign reading "Restaurante Mirador Cinchona." Check them both out if you like, but be sure to leave a tip. Here you will find a number of active hummingbird feeders hanging just off a balcony, where you can sit and enjoy a hot coffee or meal and watch the little birds as they buzz around your head. Expect to see Violet Sabrewing, Brown Violetear, Green Violetear, Coppery-headed Emerald, Green-crowned Brilliant, and if you are lucky, Green Thorntail and White-bellied Mountain-gem.

Site 1B-4: Braulio Carrillo National Park

Birding Time: 4 hours
Elevation: 500 meters
Trail Difficulty: 2
Reserve Hours: 8 a.m. to 4 p.m.
Entrance Fee: $10 (US 2008)
Trail Map Available on Site

Braulio Carrillo National Park protects 118,500 acres of mostly virgin forest along the Caribbean Slope of the Cordillera Central. In 1977 construction began on the San José-Guápiles Road, now called Rt. 32. A year later Braulio Carrillo National Park was created, in part to protect the delicate hillsides through which the highway runs. Braulio Carrillo is quite progressive in its design, protecting both high- and low-elevation habitats and thus providing a corridor for animals to move up and down the slope.

Unfortunately for birders, access to the park is limited, most of the forest being inaccessible. Previously there were more trails into the park off Rt. 32, but a series of attacks on tourists by thieves forced the park service to close these. Currently, the only trails off the main highway that the park officially maintains are located at the Quebrada Gonzalez Ranger Station. The elevation of Quebrada Gonzalez is about 500 meters, and the trails lead into beautiful primary forest. While any one trip to the area will yield only a fraction of the potential species found here, it is possible to spot many rarities. Purplish-backed Quail-Dove, Yellow-eared Toucanet, Black-crowned Antpitta, and Ashy-throated Bush-Tanager are just a few examples of species found here that are hard to find elsewhere in Costa Rica.

A visit to Quebrada Gonzalez is easy to fit into many itineraries, thanks to its convenient location along Rt. 32. It also makes for a very easy day trip from San José by either car or public bus.

Target Birds

*Black-eared Wood-Quail (13) 12:12
Black-and-white Hawk-Eagle (49) 17:13
Ornate Hawk-Eagle (49) 17:11
Purplish-backed Quail-Dove (93) 18:21
*Red-fronted Parrotlet (97) 19:15
*Lattice-tailed Trogon (145) 26:3
Lanceolated Monklet (151) 28:5
*Yellow-eared Toucanet (155) 27:20
*Rufous-winged Woodpecker (159) 28:10
Cinnamon Woodpecker (161) 28:9
Striped Woodhaunter (165) 30:4
Brown-billed Scythebill (175) 29:14
*Streak-crowned Antvireo (181) 32:7
Checker-throated Antwren (183) 32:5
Dull-mantled Antbird (181) 31:14
Immaculate Antbird (179) 31:10
*Black-crowned Antpitta (185) 30:19
Rufous-browed Tyrannulet (193) 37:12
Ruddy-tailed Flycatcher (199) 36:23
*Gray-headed Piprites (215) 33:3
White-ruffed Manakin (223) 33:9
White-crowned Manakin (223) 33:8
Sharpbill (225) 34:7
Green Shrike-Vireo (229) 40:1
Tawny-faced Gnatwren (237) 32:14
Black-headed Nightingale-Thrush (247) 38:25
Pale-vented Thrush (251) 39:5
Ashy-throated Bush-Tanager (279) 45:12
*Black-and-yellow Tanager (283) 46:14
*White-throated Shrike-Tanager (283) 47:1
*Blue-and-gold Tanager (283) 45:11
Emerald Tanager (289) 46:5
Speckled Tanager (289) 46:8
Slate-colored Grosbeak (309) 48:6

Access

From San José: Take Rt. 32 heading toward Guápiles and Puerto Limón. At the toll plaza, which is about 15 km outside San José, turn your trip odometer to zero. Proceed straight for 27.6 km to the Quebrada Gonzalez Ranger Station, which will be on your right. (Driving time from San José: 55 min)

From the intersection of Rt. 32 and Rt. 4: Head toward San José on Rt. 32 for 8.4 km to the Quebrada Gonzalez Ranger Station, which will be on your left. (Driving time from Rt. 4: 10 min)

Logistics: San José is a good base from which to access Quebrada Gonzalez because it offers a full range of accommodations. Just be sure to get an early start to avoid the morning rush hour. Alternatively, you could stay in Puerto Viejo or Guápiles on the Caribbean side. While there is no food available inside the park, there are a number of restaurants located at the intersection of Rt. 32 and Rt. 4, just 10 minutes away.

Birding Sites

Upon arriving at Quebrada Gonzalez, the first thing to do is check out the trees around the buildings and parking area. These often hold mixed flocks of tanagers, which could include Emerald Tanager, Blue-and-gold Tanager, Black-and-yellow Tanager, Golden-hooded Tanager, and Bay-headed Tanager. It is also a good place to look for Cinnamon Woodpecker. The small streambed that runs along the right side of the driveway (in between the entrance and exit of the

Sendero Las Palmas) often yields Dull-mantled Antbird and seems to consistently attract mixed-species flocks.

The Quebrada Gonzalez station has a small trail system that provides access to beautiful primary forest. Sendero Las Palmas, which leaves just to the right of the park buildings, does a 1.6 km loop and is consistently the most productive trail. Most of the time the forest at Quebrada Gonzalez is fairly quiet, although White-breasted Wood-Wren, Nightingale Wren, Black-headed Nightingale-Thrush, and Lattice-tailed Trogon can often be encountered if you are patient and observant. Also, be sure to keep an eye open for Purplish-backed Quail-Dove feeding in the trail. Diverse mixed-species flocks are the main attraction at Quebrada Gonzales and are relatively easy to find. Black-faced Grosbeak, Tawny-crested Tanager, Carmiol's Tanager, and Dusky-faced Tanager are the most common species to be leading the flocks. However, many other species are often feeding with them. The list of birds that could join them is large, but the more reliable and interesting species include Slate-colored Grosbeak, White-throated Shrike-Tanager, Striped Woodhaunter, Checker-throated Antwren, and Ashy-throated Bush-Tanager. If you are really lucky you may find a swarm of army ants, which could be accompanied by Oscillated Antbird and Black-crowned Antpitta. Be forewarned, however, that birding at this location can be difficult because of the narrow path and dense vegetation.

The other trails are located across the highway from the ranger station. Sendero El Ceibo is a 1 km loop, at the far end of which starts Sendero La Botarrama, which forms an additional 2.5 km loop, and is a good place to find Black-eared Wood-Quail. These trails are worth exploring if you have the time, although the birding is generally not as productive as on Sendero Las Palmas.

Species to Expect

Black Vulture (31) 13:4
Turkey Vulture (31) 13:3
Tiny Hawk (39) 16:2
Ornate Hawk-Eagle (49) 17:11
Short-billed Pigeon (87) 18:5
Purplish-backed Quail-Dove (93) 18:21
Ruddy Quail-Dove (93) 18:17
Brown-hooded Parrot (97) 19:9
Squirrel Cuckoo (103) 21:7
Green Hermit (121) 23:3
Long-billed Hermit (121) 23:2
Stripe-throated Hermit (121) 23:1
Violet-crowned Woodnymph (127) 24:4
Green-crowned Brilliant (123) 23:15
Purple-crowned Fairy (125) 23:14
Black-throated Trogon (141) 26:9
Slaty-tailed Trogon (145) 26:2
***Lattice-tailed Trogon (145) 26:3**
Broad-billed Motmot (147) 27:9
White-whiskered Puffbird (151) 28:6
Red-headed Barbet (153) 28:2
Collared Aracari (155) 27:15
*Yellow-eared Toucanet (155) 27:20
Chestnut-mandibled Toucan (155) 27:19
Black-cheeked Woodpecker (157) 28:14
Smoky-brown Woodpecker (159) 28:13
*Rufous-winged Woodpecker (159) 28:10
Cinnamon Woodpecker (161) 28:9
Striped Woodhaunter (165) 30:4
Buff-throated Foliage-gleaner (165) 30:3
Plain Xenops (169) 29:3
Wedge-billed Woodcreeper (171) 29:6
Northern Barred-Woodcreeper (169) 29:19
Spotted Woodcreeper (173) 29:20

Russet Antshrike (177) 31:4
*Streak-crowned Antvireo (181) 32:7
Dot-winged Antwren (183) 32:1
Chestnut-backed Antbird (179) 31:11
Dull-mantled Antbird (181) 31:14
Spotted Antbird (181) 31:12
Ochre-bellied Flycatcher (199) 36:25
Scale-crested Pygmy-Tyrant (197) 37:4
Common Tody-Flycatcher (197) 37:7
Ruddy-tailed Flycatcher (199) 36:23
Sulphur-rumped Flycatcher (199) 36:21
Tufted Flycatcher (207) 36:11
Rufous Mourner (201) 34:9
Boat-billed Flycatcher (211) 35:12
Tropical Kingbird (213) 35:1
Rufous Piha (215) 34:10
Cinnamon Becard (215) 33:11
Masked Tityra (217) 34:1
White-ruffed Manakin (223) 33:9
White-crowned Manakin (223) 33:8
Tawny-crowned Greenlet (229) 32:4
Lesser Greenlet (229) 40:7
Green Shrike-Vireo (229) 40:1
Bay Wren (241) 38:12
***Stripe-breasted Wren (241) 38:5**
White-breasted Wood-Wren (245) 38:15
Nightingale Wren (245) 38:21
Song Wren (245) 38:22
Tawny-faced Gnatwren (237) 32:14
Black-headed Nightingale-Thrush (247) 38:25
Pale-vented Thrush (251) 39:5
Golden-winged Warbler (257) 41:5
Tropical Parula (257) 41:3
Chestnut-sided Warbler (259) 43:4
Blackburnian Warbler (261) 41:8
Black-and-white Warbler (267) 41:13
Canada Warbler (269) 42:8
Buff-rumped Warbler (271) 40:23
Bananaquit (277) 40:24
Ashy-throated Bush-Tanager (279) 45:12
***Black-and-yellow Tanager (283) 46:14**
Dusky-faced Tanager (281) 45:18
Carmiol's Tanager (279) 45:16
*White-throated Shrike-Tanager (283) 47:1
White-shouldered Tanager (281) 46:18
Tawny-crested Tanager (281) 46:17
Summer Tanager (285) 47:5
Blue-gray Tanager (291) 46:15
*Blue-and-gold Tanager (283) 45:11
Emerald Tanager (289) 46:5
Silver-throated Tanager (291) 46:6
Speckled Tanager (289) 46:8
Bay-headed Tanager (289) 46:9
Golden-hooded Tanager (289) 46:13
Scarlet-thighed Dacnis (291) 46:4
Green Honeycreeper (293) 46:7
Shining Honeycreeper (293) 46:1
Orange-billed Sparrow (303) 48:17
Slate-colored Grosbeak (309) 48:6
Black-faced Grosbeak (311) 48:5
Blue-black Grosbeak (311) 48:10
Baltimore Oriole (321) 44:7
Scarlet-rumped Cacique (321) 44:11
***Tawny-capped Euphonia (325) 45:6**

Nearby Birding Opportunities

El Tapir Reserve

Another place to check is El Tapir, located 1.8 km downslope (heading away from San José) from the Quebrada Gonzalez station, on the right. The main attraction here is the hummingbirds. For $5 (US 2008) per person you can enter the grounds, which have been planted with many porterweed bushes, a favorite of hummingbirds. Species of note include Green Thorntail, Black-crested Coquette, Snowcap, White-necked Jacobin, Brown Violetear, and Violet-headed Hummingbird, although many others are possibilities. Because all the flowering bushes are at eye level, the viewing of these energetic little birds is excellent.

El Tapir also has a system of trails through beautiful forest similar to that of Quebrada Gonzalez. To walk the trails you will need to hire a guide from the reserve, although the extra cost is reasonable. Expect to see a range of species similar to those found along the Quebrada Gonzalez trails.

El Tapir opens early. You can get in to see the hummingbirds by 6 a.m., and walking the trails becomes possible by about 7 a.m. (when there is enough light to safely see the trail). Be aware, however, that because this is a local operation you are not guaranteed to find somebody operating the facility.

Aerial Tram

The Aerial Tram is another nearby location that offers good birding as well as a unique experience—being transported through the forest canopy by a giant chairlift. The habitat here is similar to that at Quebrada Gonzalez and El Tapir, but the birding can be easier because of the paved trails and knowledgeable guides. Black-crested Coquet, Snowcap, Rufous-vented Ground-Cuckoo, Black-crowned Antpitta, White-tipped Sicklebill, and Speckled Mourner are some of the rarities that are possible to see, along with the more common species. If you want to bird the area, be sure to make a reservation in advance.

Aerial Tram
Tel: (506) 2257-5961
Website: www.rainforestrams.com
E-mail: reservations5.cr@rsat.com

Site 1B-5: Rara Avis Rainforest Lodge

Birding Time: 3 days
Elevation: 700 meters
Trail Difficulty: 2–3
Reserve Hours: N/A
Entrance Fee: Included with overnight
Bird Guides Available
Trail Map Available on Site

In 1983 Rara Avis opened its doors as one of the world's first ecotourism lodges. The vision for Rara Avis, situated on a secluded 1,200-acre private reserve bordering the immense Braulio Carrillo National Park, was to create a destination that put the traveler right in the heart of the rainforest. Away from all civilization, adventurous birders and ecotourists would be able to experience the wild beauty of the undisturbed natural world, a goal that was certainly realized.

Rara Avis Lodge is found only 15 kilometers from the small town of Horquetas, yet the journey between the two takes almost four hours because of poor road conditions. After checking in at the office in Horquetas, you will load into an old army transport vehicle, which will then bounce up into the wilderness. A few kilometers from the lodge, the road becomes impassible even for this truck and you will be transferred into a large trailer, pulled by a farm tractor, for the final stretch.

When you finally arrive at the lodge you will find comfortable hotel rooms, a small lounge/classroom, and a communal dining area, where three meals are served daily. All of the buildings are rustic, built from wood taken from fallen trees in the surrounding forest. While there is a bit of hot water in the showers and the beds are comfortable, the electricity is very limited and there are no washing machines.

An extensive 16-kilometer trail system leads through the stunningly beautiful forest of Rara Avis. Be forewarned that the footing on these often-muddy trails can be difficult. Because it borders Braulio Carrillo National Park, the forest still holds its full complement of wildlife, including jaguar, puma, tapir, bushmaster, and other amazing creatures. While your chances of seeing many of these animals are quite low, it is still exciting to happen across the tracks of a large cat or simply to know that they are out there. The birding is excellent, and many rarities and Caribbean Slope specialties can be found in the Rara Avis forest.

Rara Avis is clearly not a destination for everyone, and travelers with health issues should think twice about visiting here. Its remote location, amusement park–like tractor ride, and slippery trails make this a difficult and dangerous destination for many people. However, for those who are up to it, it will be an exciting and unforgettable experience.

Target Birds

*Black-eared Wood-Quail (13) 12:12
White Hawk (41) 17:2
Great Black-Hawk (45) 13:7
Red-throated Caracara (49) 14:11
Sunbittern (29) 6:16
Olive-backed Quail-Dove (93) 18:22
Purplish-backed Quail-Dove (93) 18:21
*Red-fronted Parrotlet (97) 19:15
*Snowcap (137) 25:8
*Lattice-tailed Trogon (145) 26:3
*Yellow-eared Toucanet (155) 27:20
Striped Woodhaunter (165) 30:4
Tawny-throated Leaftosser (167) 30:11
Dull-mantled Antbird (181) 31:14
Immaculate Antbird (179) 31:10
Black-headed Antthrush (187) 30:17
*Black-crowned Antpitta (185) 30:19
*Thicket Antpitta (185) 30:14
Rufous-browed Tyrannulet (193) 37:12
*Gray-headed Piprites (215) 33:3
White-crowned Manakin (223) 33:8
Sharpbill (225) 34:7
Black-headed Nightingale-Thrush (247) 38:25
Pale-vented Thrush (251) 39:5
Ashy-throated Bush-Tanager (279) 45:12
*Black-and-yellow Tanager (283) 46:14
*Blue-and-gold Tanager (283) 45:11
Emerald Tanager (289) 46:5
*Sooty-faced Finch (301) 48:21
Slate-colored Grosbeak (309) 48:6

Access

From the intersection of Rt. 4 and Rt. 32: (Refer to site 1A-4 for a map.) Take Rt. 4 toward Puerto Viejo for 15.6 km. Turn left here on what is currently the second signed entrance to the town of Horquetas. The small Rara Avis office will be on your right in 0.2 km. (Driving time from Rt. 32: 15 min)

From Puerto Viejo de Sarapiquí: (Refer to site 1A-4 for a map.) Take Rt. 4 toward Rt. 32 and Guápiles. After 15.2 km, turn right onto a small paved road that leads into the town of Horquetas. In 0.2 km the small Rara Avis office will be on your right. (Driving time from Puerto Viejo: 15 min)

Logistics: At Horquetas there is a guarded parking area where you can park your car while at Rara Avis. The transport vehicle usually leaves Horquetas at 9 a.m., arriving at Rara Avis in time for lunch. On the return trip, the tractor leaves Rara Avis after lunch, getting you back to civilization by about 4 p.m. Aside from lodging, Rara Avis provides three meals a day as well as local guides. If you need anything else during your stay, you should be sure to bring it with you.

Rara Avis
Tel: (506) 2764-1111
Website: www.rara-avis.com
E-mail: reservations@rara-avis.com

Birding Sites

Although Rara Avis does not offer much in the way of new-growth habitats, it does offer some of the most beautiful, pristine, and secluded forests in the country, filled with interesting birds. Many of the middle-elevation Caribbean Slope specialty birds such as Lattice-tailed Trogon, Purplish-backed Quail-Dove, Ashy-throated Bush-Tanager, Blue-and-gold Tanager, and White-crowned Manakin can reliably be found at Rara Avis.

You will find second-growth and open-forest habitat along trails such as Bromelia, Danta, Azul, Guácimo, and the Corduroy Road. These are good places to look for a wide range of tanagers. Some of the more interesting species include Ashy-throated Bush-Tanager, Black-and-yellow Tanager, Tawny-crested Tanager, Hepatic Tanager, Blue-and-gold Tanager, and Emerald Tanager. These trails are also good places to find Immaculate Antbird, Thicket Antpitta, Bay Wren, and Russet Antshrike. Along Danta and Morpho trails there are also two White-crowned Manakin leks, which the guides will be pleased to show you.

Other trails, such as Catarata, Plantanilla, Carril, and Ronda Norte, lead through primary forest and can yield some very interesting birds. Nightingale Wren and Song Wren are often heard singing. Trogons such as Black-throated Trogon, Violaceous Trogon, and the rare Lattice-tailed Trogon can all be found here. The understory harbors a number of interesting species, including Slaty-breasted Tinamou, Olive-backed Quail-Dove, Purplish-Backed Quail-Dove, Tawny-throated Leaftosser, and Black-eared Wood-Quail. Other birds to look for include Dull-mantled Antbird, Streak-crowned Antvireo, Yellow-eared Toucanet, Tawny-faced Gnatwren, and Black-headed Nightingale-Thrush.

Around the buildings is a good place to hear Gray-necked Wood-Rail early in the morning. Near the *casitas* (cabins), it is easy to see the snapping displays of White-collared Manakins at their lek. There are some hummingbird feeders around the dining hall, which principally attract Violet-crowned Woodnymph, although you can also see Green Hermit, Green-crowned Brilliant, and Violet-headed Hummingbird. This is also a good place to see raptors circling overhead such as Swallow-tailed Kite, White Hawk, and Barred Hawk.

Species to Expect

Little Tinamou (3) 12:17
Slaty-breasted Tinamou (3) 12:8
Crested Guan (11) 12:4
Black Vulture (31) 13:4
Turkey Vulture (31) 13:3
Swallow-tailed Kite (37) 15:2
Sharp-shinned Hawk (39) 16:3
Barred Hawk (41) 17:1
White Hawk (41) 17:2
Barred Forest-Falcon (53) 16:5
White-throated Crake (57) 6:9
Gray-necked Wood-Rail (55) 6:13
Red-billed Pigeon (87) 18:2
Short-billed Pigeon (87) 18:5
Gray-chested Dove (91) 18:16
Olive-backed Quail-Dove (93) 18:22
Purplish-backed Quail-Dove (93) 18:21
Orange-chinned Parakeet (97) 19:14
Brown-hooded Parrot (97) 19:9
White-crowned Parrot (97) 19:7
Red-lored Parrot (99) 19:4
Squirrel Cuckoo (103) 21:7
Crested Owl (105) 20:3
Common Pauraque (111) 21:18
White-collared Swift (117) 22:1
Gray-rumped Swift (119) 22:10
Green Hermit (121) 23:3
Stripe-throated Hermit (121) 23:1
Brown Violetear (131) 23:6
Violet-headed Hummingbird (137) 25:11
Black-crested Coquette (137) 25:6
Violet-crowned Woodnymph (127) 24:4
*Black-bellied Hummingbird (133) 24:21
*Snowcap (137) 25:8
Green-crowned Brilliant (123) 23:15
Purple-crowned Fairy (125) 23:14
Violaceous Trogon (141) 26:8
Collared Trogon (143) 26:5
Black-throated Trogon (141) 26:9
Slaty-tailed Trogon (145) 26:2
***Lattice-tailed Trogon (145) 26:3**
Broad-billed Motmot (147) 27:9
Collared Aracari (155) 27:15
Keel-billed Toucan (155) 27:18
Chestnut-mandibled Toucan (155) 27:19
Black-cheeked Woodpecker (157) 28:14
Smoky-brown Woodpecker (159) 28:13
Cinnamon Woodpecker (161) 28:9
Pale-billed Woodpecker (161) 27:14
Buff-throated Foliage-gleaner (165) 30:3
Tawny-throated Leaftosser (167) 30:11
Wedge-billed Woodcreeper (171) 29:6
Spotted Woodcreeper (173) 29:20
Russet Antshrike (177) 31:4
Plain Antvireo (181) 32:6
*Streak-crowned Antvireo (181) 32:7
Dusky Antbird (177) 31:8
Chestnut-backed Antbird (179) 31:11
Dull-mantled Antbird (181) 31:14
Immaculate Antbird (179) 31:10
Black-faced Antthrush (187) 30:15
***Thicket Antpitta (185) 30:14**
Paltry Tyrannulet (191) 37:10
Scale-crested Pygmy-Tyrant (197) 37:4
Black-headed Tody-Flycatcher (197) 37:8
Tufted Flycatcher (207) 36:11
Olive-sided Flycatcher (203) 36:4

Western Wood-Pewee (203) 36:7
Tropical Pewee (203) 36:9
Bright-rumped Attila (201) 35:6
Rufous Mourner (201) 34:9
Dusky-capped Flycatcher (209) 35:21
Great Kiskadee (211) 35:13
Boat-billed Flycatcher (211) 35:12
Tropical Kingbird (213) 35:1
Thrush-like Schiffornis (215) 33:10
Rufous Piha (215) 34:10
Cinnamon Becard (215) 33:11
Masked Tityra (217) 34:1
White-collared Manakin (221) 33:1
White-ruffed Manakin (223) 33:9
White-crowned Manakin (223) 33:8
Red-eyed Vireo (227) 40:3
Lesser Greenlet (229) 40:7
Green Shrike-Vireo (229) 40:1
Bay Wren (241) 38:12
***Stripe-breasted Wren (241) 38:5**
White-breasted Wood-Wren (245) 38:15
Nightingale Wren (245) 38:21
Song Wren (245) 38:22
Tawny-faced Gnatwren (237) 32:14
Long-billed Gnatwren (237) 32:15
Black-headed Nightingale-Thrush (247) 38:25
Swainson's Thrush (249) 39:2
Pale-vented Thrush (251) 39:5
Clay-colored Thrush (251) 39:8
Chestnut-sided Warbler (259) 43:4
Canada Warbler (269) 42:8
Buff-rumped Warbler (271) 40:23
Bananaquit (277) 40:24
Common Bush-Tanager (279) 45:13
Ashy-throated Bush-Tanager (279) 45:12
***Black-and-yellow Tanager (283) 46:14**
Carmiol's Tanager (279) 45:16
White-shouldered Tanager (281) 46:18
Tawny-crested Tanager (281) 46:17
Hepatic Tanager (285) 47:6
Passerini's Tanager (287) 47:4
Blue-gray Tanager (291) 46:15
*Blue-and-gold Tanager (283) 45:11
Emerald Tanager (289) 46:5
Speckled Tanager (289) 46:8
Golden-hooded Tanager (289) 46:13
Scarlet-thighed Dacnis (291) 46:4
Green Honeycreeper (293) 46:7
Shining Honeycreeper (293) 46:1
Red-legged Honeycreeper (293) 46:2
Orange-billed Sparrow (303) 48:17
Buff-throated Saltator (309) 48:2
Slate-colored Grosbeak (309) 48:6
Black-faced Grosbeak (311) 48:5
Blue-black Grosbeak (311) 48:10
Scarlet-rumped Cacique (321) 44:11
Chestnut-headed Oropendola (323) 44:9
Olive-backed Euphonia (325) 45:3
White-vented Euphonia (327) 45:7
***Tawny-capped Euphonia (325) 45:6**

Site 1B-6: Guayabo National Monument

Birding Time: 4 hours
Elevation: 1,000 meters
Trail Difficulty: 1
Reserve Hours: 8 a.m. to 3:30 p.m.
Entrance Fee: $6 (US 2008)
4×4 Recommended
Trail Map Available on Site

Guayabo National Monument is Costa Rica's most famous archeological site. The area has been only partially excavated, but a well-maintained trail leads through an impressive ancient city, and lets you catch a glimpse of the bygone society that once occupied this beautiful countryside. Archeologists believe that the city was the center of a chiefdom roughly 1,000 years ago. Guayabo was not part of a large civilization like the Aztecs to the north, or the Incas to the south (both contemporaries). Instead, it was the nucleus of a smaller society, with its own political, commercial, and religious sovereignty. It is estimated that over 10,000 people once lived in and around the city of Guayabo, farming and trading with other chiefdoms. They were skilled masons, as evidenced by the impressive ruins they left behind. At the site you will find cobblestone roadways, burial mounds, house foundations, and a series of impressive subterranean aqueducts, two of which are still operational.

For birders Guayabo can be a pleasant stop. The habitat in the area is mostly second-growth forest, and well-maintained trails provide easy access. While it is not one of the best birding locations around, it is an interesting stop if you like archeology.

Access

From Turrialba: Take the road that leads north out of town, immediately crossing over a bridge. From the bridge proceed straight for 1.0 km until the road comes to a four-way intersection. Follow the main road, as it turns right. Go 12.6 km from this intersection and you should turn left onto a dirt road following signs for Guayabo. You will arrive at the park in 4.2 km. (Driving time from Turrialba: 30 min)

Logistics: There are a few local hotels in the town of Turrialba. If you stay in one of these, be sure to park your car in a protected parking lot for the night.

Birding Sites

One of the most productive birding areas of Guayabo can be the picnic grounds/campsite and the road at the entrance to the park. The trees around the picnic tables often hold mixed flocks of Golden-hooded Tanager, Bay-headed

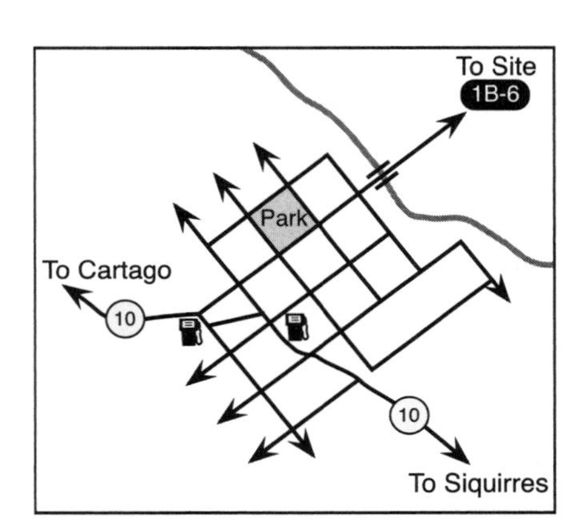

Turrialba City Map

Tanager, White-shouldered Tanager, and Golden-winged Warbler. From here, walk farther up the road, birding the edge habitat on either side. It may yield Buff-throated Foliage-gleaner, Green-fronted Lancebill, Tropical Pewee, and Brown Jay. Walking back toward Turrialba, you may find Yellow-faced Grassquit, Thick-billed Seed-Finch, and Yellow-bellied Elaenia in the scrubby fields.

Beyond the ticket building, the park maintains two trails. I suggest first walking the trail that leaves to the left, which brings you past the excavated ruins. The first half of this trail runs through second-growth forest and offers productive birding. Stripe-breasted Wrens and White-breasted Wood-Wrens call from the undergrowth. Mixed flocks of Golden-crowned Warblers move low through the forest and often include Slaty Antwren, Spotted Barbtail, and White-ruffed Manakin, as well as middle-elevation flycatchers such as Slaty-capped Flycatcher, Scale-crested Pygmy-Tyrant, and Olive-striped Flycatcher.

You will soon emerge from the forest at the site of the excavated city. The edge habitat around the ruins can hold some interesting species. This is an excellent place to look for Immaculate Antbird, although it often stays hidden in the thick tangle of undergrowth. Listen for its clear repeated whistles. Bay Wren and Band-backed Wren are also common here, and you might find Crimson-collared Tanager as well as Black-headed Saltator. If they are with fruit, the trees around the ruins can be a good place to find flocks of tanagers and warblers.

The other trail within the park completes a short loop down to a small river and does not pass any ancient ruins. The birding along this trail is generally less productive, but it is a good place to find a lek of White-collared Manakins, so keep an ear open for their distinctive snapping. Other birds you may encounter include Gray-headed Chachalaca, Keel-billed Toucan, Collared Aracari, and Orange-billed Sparrow.

Nearby Birding Opportunities

C.A.T.I.E.

C.A.T.I.E. is a research and education facility for tropical agriculture located just three kilometers east of Turrialba along Rt. 10. The facility maintains a botanical garden and manicured grounds that you can stroll. The entrance fee is $5 (US 2008).

One of the most productive places is a small pond near the entrance, where you can find Northern Jacana, Purple Gallinule, Tricolored Heron, and Yellow-crowned Night-Heron. If you scan the stand of bamboo on the small island, you can usually find a Boat-billed Heron or two, which feed at the pond during the night.

Touring the botanical gardens can be interesting, although the habitat is not as productive as you might expect. Because the facility focuses on tropical agriculture, the botanical garden feels more like a small farm than a garden. Even so, there are many birds about, including Red-billed Pigeon, Roadside Hawk, White-winged Becard, and Cinnamon Becard. It is also a reliable place to find Green Ibis feeding in the grass.

Site 1B-7: Rancho Naturalista

Birding Time: 2–5 days
Elevation: 850 meters
Trail Difficulty: 2
Reserve Hours: N/A
Entrance Fee: Included with overnight
Bird Guides Available

Rancho Naturalista is something of a birder's inn. Here you will find warm hospitality, comfortable rooms, great food, expert bird guides, and plenty of birds. The main lodge serves as the center of activity. On the first floor there is a small lounge and the dining area where three delicious meals are served daily, and which affords views of the fruit feeders in the backyard. On the second floor is an observation deck, where you get a bird's eye view of the activity at the hummingbird and fruit feeders, as well as enjoy a magnificent vista of Irazú and Turrialba volcanoes. A few guest rooms are located in the main lodge, and more can be found in three nearby buildings. This allows Rancho Naturalista to comfortably host about 30 people.

Rancho Naturalista caters to birders in a way that few other lodges do. They offer very knowledgeable bird guides to take you around the 125-acre property and help you find many exciting species. They can also arrange morning and day trips to a number of interesting nearby locations. Three of these, Silent Mountain, Tuis River, and La Mina, are described below. Other areas, such as Irazú Volcano (4B-2) and C.A.T.I.E. (1B-6), are described elsewhere in this volume. Because of the excellent guides, the many rarities, and the diversity of species and habitats found near Rancho Naturalista, many birders choose to use this well-appointed lodge as a comfortable base of operations for four or five days of birding.

Target Birds

Bicolored Hawk (39) 16:8
Sunbittern (29) 6:16
Purplish-backed Quail-Dove (93) 18:21
Mottled Owl (107) 20:6
White-necked Jacobin (125) 23:17
Green-breasted Mango (129) 23:13
Black-crested Coquette (137) 25:6
Green Thorntail (137) 25:1
*Snowcap (137) 25:8
Bronze-tailed Plumeleteer (129) 23:12
Purple-crowned Fairy (125) 23:14
Lanceolated Monklet (151) 28:5
*Rufous-winged Woodpecker (159) 28:10

Golden-olive Woodpecker (159) 28:11
Tawny-throated Leaftosser (167) 30:11
Brown-billed Scythebill (175) 29:14
Checker-throated Antwren (183) 32:5
Dull-mantled Antbird (181) 31:14
Immaculate Antbird (179) 31:10
Black-headed Tody-Flycatcher (197) 37:8
*Tawny-chested Flycatcher (199) 36:10
White-ruffed Manakin (223) 33:9
White-crowned Manakin (223) 33:8
*Black-throated Wren (243) 38:13
Scaly-breasted Wren (245) 38:20
Black-headed Nightingale-Thrush (247) 38:25
Ashy-throated Bush-Tanager (279) 45:12
White-lined Tanager (281) 46:16

Access

From Turrialba: Head east on Rt. 10 for 6.0 km and then turn right onto Rt. 232, following signs for La Suiza. (**From Siquirres:** Take Rt. 10 west for 39.6 km and then turn left onto Rt. 232, following signs for La Suiza.) Continue straight on the main paved road for 11.2 km, and the entrance to Rancho Naturalista will be on your left. (Driving time from Turrialba: 20 min; from Siquirres: 65 min)

Logistics: If you plan to bird Rancho Naturalista, you must stay at the lodge, which includes all of your meals. If you do not have a rental car, the lodge can help you to arrange transportation from San José or Turrialba.

Rancho Naturalista
Tel: (506) 2554-8101
Website: www.ranchonaturalista.net
E-mail: info@ranchonaturalista.net

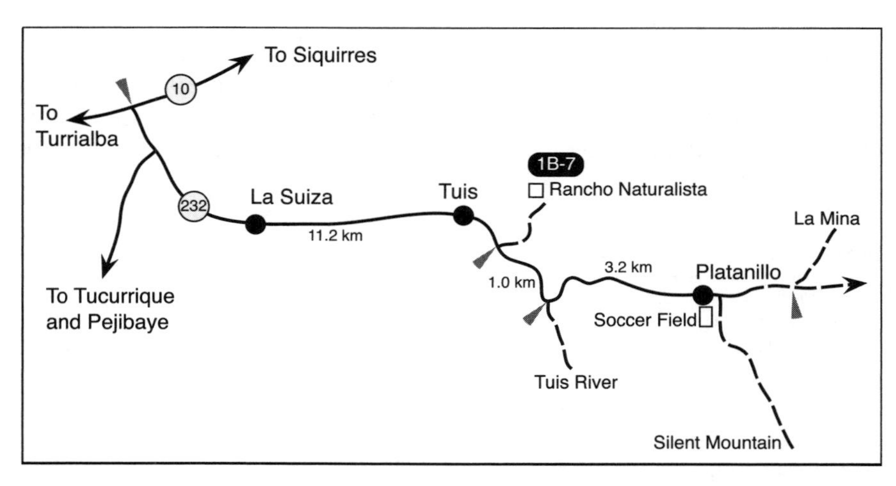

Rancho Naturalista Area Map. Tuis River, La Mina, and Silent Mountain are included.

Birding Sites

The expert guides provided for you at Rancho Naturalista can take you to all the birding locations described below. For this reason I have omitted the precise directions to these areas.

Rancho Naturalista

The grounds right around the lodge can be a productive area to look for birds. Hoffmann's Woodpecker, Montezuma Oropendola, and Scarlet-rumped Cacique are commonly found here. Gray-headed Chachalaca, White-lined Tanager, Green Honeycreeper, and Black-cheeked Woodpecker frequent the fruit feeders set up behind the lodge. On the second-floor porch you will find hummingbird feeders buzzing with activity. White-necked Jacobin, Green-breasted Mango, Violet-crowned Woodnymph, Brown Violetear, and Green-crowned Brilliant are among the more regular species to be seen at these feeders, and all can be seen right in front of your nose! A patch of porterweed flowers near the small parking area sometimes attracts smaller hummingbird species such as Green Thorntail, Violet-headed Hummingbird, and Black-crested Coquette. These are most reliably found early in the morning.

Following a small trail about 200 meters into the forest you will find a second hummingbird feeding station. While some of the regulars here are the same as those found near the lodge, also look for Bronze-tailed Plumeleteer, Snowcap, Violet Sabrewing, and Green Hermit. On the ground below the feeders you may find Orange-billed Sparrow and Chestnut-capped Brush-Finch.

Be sure not to miss the hummingbird pools in the streambed, a short walk from the lodge. In the late afternoon many birds find these naturally formed pools an enticing place to take a bath. The hummingbirds hover, then drop briefly into the water, then hover again. Other birds stand at the edge of the pools and splash about. Species to look for include Violet-crowned Woodnymph, Purple-crowned Fairy, Snowcap, Black-headed Nightingale-Thrush, and Tawny-throated Leaftosser.

Rancho Naturalista's well-developed trail system accesses a variety of habitats including pasture and secondary and primary forest. The guides know exactly where key species can be found at any particular time. Some of the more reliable birds of interest include Tawny-chested Flycatcher, Purplish-backed Quail-Dove, Scaled Antpitta, White-crowned Manakin, Black-headed Nightingale-Thrush, and Ashy-throated Bush-Tanager. At night it is easy to hear Common Paraque and Mottled Owl, and if you are lucky, you may hear a Crested Owl or Common Potoo.

Tuis River

Another interesting place to bird near Rancho Naturalista is a small trail that leads along the Tuis River, through forest fragments and pastures. This is an easy half-day walk that often produces tanagers such as Black-and-yellow Tanager,

Ashy-throated Bush-Tanager, White-shouldered Tanager, Emerald Tanager, and Speckled Tanager. This is also a good place to find water-based species including Torrent Tyrannulet, Black Phoebe, American Dipper, Buff-rumped Warbler, and Sunbittern. Even the very rare Lanceolated Monklet occasionally shows up here.

La Mina

At La Mina you will find a dirt road that proceeds through a small wooded gorge along a rushing river. Look for Green Kingfisher, Amazon Kingfisher, Buff-rumped Warbler, Black Phoebe, and Sunbittern along the river's edge. Dull-mantled Antbird is a reliable find in the undergrowth along the hillsides.

Species to Expect

Little Tinamou (3) 12:17
Gray-headed Chachalaca (11) 12:1
Cattle Egret (23) 5:13
Black Vulture (31) 13:4
Turkey Vulture (31) 13:3
King Vulture (31) 13:5
Bicolored Hawk (39) 16:8
Roadside Hawk (43) 16:12
Broad-winged Hawk (43) 16:13
Barred Forest-Falcon (53) 16:5
White-throated Crake (57) 6:9
Sunbittern (29) 6:16
Rock Pigeon (85)
Red-billed Pigeon (87) 18:2
Ruddy Pigeon (87) 18:5
Short-billed Pigeon (87) 18:5
Ruddy Ground-Dove (89) 18:7
White-tipped Dove (91) 18:14
Gray-chested Dove (91) 18:16
Purplish-backed Quail-Dove (93) 18:21
*Sulphur-winged Parakeet (95) 19:12
***Crimson-fronted Parakeet (95) 19:10**
Brown-hooded Parrot (97) 19:9
White-crowned Parrot (97) 19:7
Squirrel Cuckoo (103) 21:7
Striped Cuckoo (101) 21:5
Groove-billed Ani (103) 21:9
Mottled Owl (107) 20:6
Common Pauraque (111) 21:18
White-collared Swift (117) 22:1
Vaux's Swift (119) 22:9
Green Hermit (121) 23:3
Stripe-throated Hermit (121) 23:1
Violet Sabrewing (123) 23:9
White-necked Jacobin (125) 23:17
Brown Violetear (131) 23:6
Green-breasted Mango (129) 23:13
Violet-headed Hummingbird (137) 25:11
Black-crested Coquette (137) 25:6
Green Thorntail (137) 25:1
Violet-crowned Woodnymph (127) 24:4
Rufous-tailed Hummingbird (129) 24:10
***Snowcap (137) 25:8**
Bronze-tailed Plumeleteer (129) 23:12
Green-crowned Brilliant (123) 23:15
Purple-crowned Fairy (125) 23:14
Collared Trogon (143) 26:5
Blue-crowned Motmot (147) 27:8
Rufous Motmot (147) 27:7
Broad-billed Motmot (147) 27:9
Amazon Kingfisher (149) 27:4
Green Kingfisher (149) 27:5
Rufous-tailed Jacamar (153) 26:12
Collared Aracari (155) 27:15
Keel-billed Toucan (155) 27:18
Black-cheeked Woodpecker (157) 28:14
***Hoffmann's Woodpecker (157) 28:16**
Smoky-brown Woodpecker (159) 28:13
Golden-olive Woodpecker (159) 28:11
Slaty Spinetail (163) 32:9
Buff-throated Foliage-gleaner (165) 30:3
Plain Xenops (169) 29:3
Tawny-throated Leaftosser (167) 30:11
Olivaceous Woodcreeper (171) 29:7
Wedge-billed Woodcreeper (171) 29:6
Cocoa Woodcreeper (173) 29:17

Spotted Woodcreeper (173) 29:20
Streak-headed Woodcreeper (173) 29:8
Brown-billed Scythebill (175) 29:14
Russet Antshrike (177) 31:4
Checker-throated Antwren (183) 32:5
Slaty Antwren (183) 32:3
Dusky Antbird (177) 31:8
Dull-mantled Antbird (181) 31:14
Immaculate Antbird (179) 31:10
Spotted Antbird (181) 31:12
Yellow-bellied Elaenia (193) 37:26
Torrent Tyrannulet (201) 36:1
Olive-striped Flycatcher (195) 36:24
Ochre-bellied Flycatcher (199) 36:25
Slaty-capped Flycatcher (195) 36:26
Paltry Tyrannulet (191) 37:10
Scale-crested Pygmy-Tyrant (197) 37:4
Common Tody-Flycatcher (197) 37:7
Black-headed Tody-Flycatcher (197) 37:8
Eye-ringed Flatbill (201) 37:23
Yellow-olive Flycatcher (195) 37:16
Sulphur-rumped Flycatcher (199) 36:21
*Tawny-chested Flycatcher (199) 36:10
Western Wood-Pewee (203) 36:7
Tropical Pewee (203) 36:9
Yellow-bellied Flycatcher (205) 36:20
Black Phoebe (207) 36:5
Bright-rumped Attila (201) 35:6
Rufous Mourner (201) 34:9
Dusky-capped Flycatcher (209) 35:21
Great Kiskadee (211) 35:13
Boat-billed Flycatcher (211) 35:12
Social Flycatcher (211) 35:14
Tropical Kingbird (213) 35:1
Cinnamon Becard (215) 33:11
Masked Tityra (217) 34:1
White-collared Manakin (221) 33:1
White-ruffed Manakin (223) 33:9
Yellow-throated Vireo (225) 40:6
Tawny-crowned Greenlet (229) 32:4
Lesser Greenlet (229) 40:7
Brown Jay (231) 39:19
Blue-and-white Swallow (233) 22:20
Northern Rough-winged Swallow (235) 22:18
Southern Rough-winged Swallow (235) 22:19
Band-backed Wren (239) 38:3
***Black-throated Wren (243) 38:13**
***Stripe-breasted Wren (241) 38:5**
Plain Wren (241) 38:17
House Wren (243) 38:18
White-breasted Wood-Wren (245) 38:15
Scaly-breasted Wren (245) 38:20
Tropical Gnatcatcher (237) 41:2
*Black-faced Solitaire (249) 39:13
Black-headed Nightingale-Thrush (247) 38:25
Swainson's Thrush (249) 39:2
Wood Thrush (249) 39:4
Clay-colored Thrush (251) 39:8
White-throated Thrush (251) 39:9
Golden-winged Warbler (257) 41:5
Tennessee Warbler (255) 40:22
Tropical Parula (257) 41:3
Chestnut-sided Warbler (259) 43:4
Blackburnian Warbler (261) 41:8
Black-and-white Warbler (267) 41:13
Kentucky Warbler (271) 42:15
Olive-crowned Yellowthroat (273) 42:11
Wilson's Warbler (269) 42:4
Slate-throated Redstart (273) 42:7
Golden-crowned Warbler (275) 40:18
Rufous-capped Warbler (275) 40:16
Buff-rumped Warbler (271) 40:23
Bananaquit (277) 40:24
Ashy-throated Bush-Tanager (279) 45:12
*Black-and-yellow Tanager (283) 46:14
Carmiol's Tanager (279) 45:16
White-shouldered Tanager (281) 46:18
White-lined Tanager (281) 46:16
Red-throated Ant-Tanager (277) 47:12
Summer Tanager (285) 47:5
Crimson-collared Tanager (287) 47:3
Passerini's Tanager (287) 47:4
Blue-gray Tanager (291) 46:15
Palm Tanager (291) 45:19
Silver-throated Tanager (291) 46:6
Speckled Tanager (289) 46:8
Bay-headed Tanager (289) 46:9
Golden-hooded Tanager (289) 46:13
Scarlet-thighed Dacnis (291) 46:4
Green Honeycreeper (293) 46:7
Variable Seedeater (295) 49:3
Yellow-faced Grassquit (297) 49:6
*Sooty-faced Finch (301) 48:21
Chestnut-capped Brush-Finch (301) 48:16

Orange-billed Sparrow (303) 48:17
Black-striped Sparrow (303) 50:14
Buff-throated Saltator (309) 48:2
Black-headed Saltator (309) 48:1
Melodious Blackbird (315) 52:6
Great-tailed Grackle (317) 44:16
Baltimore Oriole (321) 44:7
Yellow-billed Cacique (321) 44:12
Scarlet-rumped Cacique (321) 44:11
Chestnut-headed Oropendola (323) 44:9
Montezuma Oropendola (323) 44:8
Olive-backed Euphonia (325) 45:3
***Tawny-capped Euphonia (325) 45:6**

Nearby Birding Opportunities

Silent Mountain

This exciting area to bird is only a 20-minute drive from Rancho Naturalista. Found at a higher elevation (1,250 meters), the forest and birds here are quite different. The major drawback to the area is access; birding here entails a strenuous full-day hike, and I only recommend it for those in good physical condition. The trail is also very difficult to find and follow, so you must visit with a guide from Rancho Naturalista. You will start the hike by climbing through a steep cow pasture, where you may find Sulfur-winged Parakeet, Acorn Woodpecker, and Golden-bellied Flycatcher. This is also the best location in the country to find the endemic Red-fronted Parrotlet. A muddy trail then ascends through beautiful forest, where you have good chances of finding many species different from those found at Rancho Naturalista. Look for Green-fronted Lancebill, White-bellied Mountain-gem, Black-bellied Hummingbird, Emerald Toucanet, and Spangle-cheeked Tanager. Along the trail you will come to a few clearings and cattle pastures, which are good places to find Dark Pewee, Elegant Euphonia, Slaty Spinetail, and Black-thighed Grosbeak, with Ornate Hawk-Eagle regularly soaring overhead.

The forest at Silent Mountain also offers a good chance of seeing some real rarities. This is the only reliable place in the country to find Rufous-rumped Antwren. You also have the possibility of seeing Rufous-browed Tyrannulet, Rough-legged Tyrannulet, Strong-billed Woodcreeper, Black-banded Woodcreeper, Rufous-breasted Antthrush, and Sharpbill (all very difficult to find in Costa Rica). If you visit the area in March, fruiting wild avocado trees often attract toucans, Resplendent Quetzal, and even the rare Lovely Cotinga.

Site 1B-8: El Copal Biological Reserve

Birding Time: 2 days
Elevation: 1,000 meters
Trail Difficulty: 2
Reserve Hours: N/A
Entrance Fee: Included with overnight
4×4 Recommended

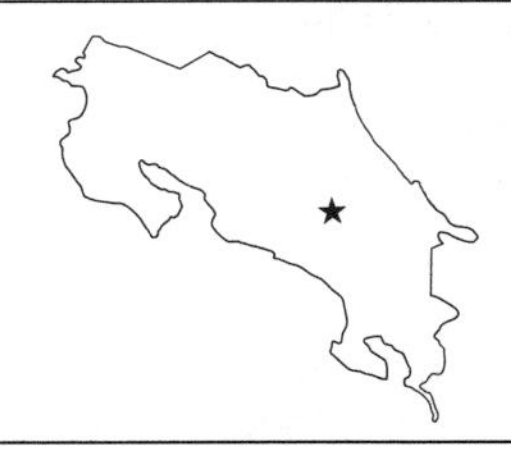

El Copal is one of Costa Rica's best-kept birding secrets. Renowned among locals as one of the most productive and exciting birding locations in the country, El Copal is virtually unheard of within the international birding community. The fact that it is a locally run operation where Spanish is the sole language spoken undoubtedly contributes to its obscurity. Also, it can be tricky to find, involving a maze of local roads. Hopefully, this write-up, and the precise directions provided, will allow more visitors to enjoy this wonderful birding spot.

What you will find when you arrive at El Copal Biological Reserve is a small lodge perched upon a hillside, with beds for 16 people within four dormitory-style rooms. There is a small dining room, shared bathrooms with cold showers, and a porch where you can relax in hammocks and survey the landscape below. Immediately behind the lodge, El Copal owns 400 acres of virgin forest that borders Tapantí National Park (4A-5). A good trail system provides hours of hiking and birding opportunities.

As for the birding, El Copal has it all. Beautifully situated at 1,000 meters on the Caribbean Slope, and providing access to a range of habitats, it will not be difficult to build a large and impressive list here. The forest, which runs continuously into Panama via Tapantí National Park and La Amistad International Park, has an extremely rich avifauna, and you have a high probability of encountering rarities. Anything from hawk-eagles to cotingas can be seen here.

Target Birds

*Black-breasted Wood-Quail (13) 12:10
Great Black-Hawk (45) 13:7
Barred Forest-Falcon (53) 16:5
*Red-fronted Parrotlet (97) 19:15
Green Thorntail (137) 25:1
*Snowcap (137) 25:8
Purple-crowned Fairy (125) 23:14
Lanceolated Monklet (151) 28:5
*Yellow-eared Toucanet (155) 27:20
Buff-fronted Foliage-gleaner (165) 30:6
Tawny-throated Leaftosser (167) 30:11
Gray-throated Leaftosser (167) 30:10
Strong-billed Woodcreeper (175) 29:18
Black-banded Woodcreeper (169) 29:15
Brown-billed Scythebill (175) 29:14
Dull-mantled Antbird (181) 31:14
Immaculate Antbird (179) 31:10
Black-headed Antthrush (187) 30:17
Scaled Antpitta (185) 30:20
Ochre-breasted Antpitta (185) 30:13
Rufous-browed Tyrannulet (193) 37:12
*Tawny-chested Flycatcher (199) 36:10
Thrush-like Schiffornis (215) 33:10
Black-and-white Becard (217) 33:15
Lovely Cotinga (219) 34:6
White-crowned Manakin (223) 33:8
Sharpbill (225) 34:7
Song Wren (245) 38:22
Black-headed Nightingale-Thrush (247) 38:25
Ashy-throated Bush-Tanager (279) 45:12
*Black-and-yellow Tanager (283) 46:14
White-lined Tanager (281) 46:16
Crimson-collared Tanager (287) 47:3
*Blue-and-gold Tanager (283) 45:11
Speckled Tanager (289) 46:8
Yellow-bellied Seedeater (295) 49:4

Access

From Paraíso: Take Rt. 224 toward Orosí for only 0.4 km and then make a left turn signed for Cachí. Follow this road 8.8 km (passing the Cachí Dam at 8.6 km) and make a left turn onto Rt. 225 toward Tucurrique. Proceed 11.2 km until you come to a T-intersection, where you will turn right. Continue straight for 4.6 km to the far end of the town of Tucurrique. Turn right here, following signs for Pejibaye. This road quickly bends left and then turns to dirt. After 2.8 km you will need to keep right at an intersection. At 1.0 km after this, keep left at a small fork in the road, and 1.2 km beyond that, turn right at a T-intersection, following signs for Pejibaye. In 0.8 km you will enter Pejibaye but will find the road blocked off in front of the soccer field. You should loop around the soccer field and then continue on the road that you had been on at the opposite end of this block. You will enter the town of Humo in 1.4 km, where you will come to a fork. Take the left road, which immediately passes over a bridge. On the other side of the bridge make a left turn and cross over a second bridge. Turn your trip odometer to zero here. In 0.2 km be sure to stay left at a fork in the road, and in 6.0 km you will arrive at the entrance to El Copal Biological Reserve on your left. (Driving time from Paraíso: 80 min)

From Site 4A-5, Tapantí National Park: El Copal Biological Reserve is extremely close to Tapantí National Park as the fruitcrow flies, but the road between the two is often in very bad condition and may not be passable even with a 4×4. There is talk of the road being improved, and I suggest asking the managers at El Copal or the park rangers at Tapantí for current driving conditions. If you are good to go, follow the road that passes in front of the Kiri Lodge (refer to site 4A-5 map), and after about 15 km El Copal Biological Reserve will be on your right.

Logistics: El Copal is quite remote, and you will find it most comfortable to stay at the reserve itself. If you speak Spanish you can make reservations directly with the reserve. Otherwise, contact the ACTUAR Travel Agency.

El Copal Biological Reserve
Tel: (506) 8880-0432 (Spanish)
Fax: (506) 2535-0047

ACTUAR Travel Agency
Tel: (506) 2248-9470
Website: www.actuarcostarica.com
E-mail: info@actuarcostarica.com

Birding Sites

I like to start my days at El Copal by walking down the Mariposa Trail just before light. Mottled Owl can sometimes be heard calling in the area, and as dawn approaches, the songs of many other species replace it. Great Tinamou, Black-headed Antthrush, Black-breasted Wood-Quail, Rufous Motmot, and

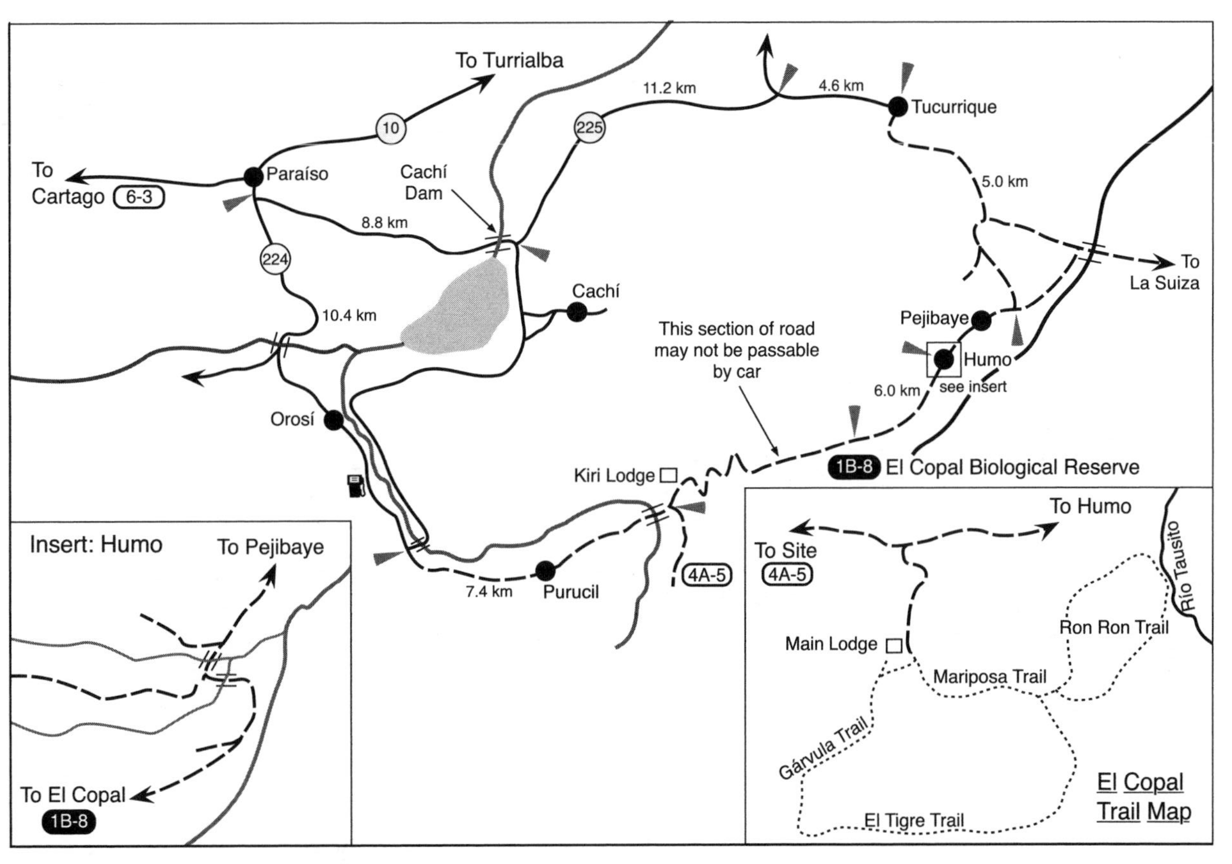

El Copal Driving Map. Tapantí National Park (4A-5) and Lankester Gardens (6-3) are included. The left insert shows details of Humo. The right insert provides a trail map for El Copal.

Barred Forest-Falcon are all regular participants in the dawn chorus. At this time you may also find Scaled Antpitta, Ochre-breasted Antpitta, or Tawny-throated Leaftosser feeding in the path. The small stream that passes the trail about halfway down is a very reliable place to find Dull-mantled Antbird. Other birds to watch for include Rufous-tailed Jacamar, Purple-crowned Fairy, Brown-billed Scythebill, Slaty Antwren, and Plain Antvireo.

Right next to the lodge there is some overgrown pasture, which can be quite productive. Expect to find Slaty Spinetail, Tropical Pewee, White-collared Manakin, Olive-crowned Yellowthroat, and Black-headed Saltator. Groups of tanagers often frequent the low trees. Regulars include Emerald Tanager, Golden-hooded Tanager, White-lined Tanager, and Black-and-yellow Tanager.

The porterweed shrubs planted around the lodge attract many hummingbirds, including Snowcap, Green Thorntail, and Brown Violetear. From the porch of the lodge you can scan the surrounding countryside. Look for Swallow-tailed Kites circling overhead. Gray-headed Chachalaca, Keel-billed Toucan, Collared Aracari, and Lineated Woodpecker often perch in the exposed trees.

The Gárvula-Tigre-Mariposa loop is another very productive forest walk that should take approximately three hours to bird. The trail proceeds up to the top of a forested ridge, then along the ridgeline before descending back down and completing the loop. Among the more interesting species you are likely to see are Song Wren, White-crowned Manakin, Thrushlike Schiffornis, Ashy-throated Bush-Tanager, Immaculate Antbird, and Black Guan. Remember that many rarities could appear along this trail, so be ready for Yellow-eared Toucanet, Lovely Cotinga, Ornate Hawk-Eagle, and Rufous-browed Tyrannulet.

The Ron Ron Trail, which takes about two hours to bird, is also worth walking. The trail leads down to the Río Tausito, which is a good place to encounter flocks of Carmiol's Tanager, Tawny-crested Tanager, and White-shouldered Tanager. Also keep an eye open for Blue-black Grosbeak, White-ruffed Manakin, Tawny-chested Flycatcher, and, if you are really lucky, a Lanceolated Monklet by the river.

While the forest trails are exciting to bird, many different species can be found by walking down the driveway and up the dirt road toward Tapantí. The habitat here is a combination of weedy overgrown fields, second growth, and sugarcane fields. Sparrows are common in the area. Variable Seedeater and Yellow-faced Grassquit are the most abundant, but the nomadic Yellow-bellied Seedeater can also be found here. Northern Rough-winged Swallow, Southern Rough-winged Swallow, and Blue-and-white Swallow commonly swoop over the fields, and White-throated Crake can often be heard calling from the thick grasses. Other birds to look for include Giant Cowbird, Black-and-white Becard, Rough-legged Tyrannulet, Black-cowled Oriole, and Scarlet-thighed Dacnis.

Species to Expect

Great Tinamou (3) 12:6
Little Tinamou (3) 12:17
Gray-headed Chachalaca (11) 12:1
Crested Guan (11) 12:4
*Black Guan (11) 12:5
*Black-breasted Wood-Quail (13) 12:10
Cattle Egret (23) 5:13
Black Vulture (31) 13:4
Turkey Vulture (31) 13:3
Swallow-tailed Kite (37) 15:2
Tiny Hawk (39) 16:2
Short-tailed Hawk (43) 16:11
Barred Forest-Falcon (53) 16:5
White-throated Crake (57) 6:9
Red-billed Pigeon (87) 18:2
Band-tailed Pigeon (87) 18:1
Ruddy Pigeon (87) 18:5
Short-billed Pigeon (87) 18:5
White-tipped Dove (91) 18:14
***Crimson-fronted Parakeet (95) 19:10**
Brown-hooded Parrot (97) 19:9
White-crowned Parrot (97) 19:7
Squirrel Cuckoo (103) 21:7
Striped Cuckoo (101) 21:5
Groove-billed Ani (103) 21:9
Mottled Owl (107) 20:6
Common Pauraque (111) 21:18
White-collared Swift (117) 22:1
Green Hermit (121) 23:3
Stripe-throated Hermit (121) 23:1
Green-fronted Lancebill (135) 23:11
Violet Sabrewing (123) 23:9
Brown Violetear (131) 23:6
Green Violetear (131) 23:7
Violet-headed Hummingbird (137) 25:11
Green Thorntail (137) 25:1
Violet-crowned Woodnymph (127) 24:4
Rufous-tailed Hummingbird (129) 24:10
Stripe-tailed Hummingbird (133) 24:17
***Snowcap (137) 25:8**
Green-crowned Brilliant (123) 23:15
Purple-crowned Fairy (125) 23:14
Violaceous Trogon (141) 26:8
Collared Trogon (143) 26:5
Black-throated Trogon (141) 26:9
Rufous Motmot (147) 27:7
Broad-billed Motmot (147) 27:9
Rufous-tailed Jacamar (153) 26:12
Red-headed Barbet (153) 28:2
*Prong-billed Barbet (153) 28:1
Collared Aracari (155) 27:15
*Yellow-eared Toucanet (155) 27:20
Keel-billed Toucan (155) 27:18
Black-cheeked Woodpecker (157) 28:14
Golden-olive Woodpecker (159) 28:11
Lineated Woodpecker (161) 27:13
Pale-billed Woodpecker (161) 27:14
Slaty Spinetail (163) 32:9
Buff-throated Foliage-gleaner (165) 30:3
Plain Xenops (169) 29:3
Tawny-throated Leaftosser (167) 30:11
Olivaceous Woodcreeper (171) 29:7
Wedge-billed Woodcreeper (171) 29:6
Northern Barred-Woodcreeper (169) 29:19
Spotted Woodcreeper (173) 29:20
Brown-billed Scythebill (175) 29:14
Fasciated Antshrike (175) 31:1
Russet Antshrike (177) 31:4
Plain Antvireo (181) 32:6
Slaty Antwren (183) 32:3
Dusky Antbird (177) 31:8
Dull-mantled Antbird (181) 31:14
Immaculate Antbird (179) 31:10
Black-headed Antthrush (187) 30:17
Scaled Antpitta (185) 30:20
*Thicket Antpitta (185) 30:14
Yellow-bellied Elaenia (193) 37:26
Olive-striped Flycatcher (195) 36:24
Ochre-bellied Flycatcher (199) 36:25
Slaty-capped Flycatcher (195) 36:26
Rough-legged Tyrannulet (191) 37:18
Paltry Tyrannulet (191) 37:10
Scale-crested Pygmy-Tyrant (197) 37:4
Common Tody-Flycatcher (197) 37:7
Yellow-olive Flycatcher (195) 37:16
White-throated Spadebill (189) 37:1
Ruddy-tailed Flycatcher (199) 36:23
Sulphur-rumped Flycatcher (199) 36:21
Olive-sided Flycatcher (203) 36:4
Western Wood-Pewee (203) 36:7

Eastern Wood-Pewee (203) 36:8
Tropical Pewee (203) 36:9
Yellow-bellied Flycatcher (205) 36:20
Bright-rumped Attila (201) 35:6
Rufous Mourner (201) 34:9
Dusky-capped Flycatcher (209) 35:21
Great Kiskadee (211) 35:13
Social Flycatcher (211) 35:14
Gray-capped Flycatcher (211) 35:15
Piratic Flycatcher (193) 35:8
Tropical Kingbird (213) 35:1
Thrush-like Schiffornis (215) 33:10
Cinnamon Becard (215) 33:11
Masked Tityra (217) 34:1
White-collared Manakin (221) 33:1
White-ruffed Manakin (223) 33:9
White-crowned Manakin (223) 33:8
Yellow-throated Vireo (225) 40:6
Red-eyed Vireo (227) 40:3
Tawny-crowned Greenlet (229) 32:4
Lesser Greenlet (229) 40:7
Brown Jay (231) 39:19
Blue-and-white Swallow (233) 22:20
Northern Rough-winged Swallow (235) 22:18
Southern Rough-winged Swallow (235) 22:19
Band-backed Wren (239) 38:3
***Black-throated Wren (243) 38:13**
***Stripe-breasted Wren (241) 38:5**
Plain Wren (241) 38:17
House Wren (243) 38:18
White-breasted Wood-Wren (245) 38:15
Song Wren (245) 38:22
Long-billed Gnatwren (237) 32:15
Tropical Gnatcatcher (237) 41:2
Black-headed Nightingale-Thrush (247) 38:25
Swainson's Thrush (249) 39:2
Pale-vented Thrush (251) 39:5
Clay-colored Thrush (251) 39:8
Tennessee Warbler (255) 40:22
Tropical Parula (257) 41:3
Chestnut-sided Warbler (259) 43:4
Blackburnian Warbler (261) 41:8
Black-and-white Warbler (267) 41:13
Olive-crowned Yellowthroat (273) 42:11
Gray-crowned Yellowthroat (273) 42:13
Wilson's Warbler (269) 42:4
Canada Warbler (269) 42:8
Slate-throated Redstart (273) 42:7
Golden-crowned Warbler (275) 40:18
Rufous-capped Warbler (275) 40:16
Buff-rumped Warbler (271) 40:23
Bananaquit (277) 40:24
Ashy-throated Bush-Tanager (279) 45:12
***Black-and-yellow Tanager (283) 46:14**
Carmiol's Tanager (279) 45:16
White-shouldered Tanager (281) 46:18
Tawny-crested Tanager (281) 46:17
White-lined Tanager (281) 46:16
Summer Tanager (285) 47:5
Scarlet Tanager (287) 47:8
Passerini's Tanager (287) 47:4
Blue-gray Tanager (291) 46:15
Palm Tanager (291) 45:19
*Blue-and-gold Tanager (283) 45:11
Emerald Tanager (289) 46:5
Silver-throated Tanager (291) 46:6
Speckled Tanager (289) 46:8
Bay-headed Tanager (289) 46:9
Golden-hooded Tanager (289) 46:13
Scarlet-thighed Dacnis (291) 46:4
Green Honeycreeper (293) 46:7
Variable Seedeater (295) 49:3
Yellow-bellied Seedeater (295) 49:4
Yellow-faced Grassquit (297) 49:6
White-naped Brush-Finch (301) 48:15
Orange-billed Sparrow (303) 48:17
Black-striped Sparrow (303) 50:14
Rufous-collared Sparrow (307) 50:13
Buff-throated Saltator (309) 48:2
Black-headed Saltator (309) 48:1
Black-faced Grosbeak (311) 48:5
Rose-breasted Grosbeak (311) 48:8
Blue-black Grosbeak (311) 48:10
Bronzed Cowbird (317) 44:15
Giant Cowbird (317) 44:10
Black-cowled Oriole (319) 44:5
Baltimore Oriole (321) 44:7
Chestnut-headed Oropendola (323) 44:9
Montezuma Oropendola (323) 44:8
Elegant Euphonia (325) 45:9
White-vented Euphonia (327) 45:7
***Tawny-capped Euphonia (325) 45:6**

Nearby Birding Opportunities

The Cachí Dam

The Cachí Dam, a hydroelectric facility near the town of Cachí, has created a large lake that attracts a variety of water birds. The main road between Paraíso and Cachí passes right over the dam and is the best place from which to scan for birds. Bat Falcon can sometimes be seen perching along the face of the dam, and Torrent Tyrannulet and Black Phoebe are both possibilities near the outflow water. The lake is a reliable place to find Least Grebe, and if you scan the shoreline you should be able to find Blue-winged Teal, Northern Jacana, Purple Gallinule, Little Blue Heron, and Ringed Kingfisher.

Region 2
The North Pacific Slope

The **Lesser Ground-Cuckoo** is a furtive species that spends most of its time on or near the ground of thickets and second growth in the dry North Pacific. Though it is much more easily heard than seen, this skulking species can often be found venturing into the open at Diriá National Park (2-5) and Rincón de la Vieja National Park (2-2).

Regional Map: The North Pacific Slope

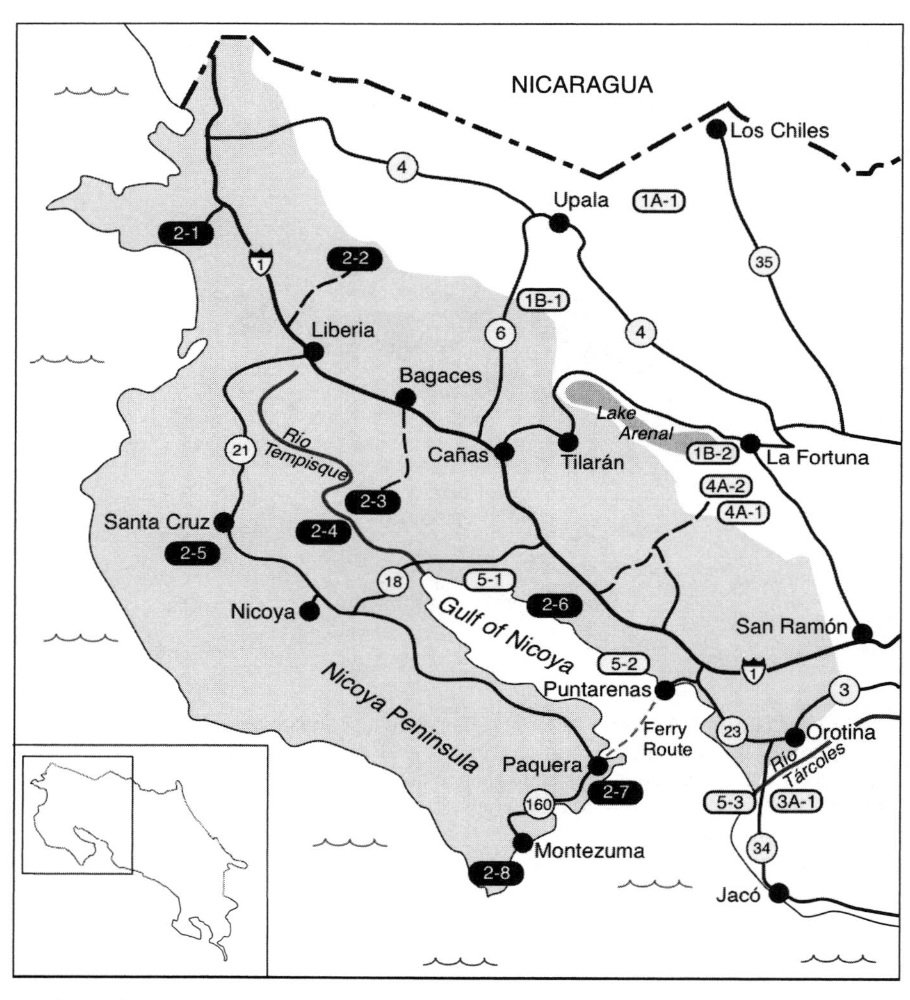

2-1: Santa Rosa National Park
2-2: Rincón de la Vieja National Park
2-3: Palo Verde National Park
2-4: Mata Redonda Marsh
2-5: Diriá National Park
2-6: La Ensenada Wildlife Refuge
2-7: Curú National Wildlife Refuge
2-8: Cabo Blanco Absolute Nature Reserve

Introduction

The North Pacific Slope, as I refer to it in this book, includes all the land between the Nicaraguan border and the Tárcoles River on the Pacific slope of the country. Somewhat coincidentally, these boundaries are close to the borders of the Guanacaste Province (which does not extend as far south). Because of this, the names "Guanacaste" and "North Pacific" are often used interchangeably. The Continental Divide, which forms the eastern border of the North Pacific Slope, is formed by two mountain ranges, the Cordillera de Tilarán and the Cordillera de Guanacaste. The Cordillera de Tilarán runs roughly from the city of San Ramón to Lake Arenal, while the Cordillera de Guanacaste runs from Lake Arenal up to the Nicaraguan border, where it ends.

The North Pacific Slope is unique in Costa Rica because of its distinctive climate. Starting in December and continuing through early May, the North Pacific region goes through an extreme dry season, where little to no rain falls. (Note that during the wet season, this region receives ample rainfall, yielding an annual average of about 1.5 meters [5 feet]. In comparison, the Caribbean and South Pacific slopes generally receive between 3 and 5 meters annually.) The causes of the distinct dry-season weather regime are complex, but a brief explanation follows.

Throughout the year Costa Rica experiences northeast trade winds due to the effects of the intertropical convergence. These winds carry moist air off the Caribbean Sea and over the Caribbean Slope. When this air hits the mountain ranges that run the length of the country, it rises, cools, and rains out its moisture. Thus by the time the air gets over the mountains and onto the Pacific side of the country, it is dry.

This account alone is not quite satisfactory because it does not explain why the North Pacific experiences such a distinct dry season and the South Pacific does not. The reason for this lies in the height and shape of the northern and southern mountain ranges. The northern mountains, Cordillera de Guanacaste and Cordillera de Tilarán, are lower and have a more gradual slope. This allows for a laminar (smooth) flow of air that closely conforms to the contours of the slope. The mountains in the south, principally the Cordillera de Talamanca, are much higher and have a steeper incline, which disrupts the laminar flow of air. With the airflow disturbed, a vortex is created and it actually sucks moisture-laden air in from the Pacific. This explains why the South Pacific Slope does not experience such a distinct dry season as the North Pacific.

The unique climate of the North Pacific region naturally leads to a unique ecosystem, the tropical dry forest. Costa Rica represents the southern extent of this ecosystem, which, before man's intervention, extended up the Pacific slope of Central America. Tropical dry forests are semi-deciduous, with two layers of trees, understory and canopy, the latter usually reaching about 20 to 30 meters high.

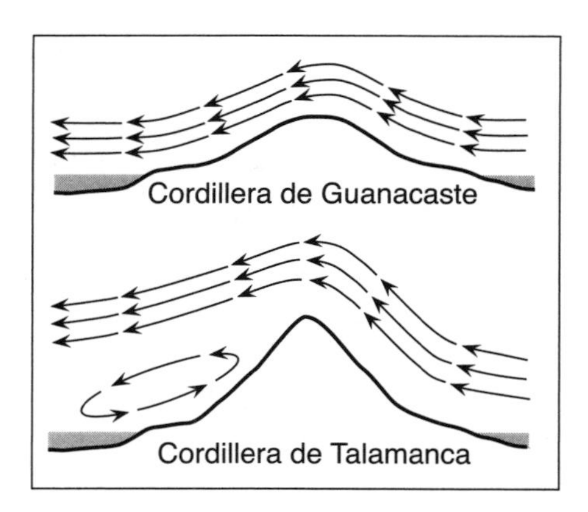

Wind Patterns over the Mountains

Ecologically, tropical dry forests are very different from their wet-forest counterparts. They are adapted to survive under the drought conditions that grip the region every year during the dry season. Since a tree loses most of its water through its leaves, many dry-forest trees are deciduous, losing their leaves during the dry season to better conserve water. Another difference between dry-forest trees and wet-forest trees is in the root system. The roots of most wet-forest trees are found just beneath the soil's surface, in an all-out effort to collect nutrients. Many dry-forest trees possess a taproot that penetrates deep into the ground to collect valuable water in times of scarcity.

If you are in a tropical dry forest during the dry season, you may notice that while many trees have dropped their leaves, they have also burst into flower. The reason is that many dry-forest plants take advantage of the constant and often high winds that sweep through the region during the dry season, using it to disperse their pollen and seeds. As a result, they are often left with leafless branches filled with colorful flowers; it is truly beautiful.

Five hundred years ago tropical dry forest was the most abundant habitat in Central America, covering most of the Pacific slope. Today, tropical dry forest is considered to be the most endangered habitat in the world. And yet, with all the great conservation effort that has focused on tropical rain forests, tropical dry forest has slipped under the radar, most people having no idea it even exists.

In Costa Rica, most of the original dry forest has been converted to cattle pastures. Cutting the forest and then burning the remains can quickly transform the land from forest to savannah, which is ideal for raising beef cattle. Guanacaste Province has always served as the foundation of the nation's cattle industry. By the late 1970s and early 1980s, beef had grown to be Costa Rica's third largest export, behind coffee and bananas. At that time, the country had about two million people and about two million head of cattle. In recent years the industry has declined somewhat, although it still dominates the land use of the North Pacific Slope and occupies an important place in the nation's economy.

Other important farming industries of the North Pacific include the production of rice, honeybees, cotton, sugarcane, corn, beans, and pigs.

This chapter covers eight sites within the North Pacific Slope. In my mind, Palo Verde National Park (2-3) is the standout, with access to dry forest as well as one of the most important wetlands in all of Central America. Each of the other sites, however, provides something unique and interesting. Middle-elevation habitats are accessible at Diriá and Rincón de la Vieja national parks (2-5, 2-2), with the latter displaying dramatic geological phenomena. If ease and accommodations are important considerations, La Ensenada Wildlife Refuge (2-6) has a comfortable lodge with productive birding right outside the door.

Regional Specialties

Thicket Tinamou (3) 12:18
Plain Chachalaca (11) 12:2
Crested Bobwhite (15) 12:13/14
Jabiru (19) 4:5
Harris's Hawk (33) 14:5
White-tailed Hawk (43) 17:6
Double-striped Thick-knee (63) 9:10
White-winged Dove (89) 18:12
Inca Dove (89) 18:11
Common Ground-Dove (89) 18:8
Orange-fronted Parakeet (95) 19:13
White-fronted Parrot (99) 19:6
Yellow-naped Parrot (99) 19:5
Lesser Ground-Cuckoo (103) 21:11
Pacific Screech-Owl (107) 20:11
Ferruginous Pygmy-Owl (109) 20:17
Northern Potoo (115) 20:4
Canivet's Emerald (127) 24:13
Cinnamon Hummingbird (129) 24:11
Plain-capped Starthroat (135) 23:19
Black-headed Trogon (141) 26:10
Elegant Trogon (143) 26:7
Turquoise-browed Motmot (147) 27:11
Ivory-billed Woodcreeper (173) 29:21
Northern Beardless-Tyrannulet (191) 37:19
Stub-tailed Spadebill (189) 37:2
Nutting's Flycatcher (209) 35:22
Brown-crested Flycatcher (209) 35:18
Scissor-tailed Flycatcher (207) 35:5
Long-tailed Manakin (223) 33:4
White-throated Magpie-Jay (231) 39:18
Rufous-naped Wren (239) 38:2
Rock Wren (239) 38:4
Rufous-and-white Wren (241) 38:9
Banded Wren (241) 38:6
White-lored Gnatcatcher (237) 41:1
Western Tanager (285) 47:2
Olive Sparrow (303) 50:15
Stripe-headed Sparrow (303) 50:7
Botteri's Sparrow (305) 50:5
Rusty Sparrow (305) 50:4
Grasshopper Sparrow (305) 50:8
Blue Grosbeak (313) 48:11
Streak-backed Oriole (319) 44:2
Spot-breasted Oriole (319) 44:1
Scrub Euphonia (327) 45:4

Common Birds to Know

Black Vulture (31) 13:4
Turkey Vulture (31) 13:3
Roadside Hawk (43) 16:12
White-winged Dove (89) 18:12
Inca Dove (89) 18:11
Common Ground-Dove (89) 18:8
White-tipped Dove (91) 18:14
Orange-fronted Parakeet (95) 19:13
Orange-chinned Parakeet (97) 19:14
White-fronted Parrot (99) 19:6
*Hoffmann's Woodpecker (157) 28:16
Yellow-olive Flycatcher (195) 37:16
Brown-crested Flycatcher (209) 35:18
Great Kiskadee (211) 35:13
Tropical Kingbird (213) 35:1
Rufous-naped Wren (239) 38:2
Baltimore Oriole (321) 44:7

Site 2-1: Santa Rosa National Park

Birding Time: 1–2 days
Elevation: 200 meters
Trail Difficulty: 1
Reserve Hours: N/A
Entrance Fee: $10 (US 2008)

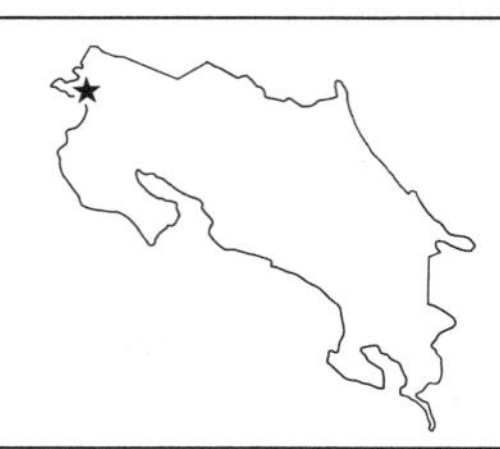

Santa Rosa National Park sits on its own small peninsula just above the Nicoya Peninsula in the northwest corner of the country. Ecologically it is an important area in that it protects 122,000 acres of tropical dry forest, including some sizable chunks of primary forest. The park shares a border with Guanacaste National Park, which continues up into the mountains, and together they provide an excellent biological corridor for animals to move up and down the slope.

As the site of three battles, Santa Rosa National Park also has important historical significance. (Costa Ricans are extremely peaceful, and the importance that these three skirmishes hold in the nation's history demonstrates this.) The first battle took place in 1856, when an American named William Walker, who had already taken control of Nicaragua, set his sites on Costa Rica. He moved a mercenary army across the border to La Casona, which was one of the most important ranching haciendas at the time. On March 20 the Costa Rican militia engaged his army there, and within 14 minutes had sent Walker retreating back to Nicaragua. In 1919 and 1955 La Casona was again the site of conflict between Costa Rican forces and invaders coming from Nicaragua. Today a replica of La Casona containing a small museum has been built where the original building stood in 1856, and a memorial to the fallen soldiers has been erected atop the hillside behind La Casona.

While Santa Rosa does not see too many visitors, it is an active center for research and education, and it is not uncommon to encounter groups of students working in the park. The facilities are good, and with prior arrangement you can take meals on site.

Target Birds

Thicket Tinamou (3) 12:18
Great Curassow (11) 12:3
Crested Bobwhite (15) 12:13/14
White-tailed Hawk (43) 17:6
Double-striped Thick-knee (63) 9:10
Yellow-naped Parrot (99) 19:5
Pacific Screech-Owl (107) 20:11
Plain-capped Starthroat (135) 23:19
Elegant Trogon (143) 26:7
Ivory-billed Woodcreeper (173) 29:21
Barred Antshrike (175) 31:3
Stub-tailed Spadebill (189) 37:2
Long-tailed Manakin (223) 33:4
White-throated Magpie-Jay (231) 39:18
Rufous-and-white Wren (241) 38:9
Banded Wren (241) 38:6
White-lored Gnatcatcher (237) 41:1
Western Tanager (285) 47:2
Scrub Euphonia (327) 45:4

Access

From Liberia: Start at the traffic light along Rt. 1 and proceed north on the highway for 34.2 km. The park entrance will be on your left. (Driving time from Liberia: 25 min)

Logistics: The national park offers dormitory-style accommodations, as well as a campsite. Meals can also be taken inside the park at the comedor, although you need to arrange these things in advance. Alternatively, there is a wide range of accommodations in the town of Liberia.

Santa Rosa National Park
Tel: (506) 2666-5051

Birding Sites

Along Rt. 1, as well as the entrance road to the park, are some fields where you will often see Double-striped Thick-knee standing in the short grass. The first half of the entrance road is also good for Thicket Tinamou and Crested Bobwhite.

About five kilometers in from the ticketing booth, the road takes some hard turns, and the forest around you becomes taller and greener. This is a small patch of primary tropical dry forest. The trees are large and ancient, and the whole area has a cooler microclimate than the surrounding forest. Not only is the forest itself worth stopping to appreciate, but you can usually find some species here much more easily than in other parts of the park. Long-tailed Manakin is present, as well as Northern Bentbill and Stub-tailed Spadebill, although all of these small birds are often difficult to spot. Olivaceous Woodcreeper, Ruddy Woodcreeper, Blue-crowned Motmot, and Rufous-and-white Wren are among the other birds that can be found here. Most birding can be done from the road, although there are a few informal trails that lead through the area. They wind and weave, so don't get lost.

Just beyond the section of primary forest a small dirt road turns sharply off to the right. Walking this short distance will provide you with one of your best chances of finding Barred Antshrike, which should come in close in response to some soft *pishing*, along with Yellow-green Vireo, Streak-backed Oriole, and Yellow-olive Flycatcher.

Another good area to bird is around La Casona. At the far side of the hacienda a trail leads up a long set of stairs to a lookout tower, from which you can get a beautiful panoramic view. During fall migration, scores of swallows can often be seen from the lookout. The majority of these are usually Barn Swallows, but you can often pick out some other species as well. A small system of trails leads out into the forest behind La Casona. The trails are called the Naked Indian Trails because of a prominent tree in the forest with the same common name

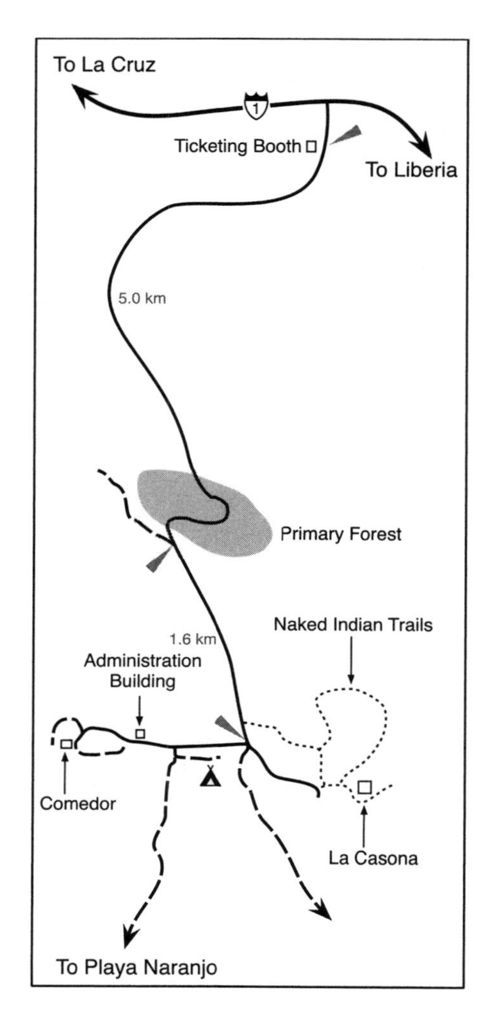

Santa Rosa National Park Area Map

whose thin bark constantly flakes off. Birds to look for here include Banded Wren, Elegant Trogon, White-lored Gnatcatcher, Turquoise-browed Motmot, and Olive Sparrow.

Many of the same species can also be encountered just walking along the small roads that lead between the buildings. The main road that runs between La Casona and the comedor provides a good chance to find Red-billed Pigeon, Roadside Hawk, Black-headed Trogon, Hoffmann's Woodpecker, and Brown-crested Flycatcher. The campsite usually holds some goodies and is a reliable place to find White-throated Magpie-Jay and Scrub Euphonia. A walk around this area at night will usually yield a calling Pacific Screech-Owl and Common Pauraque.

If you follow the road that leaves from the campsite and heads to Playa Naranjo, you will come to some overgrown fields and scrubland in less than half a kilometer. This is another interesting place to explore, and it often yields Crested Bobwhite, Stripe-headed Sparrow, Olive Sparrow, and Blue Grosbeak. The dirt road is practically impassable, even for 4×4's, so I suggest walking.

The road continues all the way to the coastline, where there are *salinas* (salt ponds), mangroves, and beach. Along the way you will pass through sections of primary forest that can offer great birding. I do not go into detail about these areas because of their inaccessibility. It is about 11 kilometers to the coast, where you will find only a rustic campsite, so you will need to carry in your own food and water. However, if you decide to make the trek, you should find some interesting birds.

Species to Expect

Thicket Tinamou (3) 12:18
Crested Guan (11) 12:4
Crested Bobwhite (15) 12:13/14
Bare-throated Tiger-Heron (21) 5:16
Black Vulture (31) 13:4
Turkey Vulture (31) 13:3
King Vulture (31) 13:5
Roadside Hawk (43) 16:12
Gray Hawk (43) 16:14
Collared Forest-Falcon (55) 16:7
Double-striped Thick-knee (63) 9:10
Red-billed Pigeon (87) 18:2
White-winged Dove (89) 18:12
Inca Dove (89) 18:11
Common Ground-Dove (89) 18:8
White-tipped Dove (91) 18:14
Orange-fronted Parakeet (95) 19:13
Orange-chinned Parakeet (97) 19:14
White-fronted Parrot (99) 19:6
Yellow-naped Parrot (99) 19:5
Squirrel Cuckoo (103) 21:7
Groove-billed Ani (103) 21:9
Pacific Screech-Owl (107) 20:11
Ferruginous Pygmy-Owl (109) 20:17
Common Pauraque (111) 21:18
Stripe-throated Hermit (121) 23:1
Canivet's Emerald (127) 24:13
Steely-vented Hummingbird (127) 24:15
Cinnamon Hummingbird (129) 24:11
Ruby-throated Hummingbird (131) 25:9
Black-headed Trogon (141) 26:10
Elegant Trogon (143) 26:7
Blue-crowned Motmot (147) 27:8
Turquoise-browed Motmot (147) 27:11
Green Kingfisher (149) 27:5
***Hoffmann's Woodpecker (157) 28:16**
Lineated Woodpecker (161) 27:13
Pale-billed Woodpecker (161) 27:14
Olivaceous Woodcreeper (171) 29:7
Northern Barred-Woodcreeper (169) 29:19
Ivory-billed Woodcreeper (173) 29:21
Streak-headed Woodcreeper (173) 29:8
Barred Antshrike (175) 31:3
Northern Bentbill (197) 37:6
Common Tody-Flycatcher (197) 37:7
Yellow-olive Flycatcher (195) 37:16
Bright-rumped Attila (201) 35:6
Dusky-capped Flycatcher (209) 35:21
Brown-crested Flycatcher (209) 35:18
Great Kiskadee (211) 35:13
Boat-billed Flycatcher (211) 35:12
Sulphur-bellied Flycatcher (213) 35:10
Tropical Kingbird (213) 35:1
Long-tailed Manakin (223) 33:4
Yellow-green Vireo (227) 40:5
Lesser Greenlet (229) 40:7
White-throated Magpie-Jay (231) 39:18
Barn Swallow (237) 22:12
Rufous-naped Wren (239) 38:2
Rufous-and-white Wren (241) 38:9
Banded Wren (241) 38:6

Long-billed Gnatwren (237) 32:15
White-lored Gnatcatcher (237) 41:1
Clay-colored Thrush (251) 39:8
Tennessee Warbler (255) 40:22
Yellow Warbler (259) 42:2
Rufous-capped Warbler (275) 40:16
Summer Tanager (285) 47:5
Western Tanager (285) 47:2
White-collared Seedeater (295) 49:2
Olive Sparrow (303) 50:15
Stripe-headed Sparrow (303) 50:7
Blue Grosbeak (313) 48:11
Streak-backed Oriole (319) 44:2
Baltimore Oriole (321) 44:7
Scrub Euphonia (327) 45:4

Site 2-2: Rincón de la Vieja National Park

Birding Time: 1–2 days
Elevation: 800 meters
Trail Difficulty: 2–3
Reserve Hours: 7 a.m. to 3 p.m. (closed on Monday)
Entrance Fee: $10 (US 2008)
4×4 Recommended
Trail Map Available on Site

Rincón de la Vieja National Park offers both interesting birding and spectacular geology, making it a great place to visit. The 34,785-acre park protects one of Costa Rica's seven active volcanoes, and while the last eruption was in 1995, impressive signs of activity are readily visible on a daily basis. Sendero Las Pailas takes you past several billowing steam vents, boiling pools of water, and bubbling mud holes, vividly reminding you of the power that lies just below your feet. In this landscape it doesn't take very much imagination to feel like you have stepped back into the time of the dinosaurs. It is also possible to hike to the crater of the volcano at the very top of the mountain, although this is quite a climb.

The reserve gets its name, Rincón de la Vieja, from an old indigenous medicine woman who once lived on the mountain with her tribe. Her powers as a healer were renowned, and people would travel great distances to see her. Thus the mountain became known as Rincón de la Vieja, or "corner of the old woman."

For birders, Rincón de la Vieja offers a range of habitats, including some that are difficult to find anywhere else in the country. Dry-forest scrub, wind-swept savannahs, and tropical humid forest are all available at Rincón. It is the only site described in this book that gives access to the Pacific slopes of the Cordillera de Guanacaste, and is thus the only place to find the few birds that are restricted to this area, including Rusty Sparrow, Botteri's Sparrow, and Rock Wren.

The facilities at the park are basic, offering little more than a bathroom and picnic tables. The park maintains four trails that provide access to the habitats in the area, although they are often steep.

Target Birds

Thicket Tinamou (3) 12:18
Violaceous Quail-Dove (93) 18:20
Lesser Ground-Cuckoo (103) 21:11
Rufous-vented Ground-Cuckoo (103) 21:10
Elegant Trogon (143) 26:7
Tody Motmot (147) 27:12
*Yellow-eared Toucanet (155) 27:20
Ruddy Woodcreeper (171) 29:13
Ivory-billed Woodcreeper (173) 29:21
Spotted Antbird (181) 31:12
Stub-tailed Spadebill (189) 37:2
Long-tailed Manakin (223) 33:4
White-throated Magpie-Jay (231) 39:18
Rock Wren (239) 38:4
Black-headed Nightingale-Thrush (247) 38:25
Botteri's Sparrow (305) 50:5
Rusty Sparrow (305) 50:4
Grasshopper Sparrow (305) 50:8

Access

From Liberia: Proceed 4.8 km north from the traffic light along Rt. 1 and then take a hard-to-see right turn onto a dirt road. **(From Santa Rosa National Park:** Head south on Rt. 1 for 29.4 km and then turn left onto a dirt road.) Proceed straight for 16.0 km to the Hacienda Guachipelín Hotel, where you will need to fork left. After another 5.0 km the road terminates at the park entrance. (Driving time from Liberia: 45 min; from Santa Rosa National Park: 60 min)

Logistics: Liberia is the nearest major town and there are plenty of hotel options there. Hacienda Guachipelín Hotel is an especially nice option, offering upscale lodging very close to the park.

Hacienda Guachipelín Hotel
Tel: (506) 2256-3600
Website: www.guachipelin.com
E-mail: info@guachipelin.com

Birding Sites

Many dry-forest species can be easily found along the road that leads to the park, as well as around the park entrance. You may encounter Crested Bobwhite feeding by the roadside, as well as Cinnamon Hummingbird, Canivet's Emerald, Blue Grosbeak, and Stripe-headed Sparrow. The telephone wires are a favorite perch of Turquoise-browed Motmot. The low, scrubby forest on either side of the road just before the park is home to the elusive Lesser Ground-Cuckoo, which sometimes can be seen in the road. Right around the park headquarters look for Keel-billed Toucan, White-throated Magpie-Jay, and Gray-crowned Yellowthroat.

Sendero Catarata La Cangreja proceeds through beautiful old-growth forest. The trail is often quite dark and quiet, but there is plenty to see if you have patience and are a keen observer. Long-tailed Manakin, Rufous-and-white Wren, and Nightingale Wren can be found along the trail, although they are more often heard than seen. Lesser Greenlet, Red-crowned Ant-Tanager, and Rufous-capped Warbler are abundant, and if you get lucky you might find an Ivory-billed Woodcreeper, Stub-tailed Spadebill, or a Tody Motmot.

About midway down Sendero Catarata La Cangreja, a path forks off to the right, leading to Catarata Escondida (the "hidden waterfall"). This trail leaves the forest and proceeds uphill through some scrubby fields, where you can often find Lesser Ground-Cuckoo. Further on, the trail rises more steeply and leads into a wind-swept savannah, where the grass becomes short with scattered small shrubs. This is the area where it is possible to find three species with very limited ranges in Costa Rica: Rusty Sparrow, Botteri's Sparrow, and Rock Wren. In addition to interesting birding, the views from the trail are spectacular, although the hike is very steep.

Sendero Las Pailas is the shortest and easiest of the hikes at Rincón, taking about two and a half hours to bird the loop. The habitat along the trail starts in forest similar to that along the Catarata La Cangreja, and a similar range of species can be expected. The second half of Sendero Las Pailas proceeds through sparser, drier forest, which holds a different range of species. Elegant Trogon is a possibility here, as well as Turquoise-browed Motmot, Brown-crested Flycatcher, White-lored Gnatcatcher, and Hoffmann's Woodpecker. This trail also displays some impressive geological features, including steam vents, boiling ponds, and bubbling mud pits. It is certainly a fun hike.

The Crater Trail heads straight up the mountain and proceeds through a variety of habitats. Starting off in tropical humid forest similar to that found on Catarata La Cangreja and Las Pailas, it climbs through premontane and montane forest all the way up past the timberline. Be forewarned that this is a demanding hike that will take about a full day. Because of the change in elevation, it is possible to see a number of species along this trail that are not found in the lower sections of the park. As you climb, look for Violet Sabrewing, Stripe-tailed Hummingbird, Purple-throated Mountain-gem, Orange-bellied Trogon, Orange-billed Nightingale-Thrush, and Golden-crowned Warbler. When you pass the tree line, you should also be able to find Mountain Elaenia and Slaty Flowerpiercer.

Swarms of army ants are common around the lower trails, and there is a good chance that you will encounter one. Look for Ruddy Woodcreeper, Northern Barred-Woodcreeper, Gray-headed Tanager, and Spotted Antbird to be feeding at the head of the swarm. The very rare Rufous-vented Ground-Cuckoo is also a possibility.

Species to Expect

Thicket Tinamou (3) 12:18
Crested Guan (11) 12:4
Crested Bobwhite (15) 12:13/14
Black Vulture (31) 13:4
Turkey Vulture (31) 13:3
King Vulture (31) 13:5
Roadside Hawk (43) 16:12
Gray Hawk (43) 16:14

Collared Forest-Falcon (55) 16:7
Crested Caracara (53) 14:12
Laughing Falcon (53) 15:8
Red-billed Pigeon (87) 18:2
White-winged Dove (89) 18:12
Inca Dove (89) 18:11
Common Ground-Dove (89) 18:8
White-tipped Dove (91) 18:14
Ruddy Quail-Dove (93) 18:17
Orange-fronted Parakeet (95) 19:13
Orange-chinned Parakeet (97) 19:14
White-fronted Parrot (99) 19:6
Yellow-naped Parrot (99) 19:5
Squirrel Cuckoo (103) 21:7
Lesser Ground-Cuckoo (103) 21:11
Groove-billed Ani (103) 21:9
Common Pauraque (111) 21:18
White-collared Swift (117) 22:1
Vaux's Swift (119) 22:9
Violet Sabrewing (123) 23:9
Canivet's Emerald (127) 24:13
Steely-vented Hummingbird (127) 24:15
Cinnamon Hummingbird (129) 24:11
Stripe-tailed Hummingbird (133) 24:17
*Purple-throated Mountain-gem (135) 24:7
Ruby-throated Hummingbird (131) 25:9
Black-headed Trogon (141) 26:10
Elegant Trogon (143) 26:7
*Orange-bellied Trogon (143) 26:4
Tody Motmot (147) 27:12
Blue-crowned Motmot (147) 27:8
Turquoise-browed Motmot (147) 27:11
Collared Aracari (155) 27:15
Keel-billed Toucan (155) 27:18
***Hoffmann's Woodpecker (157) 28:16**
Lineated Woodpecker (161) 27:13
Pale-billed Woodpecker (161) 27:14
Ruddy Woodcreeper (171) 29:13
Olivaceous Woodcreeper (171) 29:7
Northern Barred-Woodcreeper (169) 29:19
Ivory-billed Woodcreeper (173) 29:21
Streak-headed Woodcreeper (173) 29:8
Barred Antshrike (175) 31:3
Spotted Antbird (181) 31:12
Yellow-bellied Elaenia (193) 37:26
Mountain Elaenia (193) 37:24
Scale-crested Pygmy-Tyrant (197) 37:4
Northern Bentbill (197) 37:6
Common Tody-Flycatcher (197) 37:7
Yellow-olive Flycatcher (195) 37:16
Stub-tailed Spadebill (189) 37:2
Yellowish Flycatcher (207) 36:19
Bright-rumped Attila (201) 35:6
Brown-crested Flycatcher (209) 35:18
Great Kiskadee (211) 35:13
Boat-billed Flycatcher (211) 35:12
Social Flycatcher (211) 35:14
Sulphur-bellied Flycatcher (213) 35:10
Tropical Kingbird (213) 35:1
Rose-throated Becard (217) 33:12
Long-tailed Manakin (223) 33:4
Lesser Greenlet (229) 40:7
White-throated Magpie-Jay (231) 39:18
Brown Jay (231) 39:19
Rufous-naped Wren (239) 38:2
Rufous-and-white Wren (241) 38:9
Banded Wren (241) 38:6
Plain Wren (241) 38:17
Nightingale Wren (245) 38:21
White-lored Gnatcatcher (237) 41:1
Orange-billed Nightingale-Thrush (247) 38:26
Clay-colored Thrush (251) 39:8
White-throated Thrush (251) 39:9
Tennessee Warbler (255) 40:22
Gray-crowned Yellowthroat (273) 42:13
Slate-throated Redstart (273) 42:7
Golden-crowned Warbler (275) 40:18
Rufous-capped Warbler (275) 40:16
Gray-headed Tanager (283) 45:17
Red-crowned Ant-Tanager (277) 47:10
Blue-gray Tanager (291) 46:15
Red-legged Honeycreeper (293) 46:2
Blue-black Grassquit (297) 49:7
Olive Sparrow (303) 50:15
Stripe-headed Sparrow (303) 50:7
Rusty Sparrow (305) 50:4
Blue Grosbeak (313) 48:11
Painted Bunting (313) 48:13
Baltimore Oriole (321) 44:7
Yellow-throated Euphonia (327) 45:5

Site 2-3: Palo Verde National Park

Birding Time: 2 days
Elevation: Sea level
Trail Difficulty: 1
Reserve Hours: 8 a.m. to 4 p.m.
Entrance Fee: $10 (US 2008)
4×4 Recommended

The Tempisque River Valley serves as the main watershed for most of the North Pacific region of Costa Rica, and Palo Verde National Park sits on the edge of this important waterway just before it empties into the Gulf of Nicoya. The park protects about 32,000 acres of land, almost half of which is flooded with water during the height of the rainy season. The area, one of the most important wetlands in all of Central America, provides essential habitat for thousands of resident and migrant birds.

In 1978 Costa Rica realized the importance of the Palo Verde Marsh and established the national park, procuring most of the land from cattle ranchers. In the years that followed, however, the marsh began to become choked by aggressive plants. Invasive species, principally the grass *Typha domingensis*, were drastically changing the habitat, making it much less desirable for most of the water birds that depend on Palo Verde. In an effort to fight the invasive species, the park has tried many different management techniques. One of the most successful has been to use a large tractor, when the lands are wet, to till up the grass. The park has also reintroduced cattle, which are allowed to graze in the marsh. Whether or not the cattle are having any affect is still being debated, and it may be that in years to come they will be removed again. But for now, the strange situation of cattle being ranched inside a national park still exists.

The park itself is a magical place, with a feeling of peace and isolation. Wildlife abounds, and not just birds. Mantled Howler Monkeys and White-faced Capuchins laze in the trees on hot afternoons, White-tailed Deer bound through the forest, and American Crocodiles bask in the sun along the riverbanks. Unfortunately mosquitoes are also all too common, at least in the wet season, so come well prepared. The landscape of the park changes drastically between the dry season and the wet season. When it is wet, the marsh is filled with water and marsh plants. At this time the marsh birds are more dispersed and harder to spot, although the land birds seem to be more plentiful. When the water dries up, the marsh birds congregate into the remaining water holes and are much easier to observe. Regardless of time of year, good birding can always be found, making this one of my personal favorite places to bird in Costa Rica.

Target Birds

Thicket Tinamou (3) 12:18
Fulvous Whistling-Duck (5) 8:9
Muscovy Duck (5) 8:10
Northern Shoveler (7) 8:2
Masked Duck (5) 7:8
Great Curassow (11) 12:3
Least Bittern (25) 6:3
Black-crowned Night-Heron (27) 5:4
Glossy Ibis (29) 4:9
Jabiru (19) 4:5
Hook-billed Kite (35) 16:9
Snail Kite (35) 14:6
Northern Harrier (41) 15:7
Crane Hawk (41) 14:4
Great Black-Hawk (45) 13:7
Harris's Hawk (33) 14:5
White-tailed Hawk (43) 17:6
Zone-tailed Hawk (45) 14:2
Collared Forest-Falcon (55) 16:7
Peregrine Falcon (51) 15:5
Sora (59) 6:12
Limpkin (63) 5:5
Double-striped Thick-knee (63) 9:10
Plain-breasted Ground-Dove (89) 18:9
Scarlet Macaw (99) 19:1
Yellow-naped Parrot (99) 19:5
Mangrove Cuckoo (101) 21:4
Pacific Screech-Owl (107) 20:11
Spectacled Owl (105) 20:8
Lesser Nighthawk (113) 21:12
Plain-capped Starthroat (135) 23:19
White-lored Gnatcatcher (237) 41:1
Painted Bunting (313) 48:13
Spot-breasted Oriole (319) 44:1
Scrub Euphonia (327) 45:4

Access

From Bagaces: Turn west off Rt. 1 onto a dirt road at the Bagaces gas station. You will quickly come to a fork, where you should keep right. Go straight for 13.0 km to another fork and bear right. You will come to a third fork after another 4.0 km, where you should keep left. You will arrive at the entrance to the park in another 3.2 km (20.2 km from Rt. 1). (Driving time from Bagaces: 45 min)

Logistics: Because of its distance from civilization, and because the best birding time is early in the morning, I highly suggest staying inside the reserve at either the ranger station or the Organization for Tropical Studies (OTS) Field Station. Both are fairly rustic, although OTS offers better facilities at a higher price. Camping is also possible.

OTS Field Station
Tel: (506) 2524-0628
Website: www.ots.ac.cr
E-mail: nat-hist@ots.ac.cr

Ranger Station
Tel: (506) 2200-0125

Birding Sites

Agricultural Land and the Park Entrance

The dirt road that leads from Bagaces to the national park consistently produces interesting birds. Raptors such as White-tailed Hawk, Swainson's Hawk, and American Kestrel are possibly flying over the fields. Scissor-tailed Flycatcher likes to perch on the fence posts, and Black-striped Sparrow moves through the bushes.

Just outside the park entrance, on either side of the road, are rice patties, which can be filled with Wood Stork, Great Egret, and Black-necked Stilt depending on the water level. Jabiru, Solitary Sandpiper, and Pectoral Sandpiper are also possible findings in the area. As you drive up to the ticketing booth, it is usually worth getting out of the car to check the trees by the side of the road. You may find Yellow-naped Parrot, Orange-fronted Parakeet, White-lored Gnatcatcher, Steely-vented Hummingbird, and White-throated Magpie-Jay. This is also one of the best places in Palo Verde to find Lesser Ground-Cuckoo. The grassy fields around the buildings beyond the ticket booth consistently produce Double-striped Thick-knee.

OTS and Ranger Station Vicinity

Conveniently some of the best birding in the park is along the main road as it runs between the OTS Field Station and the ranger station. The old airstrip that starts in front of the OTS buildings can yield a wide variety of interesting species, including Laughing Falcon, Crested Caracara, and Yellow-naped Parrot. The scrubby plants that border the field often attract hummingbirds such as Canivet's Emerald, Steely-vented Hummingbird, Ruby-throated Hummingbird, and Green-breasted Mango.

The old airstrip also provides the best views of the marsh, with two especially good vantage points. The first is an old metal tower that offers an elevated look over the entire Palo Verde Marsh, although it cannot really hold more than two people at a time. (It might be a good idea to ask at the OTS station if the tower is still safe before you venture up.) The second option is a boardwalk, which extends into the marsh a bit farther down the shoreline. Either of these places is great for spending time scanning the marsh. In the wet season, dawn and dusk are the best times to observe, as birds are busy flying between their foraging and roosting sites. In the dry season timing isn't as important, as most species spend all day foraging right in front of the observation areas, and any time of day will yield a large variety. Look for Fulvous Whistling-Duck, Northern Shoveler, Glossy Ibis, Bare-throated Tiger Heron, Limpkin, and Roseate Spoonbill. Many species of raptors constantly patrol the area, including Snail Kite, Peregrine Falcon, Harris's Hawk, and Zone-tailed Hawk. In the dry season you can often find Jabiru in this area as well.

The road that continues between the airfield and the ranger station is productive for land birds, and you might find Streak-backed Oriole, Brown-crested Flycatcher, Mangrove Cuckoo, and Rose-throated Becard. This area is also good for raptors such as Common Black-Hawk, Crested Caracara, Crane Hawk, and Ferruginous Pygmy-Owl. Just to the right of the ranger station a small spring bubbles up. Especially in the dry season this can be a worthwhile place to check for birds that might be attracted to the precious water.

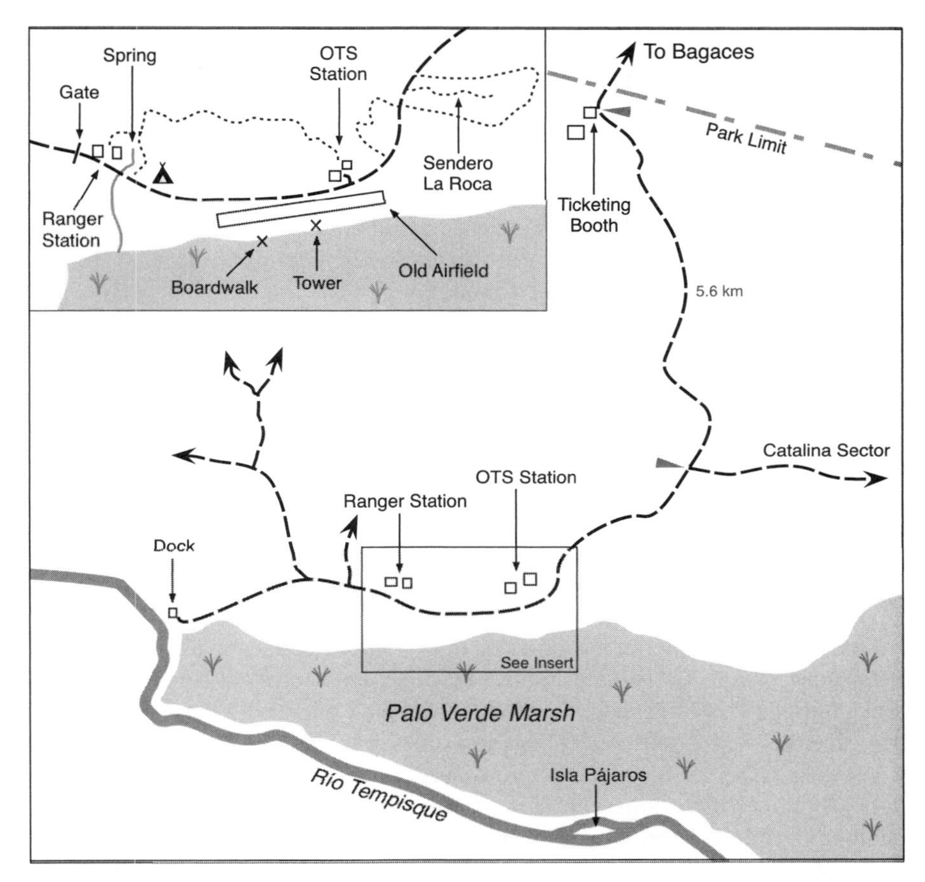

Palo Verde National Park Area Map. The insert shows details around the Organization for Tropical Studies (OTS) station and the ranger station.

The main road continues past the ranger station for about two kilometers to the Tempisque River, where it terminates. This stretch of road provides access to some of the most productive birding in the park, and I highly recommend taking the time to walk it. The first half of the road passes through scrubby forest. Look for Indigo Bunting, Painted Bunting, Black-headed Trogon, and White-throated Magpie-Jay among others. Flowering plants along the roadside could attract Canivet's Emerald, Green-breasted Mango, Plain-capped Starthroat, and Ruby-throated Hummingbird.

In the wet season, the terrain along the roadside becomes flooded as you get closer to the river. Water-associated birds, such as Bare-throated Tiger-heron, White Ibis, and Belted Kingfisher, can be found in this area, as well as Blue-throated Goldentail and Spot-breasted Oriole. This area also provides the best chance of seeing Scarlet Macaws, which often fly overhead. The macaws at Palo

Verde are the only wild population of Scarlet Macaws left living in tropical dry forest. Once at the river, look for Neotropic Cormorant, Ringed Kingfisher, and Osprey. If the tide is low, you will also probably see American Crocodiles sunning along the riverbank.

Another road worth exploring forks right off the road to the Tempisque River, about 400 meters beyond the ranger station. (This is the second small right fork after the station.) This relatively long road leads into dry forest, and walking it may yield Streaked Flycatcher, Turquoise-browed Motmot, Elegant Trogon, Long-tailed Manakin, and Thicket Tinamou.

The national park also maintains a number of trails through the forest that abuts the two field stations. The birding along these trails usually is not as productive as along the roads discussed earlier, and I do not suggest taking the time to walk them. The one exception is the short trail called Sendero La Roca. This leaves the main road before the OTS station and climbs up to the top of a hill. While the birding is nothing special, the panoramic vista of the marsh from this lookout is quite breathtaking.

Isla Pájaros Boat Trip

One enjoyable outing that is possible from Palo Verde is a boat ride down the Tempisque River. You should inquire at either the ranger station or the OTS Field Station, where the staff will be able to arrange a trip with local guides. The trip generally lasts about three hours and starts at the dock along the Tempisque River. The boat ride is a good place to observe a variety of wildlife, including large American Crocodiles. Scarlet Macaw, Spot-breasted Oriole, and Lesser Nighthawk are also regularly seen from the boat. The ride takes you to the famous Isla Pájaros, where hundreds of herons and other birds nest. It is one of the only known nesting sites of Glossy Ibis in Costa Rica.

Species to Expect

Thicket Tinamou (3) 12:18
Black-bellied Whistling-Duck (5) 8:7
Fulvous Whistling-Duck (5) 8:9
Muscovy Duck (5) 8:10
Blue-winged Teal (7) 8:5
Northern Shoveler (7) 8:2
Great Curassow (11) 12:3
Least Grebe (9) 7:4
Neotropic Cormorant (17) 4:4
Anhinga (17) 4:3
Bare-throated Tiger-Heron (21) 5:16
Great Blue Heron (23) 5:6
Great Egret (23) 5:14
Little Blue Heron (23) 5:9
Tricolored Heron (25) 5:8
Cattle Egret (23) 5:13
Green Heron (25) 6:2
Black-crowned Night-Heron (27) 5:4
White Ibis (29) 4:8
Glossy Ibis (29) 4:9
Roseate Spoonbill (27) 4:7
Jabiru (19) 4:5
Wood Stork (19) 4:6
Black Vulture (31) 13:4
Turkey Vulture (31) 13:3
Osprey (33) 17:14
Hook-billed Kite (35) 16:9
Snail Kite (35) 14:6

Common Black-Hawk (45) 13:6
Harris's Hawk (33) 14:5
Roadside Hawk (43) 16:12
Broad-winged Hawk (43) 16:13
Gray Hawk (43) 16:14
Short-tailed Hawk (43) 16:11
White-tailed Hawk (43) 17:6
Zone-tailed Hawk (45) 14:2
Crested Caracara (53) 14:12
Laughing Falcon (53) 15:8
Peregrine Falcon (51) 15:5
Purple Gallinule (59) 6:15
Common Moorhen (61) 7:2
American Coot (61) 7:1
Limpkin (63) 5:5
Double-striped Thick-knee (63) 9:10
Black-necked Stilt (69) 9:12
Northern Jacana (61) 6:18
Spotted Sandpiper (73) 11:8
Red-billed Pigeon (87) 18:2
White-winged Dove (89) 18:12
Inca Dove (89) 18:11
Common Ground-Dove (89) 18:8
White-tipped Dove (91) 18:14
Orange-fronted Parakeet (95) 19:13
Scarlet Macaw (99) 19:1
Orange-chinned Parakeet (97) 19:14
White-fronted Parrot (99) 19:6
Yellow-naped Parrot (99) 19:5
Squirrel Cuckoo (103) 21:7
Mangrove Cuckoo (101) 21:4
Groove-billed Ani (103) 21:9
Pacific Screech-Owl (107) 20:11
Ferruginous Pygmy-Owl (109) 20:17
Lesser Nighthawk (113) 21:12
Common Pauraque (111) 21:18
Stripe-throated Hermit (121) 23:1
Green-breasted Mango (129) 23:13
Canivet's Emerald (127) 24:13
Blue-throated Goldentail (129) 24:9
Steely-vented Hummingbird (127) 24:15
Rufous-tailed Hummingbird (129) 24:10
Cinnamon Hummingbird (129) 24:11
Plain-capped Starthroat (135) 23:19
Ruby-throated Hummingbird (131) 25:9
Black-headed Trogon (141) 26:10
Turquoise-browed Motmot (147) 27:11
Ringed Kingfisher (149) 27:1
Amazon Kingfisher (149) 27:4
Green Kingfisher (149) 27:5
***Hoffmann's Woodpecker (157) 28:16**
Pale-billed Woodpecker (161) 27:14
Streak-headed Woodcreeper (173) 29:8
Common Tody-Flycatcher (197) 37:7
Yellow-olive Flycatcher (195) 37:16
Dusky-capped Flycatcher (209) 35:21
Great Crested Flycatcher (209) 35:17
Brown-crested Flycatcher (209) 35:18
Great Kiskadee (211) 35:13
Social Flycatcher (211) 35:14
Streaked Flycatcher (213) 35:11
Sulphur-bellied Flycatcher (213) 35:10
Piratic Flycatcher (193) 35:8
Tropical Kingbird (213) 35:1
Rose-throated Becard (217) 33:12
Masked Tityra (217) 34:1
Long-tailed Manakin (223) 33:4
Yellow-green Vireo (227) 40:5
White-throated Magpie-Jay (231) 39:18
Barn Swallow (237) 22:12
Rufous-naped Wren (239) 38:2
Rufous-and-white Wren (241) 38:9
Banded Wren (241) 38:6
Long-billed Gnatwren (237) 32:15
Tropical Gnatcatcher (237) 41:2
Tennessee Warbler (255) 40:22
Yellow Warbler (259) 42:2
Northern Waterthrush (269) 43:14
Summer Tanager (285) 47:5
Blue-black Grassquit (297) 49:7
White-collared Seedeater (295) 49:2
Olive Sparrow (303) 50:15
Stripe-headed Sparrow (303) 50:7
Indigo Bunting (313) 48:12
Painted Bunting (313) 48:13
Great-tailed Grackle (317) 44:16
Streak-backed Oriole (319) 44:2
Spot-breasted Oriole (319) 44:1
Baltimore Oriole (321) 44:7
Scrub Euphonia (327) 45:4

Site 2-4: Mata Redonda Marsh

Birding Time: 4 hours
Elevation: Sea level
Trail Difficulty: 1
Reserve Hours: N/A
Entrance Fee: N/A
4×4 Recommended

Mata Redonda is a vast wetland along the western side of the Tempisque River Basin that provides an important habitat for many water-dependent species of birds. This marsh is one of the largest in the country, and one of the only areas in the North Pacific region to remain wet throughout the year. The birding here is best during the dry season when the parched conditions of the surrounding countryside push thousands of birds into the water that remains at the Mata Redonda Marsh. The site is especially noteworthy because of its importance to Jabiru. In recent years more than 60 individuals have been seen frequenting the area, and it is the most reliable site in Costa Rica to find this impressive species. Many of the other water-based species that are "Target Birds" for Palo Verde National Park (2-3) should also be looked for here, including Snail Kite, Harris's Hawk, Limpkin, and Muscovy Duck.

Please note that Mata Redonda is quite remote and is only recommended for seasoned travelers to Latin America. To access the marshes you must pass through a private cattle ranch, and basic Spanish is helpful to gain permission. Also, the roads around the marsh get rough and you will certainly want to be driving a 4×4. That being said, it is a fun place to explore, the scenery is beautiful, and the area will give you a real flavor for rural life in Guanacaste.

Access

From Nicoya: Start at the major intersection along Rt. 21 near the city of Nicoya. Turn east off Rt. 21 onto a dirt road, heading in the opposite direction of downtown Nicoya. Follow this for 4.4 km, bearing left after 3.6 km, and then turn right. Proceed for 8.0 km and turn right again. Continue 18.4 km to the town of Rosario, passing through Pozo de Agua and Puerto Humo along the way. In Rosario you will come to a T-intersection. Turn right, and the road will immediately bend left, passing around a soccer field. Follow the road 2.0 km out of town to a small farmhouse on your left opposite expansive cattle fields. (This last stretch of road is in poor condition, and you will want a 4×4.) Directions to the birding sites start at this farmhouse. (Driving time from Nicoya: 70 min)

Logistics: Lodging in the area is difficult to find. The city of Santa Cruz could serve as a base from which to bird both this site and Diriá National Park (2-5). There are also some small lodges near Barra Honda National Park that offer

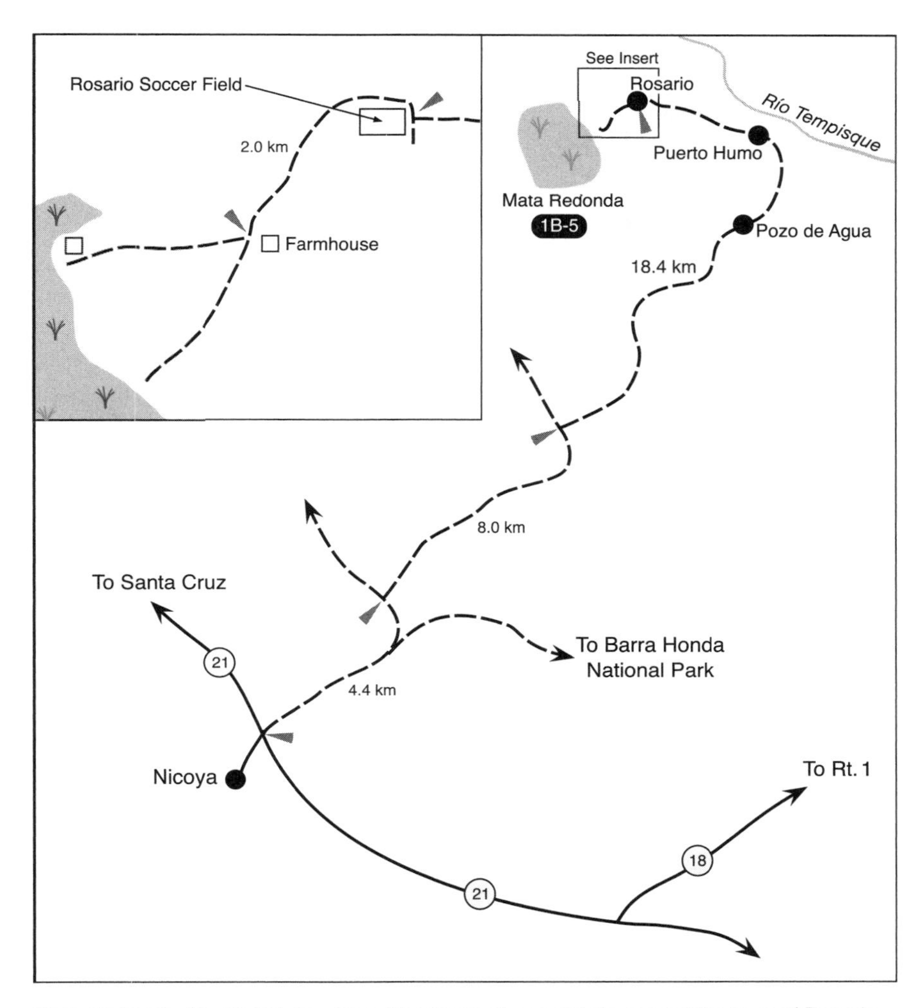

Mata Redonda Marsh Driving Map. The insert shows details around the town of Rosario.

basic accommodations and are closer to the marsh. Alternatively, it is possible to visit the marsh from Palo Verde National Park (2-3). This day trip must be arranged with the staff at Palo Verde, and travel requires both boat and flatbed truck. The birding all along this journey is good, and it should make for a pleasant excursion.

Birding Sites

Across from the small farmhouse described in "Access" is a gate that blocks a little-used road leading through cattle fields. To gain access to this road, which leads to the Mata Redonda Marsh, you must ask the owners of the small farmhouse for permission to pass through their land. This should not be a problem, and they will open the gate for you. The road proceeds for 2.6 km through cattle pastures to another small house set at the edge of the marsh. If you encounter any closed gates along the way, you can open them to drive through, but be sure to close them behind you. As you pass through the cattle fields, keep your eyes open for Double-striped Thick-knee and Crested Caracara.

When you arrive at the marsh, park near the small house in front of the extensive wetland. During the dry season I suggest walking along the water, where shorebirds, including Black-necked Stilt, Lesser Yellowlegs, and Stilt Sandpiper, can often be found along the muddy edges. Jabiru is regularly seen feeding among the tall grasses along with Roseate Spoonbill, Wood Stork, and Limpkin. Thousands of Blue-winged Teal winter in the marsh, and you might find American Coot, Northern Shoveler, or something else interesting among them. You should also expect to see a few species of raptors, perhaps Zone-tailed Hawk or Snail Kite, gliding past.

If you are interested in further exploring Mata Redonda, there is a second major access point. Return to the farmhouse with the gate and make a hard right. This road ends in about two kilometers, at a different section of the marsh. Anything you may have missed at the first location might be found here.

Site 2-5: Diriá National Park

Birding Time: 1 day
Elevation: 100 meters
Trail Difficulty: 1–2
Reserve Hours: N/A
Entrance Fee: N/A
4×4 Recommended

Diriá National Park offers excellent birding, thanks to the range of habitats that are accessible in the area. Low-elevation dry forest and gallery forest can be found at the La Casona sector of the park, while up higher, around Vista del Mar, is a very interesting habitat that feels like a cross between a dry forest and a cloud forest. Diriá is the best place to find Plain Chachalaca, a Nicoya Peninsula specialty, and also offers many other interesting birds, such as Nutting's Flycatcher and Lesser Ground-Cuckoo.

The reserve was originally created in 1991 as a national forest. Its status has been upgraded twice since its creation, once in 1994, becoming a national wildlife refuge, and again in 2004, becoming a national park. Diriá currently protects 13,400 acres, although the infrastructure is minimal. The park does not receive many visitors and currently does not even charge an entrance fee. There is one ranger station called La Casona, next to which is a short trail. Even without a large trail system, however, many good birds can be found along the dirt roads, especially up near the radio towers at Vista del Mar.

Target Birds

Thicket Tinamou (3) 12:18	Ivory-billed Woodcreeper (173) 29:21
Plain Chachalaca (11) 12:2	Stub-tailed Spadebill (189) 37:2
King Vulture (31) 13:5	Nutting's Flycatcher (209) 35:22
Gray-headed Dove (91) 18:15	Long-tailed Manakin (223) 33:4
Lesser Ground-Cuckoo (103) 21:11	Rufous-browed Peppershrike (229) 40:2
Northern Potoo (115) 20:4	White-throated Magpie-Jay (231) 39:18
Plain-capped Starthroat (135) 23:19	Western Tanager (285) 47:2
Elegant Trogon (143) 26:7	

Access

From Santa Cruz: Turn off Rt. 21 toward Santa Cruz at the intersection with the stoplight. Proceed straight through town for 10 city blocks (0.8 km) until you come to the town park on your right. Continue 1 block beyond the park and turn right. Now proceed 6 blocks (0.5 km) to where the road in front of you turns to dirt, and turn left onto a paved road. Follow this straight for 6.0 km, passing through the town of Arado, to a fork. If you take the left fork you will arrive at La Casona Ranger Station in another 5.8 km. If you take the right fork, the road climbs, bringing you to Vista del Mar in 11.6 km. (Driving time from Santa Cruz to La Casona: 20 min)

Logistics: There are a number of local hotels in Santa Cruz encompassing a range of comfort levels. Business travel tends to book up many of the rooms during the week, so it is advisable to make a reservation in advance. Arado is the closest town to Diriá with food; if you plan to stay in the field for most of the day, be sure to bring along a bag lunch.

Birding Sites

La Casona Sector

The road to La Casona proceeds through patchy forests and fields. This can be an excellent area to bird, and should produce a number of species, including Rose-throated Becard, Plain Chachalaca, Elegant Trogon, Thicket Tinamou, and Orange-fronted Parakeet.

After about 5.5 km the road forks just before you reach La Casona Ranger Station. To get to the station take the left fork, descending a small hill and crossing a small river. (Be sure to check the water level of the river before you drive your car through. You can always wade across if you don't want to risk your car.) Around the buildings you should be able to find Squirrel Cuckoo, Banded Wren, White-throated Magpie-Jay, Lesser Greenlet, and Rufous-capped Warbler.

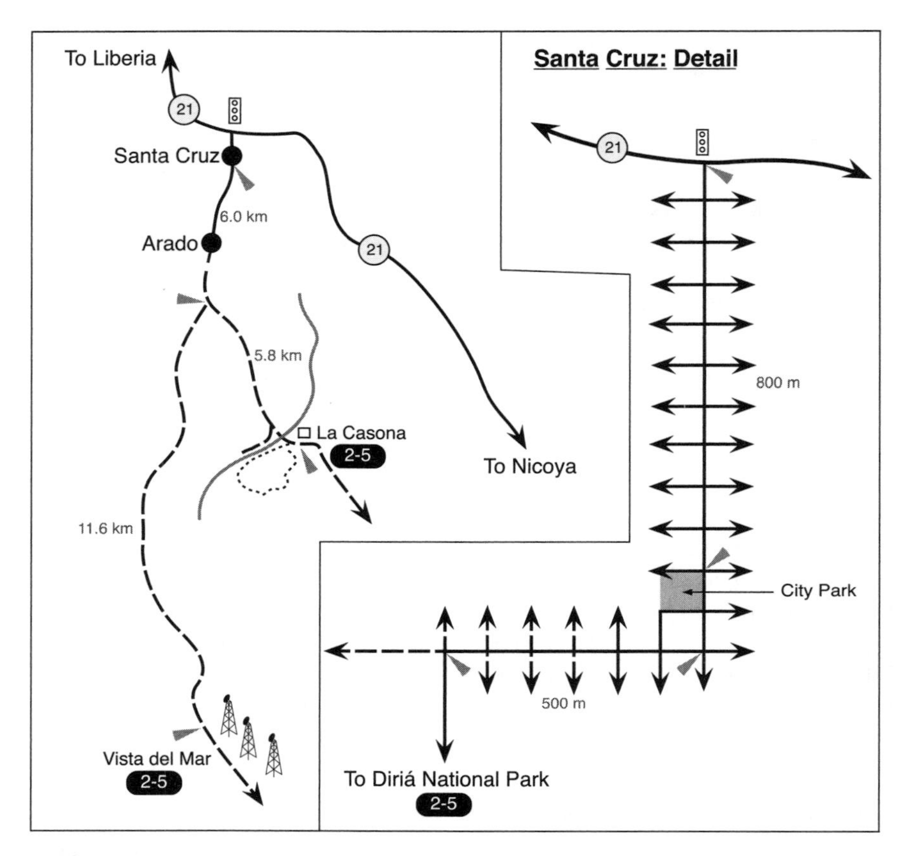

Diriá National Park Driving Map. The right side of the map depicts roads in downtown Santa Cruz.

The small river that runs past the station sustains an evergreen forest along its banks that holds some interesting species. The best access to this forest is a trail that leads upstream from the ranger station, and does a small one-kilometer loop. This is a great place to find Stub-tailed Spadebill, and you should also look for Stripe-throated Hermit, Violaceous Trogon, Bright-rumped Attila, Turquoise-browed Motmot, and Long-tailed Manakin. I also suggest walking along a small road that proceeds through the forest on the opposite side of the river. To get to this area you must cross the river and take the right fork where you had previously gone left.

Vista del Mar Sector

Vista del Mar is a very interesting area, and a mandatory stop if you are birding Diriá, although a 4×4 is suggested. When you arrive at the top of the hills, which reach about 900 meters in elevation, you will see a number of radio towers and some small farms. If the weather is clear you will have a beautiful view of the Pacific coastline. There are, unfortunately, no trails in the area so you are confined to the road. From the first tower there is good birding habitat for the next three kilometers, and you can either walk or bird from the car.

The area holds a strange conglomerate of species, which makes it especially fun to bird. You will undoubtedly run into a number of species that prefer higher elevations such as Orange-billed Nightingale-Thrush, Rufous-browed Peppershrike, Yellow-faced Grassquit, and Mountain Elaenia. However, you should expect to see many dry-forest species as well. Look for Thicket Tinamou, Long-tailed Manakin, Rufous-and-white Wren, and Olive Sparrow. In particular, keep an eye out for Nutting's Flycatcher and Lesser Ground-Cuckoo, the latter being particularly easy to see along the road. During the last months of the year Three-wattled Bellbird is occasionally found here as well.

Species to Expect

Thicket Tinamou (3) 12:18
Plain Chachalaca (11) 12:2
Cattle Egret (23) 5:13
Boat-billed Heron (27) 5:2
White Ibis (29) 4:8
Black Vulture (31) 13:4
Turkey Vulture (31) 13:3
King Vulture (31) 13:5
Roadside Hawk (43) 16:12
Gray Hawk (43) 16:14
Collared Forest-Falcon (55) 16:7
Red-billed Pigeon (87) 18:2
White-winged Dove (89) 18:12
Inca Dove (89) 18:11
Common Ground-Dove (89) 18:8
Ruddy Ground-Dove (89) 18:7
White-tipped Dove (91) 18:14
Orange-fronted Parakeet (95) 19:13
Orange-chinned Parakeet (97) 19:14
White-fronted Parrot (99) 19:6
Squirrel Cuckoo (103) 21:7
Lesser Ground-Cuckoo (103) 21:11
Groove-billed Ani (103) 21:9
Pacific Screech-Owl (107) 20:11
Ferruginous Pygmy-Owl (109) 20:17
Common Pauraque (111) 21:18
Canivet's Emerald (127) 24:13
Steely-vented Hummingbird (127) 24:15
Rufous-tailed Hummingbird (129) 24:10
Cinnamon Hummingbird (129) 24:11
Plain-capped Starthroat (135) 23:19
Black-headed Trogon (141) 26:10
Violaceous Trogon (141) 26:8
Elegant Trogon (143) 26:7
Turquoise-browed Motmot (147) 27:11
Green Kingfisher (149) 27:5
***Hoffmann's Woodpecker (157) 28:16**
Pale-billed Woodpecker (161) 27:14
Northern Barred-Woodcreeper (169) 29:19
Streak-headed Woodcreeper (173) 29:8
Barred Antshrike (175) 31:3
Mountain Elaenia (193) 37:24
Ochre-bellied Flycatcher (199) 36:25
Yellow-olive Flycatcher (195) 37:16
Stub-tailed Spadebill (189) 37:2
Bright-rumped Attila (201) 35:6
Dusky-capped Flycatcher (209) 35:21
Nutting's Flycatcher (209) 35:22
Great Crested Flycatcher (209) 35:17
Brown-crested Flycatcher (209) 35:18
Great Kiskadee (211) 35:13
Boat-billed Flycatcher (211) 35:12
Social Flycatcher (211) 35:14
Streaked Flycatcher (213) 35:11
Sulphur-bellied Flycatcher (213) 35:10
Tropical Kingbird (213) 35:1
Rose-throated Becard (217) 33:12
Masked Tityra (217) 34:1
Long-tailed Manakin (223) 33:4
Yellow-green Vireo (227) 40:5

Lesser Greenlet (229) 40:7
Rufous-browed Peppershrike (229) 40:2
White-throated Magpie-Jay (231) 39:18
Rufous-naped Wren (239) 38:2
Rufous-and-white Wren (241) 38:9
Banded Wren (241) 38:6
Plain Wren (241) 38:17
House Wren (243) 38:18
Long-billed Gnatwren (237) 32:15
White-lored Gnatcatcher (237) 41:1
Orange-billed Nightingale-Thrush (247) 38:26
Clay-colored Thrush (251) 39:8
Tennessee Warbler (255) 40:22
Yellow Warbler (259) 42:2
American Redstart (267) 41:10
Northern Waterthrush (269) 43:14
Rufous-capped Warbler (275) 40:16
Summer Tanager (285) 47:5
Western Tanager (285) 47:2
Blue-gray Tanager (291) 46:15
Red-legged Honeycreeper (293) 46:2
Variable Seedeater (295) 49:3
Yellow-faced Grassquit (297) 49:6
Olive Sparrow (303) 50:15
Stripe-headed Sparrow (303) 50:7
Grayish Saltator (309) 48:3
Blue-black Grosbeak (311) 48:10
Indigo Bunting (313) 48:12
Baltimore Oriole (321) 44:7
Yellow-billed Cacique (321) 44:12
Yellow-throated Euphonia (327) 45:5

Site 2-6: La Ensenada Wildlife Refuge

Birding Time: 1 day
Elevation: Sea level
Trail Difficulty: 2
Reserve Hours: 6 a.m. to 8 p.m.
Entrance Fee: Included with overnight
Trail Map Available on Site

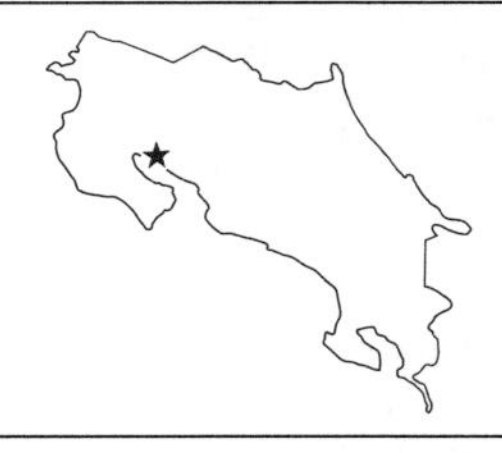

Situated on the mainland coast of the Gulf of Nicoya, La Ensenada Wildlife Refuge offers a comfortable and relaxed setting for birding the North Pacific Slope. La Ensenada is a seemingly paradoxical combination of a working cattle ranch, private hotel, and national wildlife refuge; and yet it manages to perform all these roles well. The owners of La Ensenada originally bought the land in 1977 as a cattle ranch. The hotel was established in 1990, and eight years later the land was officially made a national wildlife refuge. The cattle farming has continued through these changes, and today La Ensenada maintains 300 head, mostly bulls, which are farmed for meat. There is also a *salina* (salt pond) on the property, where sea salt is harvested during the dry season.

The hotel is comfortable, with 22 cabins looking out onto the Gulf of Nicoya and the mountains of the Nicoya Peninsula rising in the distance. Three delicious meals are served buffet style, daily, in the open-air restaurant. This is definitely one of the most comfortable sites to stay at in the North Pacific region. Non-birders will also enjoy La Ensenada, as they can take advantage of the swimming pool, horseback riding, and boat rides.

The birding in the area is rewarding, and it is easy to build a good-sized list. Most of the habitat within the reserve is patchy forest and cow pasture, which will produce most of the dry-forest species. In particular, it is the most reliable place to find Spot-breasted Oriole, and Three-wattled Bellbird is easily found during January and the beginning of February. La Ensenada also offers excellent coastal birding at the *salina* and in the mangrove that surrounds it. In fact, this site could easily have been included in the Coastline region.

Target Birds

Crested Bobwhite (15) 12:13/14
Hook-billed Kite (35) 16:9
Common Black-Hawk (45) 13:6
White-tailed Hawk (43) 17:6
Rufous-necked Wood-Rail (55) 6:14
Double-striped Thick-knee (63) 9:10
Wilson's Plover (65) 10:2
Marbled Godwit (69) 9:8
Surfbird (73) 10:7
Red Knot (77) 11:19
Stilt Sandpiper (77) 11:6
Wilson's Phalarope (79) 10:9
Black Skimmer (85) 3:11
Plain-breasted Ground-Dove (89) 18:9
Yellow-naped Parrot (99) 19:5
Lesser Ground-Cuckoo (103) 21:11
Pacific Screech-Owl (107) 20:11
Northern Potoo (115) 20:4
*Mangrove Hummingbird (131) 24:2
Northern Scrub-Flycatcher (193) 37:22
Panama Flycatcher (209) 35:20
*Three-wattled Bellbird (221) 34:12
Mangrove Vireo (225) 40:11
White-throated Magpie-Jay (231) 39:18
White-lored Gnatcatcher (237) 41:1
Yellow (Mangrove) Warbler (259) 42:3
Painted Bunting (313) 48:13
Spot-breasted Oriole (319) 44:1
Scrub Euphonia (327) 45:4

Access

From the intersection of Rt. 1 and Rt. 23: Take Rt. 1 north for 26.6 km and then fork left onto Rt. 132, following signs for Venegas. Go 3.4 km and turn right onto a dirt road just after crossing a bridge. Follow this road for 11.8 km (keep left at a fork after 9.0 km) until you come to a three-way intersection. Turn left here, and then take a right almost immediately. In another 4.6 km you will come to the town of Abangaritos. Proceed straight and you will come to the entrance of La Ensenada at the other side of town. (Driving time from Rt. 23: 90 min)

From the intersection of Rt. 1 and Rt. 18: Take Rt. 1 south for 11.6 km and then turn right onto a dirt road, following signs for La Ensenada. (Ponds on the right side of the road just before this turn often hold Least Grebe.) Proceed 8.2 km until the road forks. Take the right fork and then almost immediately turn right. In 4.6 km you will come to the town of Abangaritos. Proceed straight and you will come to the entrance of La Ensenada at the other side of town. (Driving time from Rt. 18: 60 min)

Logistics: I suggest staying at La Ensenada Lodge if you plan to bird the area. The accommodations are comfortable, the food is good, and the prices are reasonable.

La Ensenada Lodge
Tel: (506) 2289-6655
Website: www.laensenada.net
E-mail: refugioensenada@gmail.com

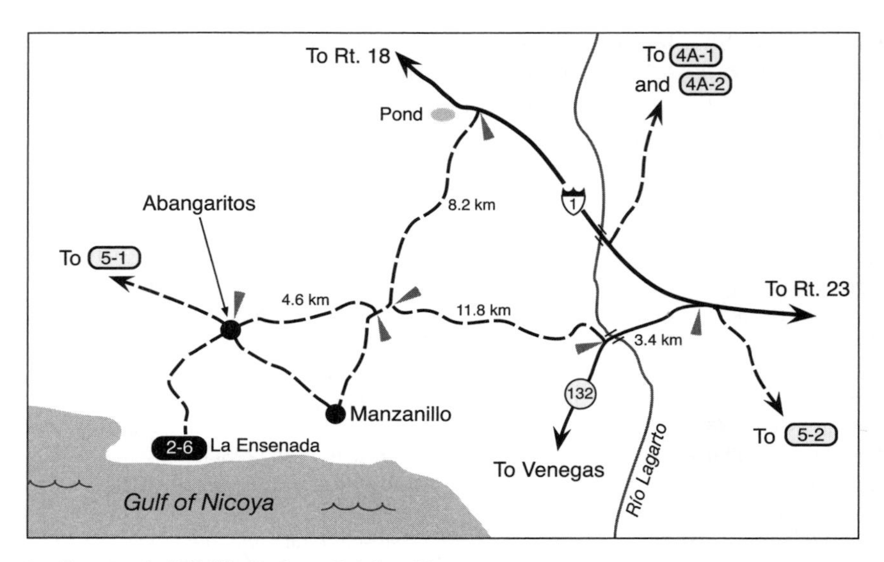

La Ensenada Wildlife Refuge Driving Map

Birding Sites

Around the grounds of the hotel you will probably encounter White-throated Magpie-Jay, Crested Bobwhite, and Cinnamon Hummingbird. Mangrove Swallows fly over the fields behind the hotel, and Lesser Ground-Cuckoo can be heard calling from the dense forest edge. At night Common Pauraque, Ferruginous Pygmy-Owl, and Pacific Screech-Owl can be heard in the area.

The refuge maintains a system of trails that meander through the grounds. Most of these access a mix of cow pasture and fragmented forest, and while the habitat may not seem ideal, many interesting species can be found. The trails that head south from the lodge (currently labeled #2 and #6) pass through some of the best habitat. Eastern Meadowlark, Orange-chinned Parakeet, and Orange-fronted Parakeet often sit in the tops of the trees. Black-striped Sparrow, Brown-crested Flycatcher, Scrub Euphonia, Rose-throated Becard, and White-lored Gnatcatcher are likely to be spotted in the trees and shrubs. Banded Wren and Lesser Ground-Cuckoo can be found in the dense undergrowth. This is also a likely place to encounter orioles. Streak-backed Oriole is the more abundant, but Spot-breasted Oriole is regular as well. Long-tailed Manakin can sometimes be found in the larger forest fragment along trail #6.

The salina usually yields some of the best birding on site and should be thoroughly checked. To get to the salina, fork right off the main entrance road about halfway down. In 200 meters or so you will come to a small pond that is worth a quick stop. Muscovy Duck, Northern Jacana, and Green Kingfisher are regulars. Laughing Falcon and Roadside Hawk are also to be expected in the area. Soon the road passes a small house and opens up to the salina. Many shorebirds, including Black-necked Stilt, Whimbrel, Least Sandpiper, Marbled Godwit, and Wilson's Plover, can be found here. Larger waders such as Wood Stork, Roseate Spoonbill, and Yellow-crowned Night-Heron also feed in the pools. Common Black-Hawk and Osprey are two of the more likely species of raptors that can be found in the area.

La Ensenada also offers access to some nice mangroves near the salina. The best trail starts by the far end of the salina, and while not currently marked, it should appear well trodden. (Be aware that at high tide this trail may not be passable.) Within the mangrove, Yellow (Mangrove) Warbler is abundant, and you have chances to see some of the other specialty birds such as Mangrove Hummingbird, Northern Scrub-Flycatcher, Panama Flycatcher, and Mangrove Vireo. Trail #1, which runs north from the hotel along the beach, provides access to another portion of mangrove. Look for many of the same species listed above.

Species to Expect

Thicket Tinamou (3) 12:18
Black-bellied Whistling-Duck (5) 8:7
Muscovy Duck (5) 8:10
Blue-winged Teal (7) 8:5
Crested Bobwhite (15) 12:13/14
Brown Pelican (17) 4:1
Neotropic Cormorant (17) 4:4
Magnificent Frigatebird (19) 1:6
Bare-throated Tiger-Heron (21) 5:16
Great Blue Heron (23) 5:6
Great Egret (23) 5:14
Little Blue Heron (23) 5:9
Cattle Egret (23) 5:13
Green Heron (25) 6:2
Yellow-crowned Night-Heron (27) 5:3
White Ibis (29) 4:8
Roseate Spoonbill (27) 4:7
Wood Stork (19) 4:6
Black Vulture (31) 13:4
Turkey Vulture (31) 13:3
Osprey (33) 17:14
Hook-billed Kite (35) 16:9
Common Black-Hawk (45) 13:6
Roadside Hawk (43) 16:12
Crested Caracara (53) 14:12
Yellow-headed Caracara (53) 15:9
Laughing Falcon (53) 15:8
Black-bellied Plover (67) 9:1
Wilson's Plover (65) 10:2
Semipalmated Plover (65) 10:5
Black-necked Stilt (69) 9:12
Northern Jacana (61) 6:18
Willet (71) 9:6
Whimbrel (71) 9:14
Marbled Godwit (69) 9:8
Semipalmated Sandpiper (75) 11:13
Western Sandpiper (75) 11:12
Least Sandpiper (75) 11:14
Short-billed Dowitcher (79) 9:3
Red-billed Pigeon (87) 18:2
White-winged Dove (89) 18:12
Inca Dove (89) 18:11
Common Ground-Dove (89) 18:8
Ruddy Ground-Dove (89) 18:7
White-tipped Dove (91) 18:14
Orange-fronted Parakeet (95) 19:13
Orange-chinned Parakeet (97) 19:14
White-fronted Parrot (99) 19:6
Yellow-naped Parrot (99) 19:5
Squirrel Cuckoo (103) 21:7
Lesser Ground-Cuckoo (103) 21:11
Groove-billed Ani (103) 21:9
Pacific Screech-Owl (107) 20:11
Ferruginous Pygmy-Owl (109) 20:17
Common Pauraque (111) 21:18
Scaly-breasted Hummingbird (125) 23:10
*Mangrove Hummingbird (131) 24:2
Cinnamon Hummingbird (129) 24:11
Ruby-throated Hummingbird (131) 25:9
Black-headed Trogon (141) 26:10
Violaceous Trogon (141) 26:8
Turquoise-browed Motmot (147) 27:11

Green Kingfisher (149) 27:5
***Hoffmann's Woodpecker (157) 28:16**
Lineated Woodpecker (161) 27:13
Streak-headed Woodcreeper (173) 29:8
Barred Antshrike (175) 31:3
Northern Scrub-Flycatcher (193) 37:22
Northern Bentbill (197) 37:6
Yellow-olive Flycatcher (195) 37:16
Dusky-capped Flycatcher (209) 35:21
Panama Flycatcher (209) 35:20
Brown-crested Flycatcher (209) 35:18
Great Kiskadee (211) 35:13
Boat-billed Flycatcher (211) 35:12
Social Flycatcher (211) 35:14
Streaked Flycatcher (213) 35:11
Sulphur-bellied Flycatcher (213) 35:10
Tropical Kingbird (213) 35:1
Scissor-tailed Flycatcher (207) 35:5
Rose-throated Becard (217) 33:12
Masked Tityra (217) 34:1
Long-tailed Manakin (223) 33:4
Yellow-green Vireo (227) 40:5
White-throated Magpie-Jay (231) 39:18
Gray-breasted Martin (235) 22:15
Mangrove Swallow (233) 22:23
Southern Rough-winged Swallow (235) 22:19
Rufous-naped Wren (239) 38:2
Banded Wren (241) 38:6
Long-billed Gnatwren (237) 32:15
White-lored Gnatcatcher (237) 41:1
Clay-colored Thrush (251) 39:8
Yellow (Mangrove) Warbler (259) 42:3
Prothonotary Warbler (267) 42:1
Gray-crowned Yellowthroat (273) 42:13
Rufous-capped Warbler (275) 40:16
Blue-black Grassquit (297) 49:7
White-collared Seedeater (295) 49:2
Olive Sparrow (303) 50:15
Stripe-headed Sparrow (303) 50:7
Blue Grosbeak (313) 48:11
Painted Bunting (313) 48:13
Eastern Meadowlark (315) 50:16
Great-tailed Grackle (317) 44:16
Streak-backed Oriole (319) 44:2
Spot-breasted Oriole (319) 44:1
Scrub Euphonia (327) 45:4

Site 2-7: Curú National Wildlife Refuge

Birding Time: 1 day
Elevation: Sea level
Trail Difficulty: 1–3
Reserve Hours: 7 a.m. to 3 p.m.
Entrance Fee: $10 (US 2008)
Trail Map Available on Site

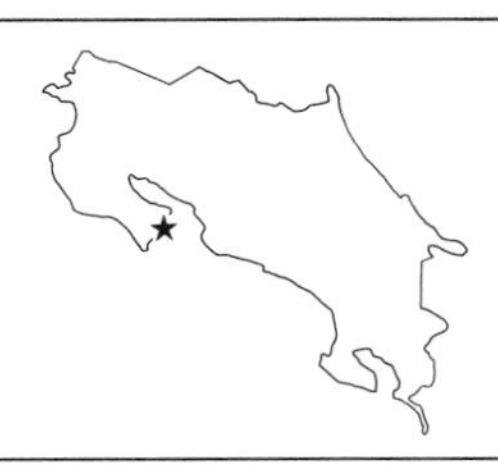

The land that is now Curú National Wildlife Refuge has been run by the Shutt family since 1933. During the early years, the Shutts' principle concern was farming, although the family always tried to practice conservation on their 3,700-acre property as well. After squatters settled part of the estate in 1974, the family began to look for better ways to protect the land that they loved. By 1983 Curú became Costa Rica's first private national wildlife refuge. Since then, Curú's focus has continued to shift away from farming, and today only 20% of the property is still worked as a farm. Ecotourism, research, and education are currently the primary goals, and 75% of the land is protected.

Curú contains a mix of habitats, including mangroves, pastures, and deciduous and semi-deciduous forests, which keeps the birding interesting. Many dry-forest species can be easily found around the grounds. Other species that have more restricted ranges in the North Pacific, such as Stripe-throated Hermit and Cocoa Woodcreeper, are also easily found here because some of the forest at the Curú reserve is relatively moist.

Curú offers room and board, although the facilities on site are rustic. There is only limited electricity, the buildings are run down, and the food served is quite basic. Some of the trails are well maintained, while others are hard to follow and quite strenuous.

Because Curú is isolated on the Nicoya Peninsula, I would not recommend including it in a serious birding itinerary for logistical reasons. However, if you happen to be staying at one of the many tourist destinations along the southern coast of the peninsula, Curú and the nearby Cabo Blanco Absolute Nature Reserve (2-8), are good places to bird.

Access

From Paquera: Head south on Rt.160 toward Tambor and Montezuma. After 2.0 km be sure to follow the main road, taking a hard left turn. At 3.2 km after this turn you will come to the entrance to Curú refuge on your left. (Driving time from Paquera: 15 min)

Note: Paquera can be reached via car ferry, which runs every two hours from Puntarenas. There are two ferry routes leaving Puntarenas, Paquera and Playa Naranjo, so be sure to get on the Paquera ferry. There are also currently two ferry companies running the route, which can make things more confusing. Expect to spend two to five hours crossing to Paquera, depending on the wait in Puntarenas.

Logistics: It is possible to stay at the Curú National Wildlife Refuge, and if you are traveling on a low budget this is an excellent option. If you want more comfortable accommodations, I suggest looking in Tambor or some of the other beach communities farther south. Wherever you stay, having lunch at Curú is extremely convenient.

Curú National Wildlife Refuge
Tel: (506) 2641-0100
Website: www.curuwildliferefuge.com
E-mail: refugiocuru@yahoo.com

Birding Sites

The dirt entrance road can yield some interesting birds and is generally worth walking. As the road passes through forest and pasture, look for Turquoise-browed Motmot, Rose-throated Becard, Orange-fronted Parakeet, and Yellow-naped Parrot. You might also take the time to walk the short Laguna Loop, which passes near a small seasonal lagoon. Gray-necked Wood-Rail,

Muscovy Duck, and White Ibis are possibilities. Scarlet Macaws are often seen flying overhead, although the population at this site is composed of released birds.

One pleasant and fairly easy set of trails to hike is Sendero Toledo to Sendero Finca de Monos. Sendero Toledo starts at the right side of the building complex and leads south, parallel to the beach. It passes through some mangroves, which could produce Scaly-breasted Hummingbird, Common Black-Hawk, and Northern Scrub-Flycatcher. After crossing a bridge, the path forks. To your left, Sendero Toledo does a short loop through the forest and returns back to this point. To your right, Sendero Finca de Monos winds its way through relatively wet forest back to the entrance road. (Be sure to stay right at any subsequent intersections.) Throughout this area keep an eye open for species that prefer evergreen forests such as Gray-headed Tanager, Cocoa Woodcreeper, and Blue-crowned Motmot. Rufous-and-white Wren and Long-tailed Manakin are also likely.

One of the best birding areas at Curú is the first part of Sendero Quesera. Follow the dirt road that parallels the beach heading north. Pass through a gate and continue on the dirt road. For the next 800 meters the road climbs up and over a small hill. The forest around this section of road is distinctly drier than what you find at some of the other trails, and it will produce many different species. Banded Wren, Barred Antshrike, Turquoise-browed Motmot, Black-headed Trogon, and Rose-throated Becard are all good possibilities.

At the far side of the hill the habitat opens up into pasture and forest edge. This can be an excellent place to find Plain Chachalaca, which is only found on the Nicoya Peninsula. Also look for Crested Caracara and White-throated Magpie-Jay.

Site 2-8: Cabo Blanco Absolute Nature Reserve

Birding Time: 1 day
Elevation: Sea level
Trail Difficulty: 2–3
Reserve Hours: 8 a.m. to 4 p.m. (Closed on Monday and Tuesday)
Entrance Fee: $8 (US 2008)
4×4 Recommended
Trail Map Available on Site

With more than 50% of its land protected, Costa Rica stands as one of the most progressive countries in the world with regard to conservation. But it was not always this way. It was not until 1963, in fact, that the Costa Rican government created the country's first nature reserve, Cabo Blanco Absolute Nature Reserve. The 3,100-acre park was established at the tip of the Nicoya Peninsula, and today is a shining example of successful reforestation. Eighty-five percent of the reserve is second-growth forest, which has developed from pastureland since the park's creation.

The southern part of the Nicoya Peninsula, with some of the best surfing waves in the country striking its shores, is a popular tourist destination among beach-goers and surfers. Tambor, Montezuma, Carmen, and Malpaís are nearby beach towns with an abundance of accommodations.

The birding at Cabo Blanco is quite productive. The forest is wetter than most forests in the North Pacific Slope, which makes for an interesting mix of birds. A number of key species, such as Stub-tailed Spadebill, Ivory-billed Wood-creeper, and Boat-billed Heron, are easily found here. However, because of Cabo Blanco's isolation, it is a difficult site to fit into serious birding itineraries. If you are looking to mix beaching and birding, though, this is an ideal destination.

Target Birds

Plain Chachalaca (11) 12:2
Brown Booby (15) 1:1
Boat-billed Heron (27) 5:2
Yellow-naped Parrot (99) 19:5
Northern Potoo (115) 20:4
Ivory-billed Woodcreeper (173) 29:21
Stub-tailed Spadebill (189) 37:2
White-throated Magpie-Jay (231) 39:18
Gray-headed Tanager (283) 45:17
Red-crowned Ant-Tanager (277) 47:10

Access

From Montezuma: Drive south along the coast for 8.8 km, passing through the village of Cabuya, and you will arrive at the entrance to the Cabo Blanco Reserve. (Driving time from Montezuma: 25 min)

Note: The southern part of the Nicoya Peninsula is most easily reached via car ferry, which leaves from Puntarenas. Refer to site 2-7 for details.

Logistics: Cabo Blanco is located near the beach towns of Montezuma, Malpaís, and Tambor. These towns have a range of hotels, restaurants, and other amenities.

Birding Sites

Cabo Blanco Absolute Nature Reserve

Two hiking trails are available at Cabo Blanco. The first, Sendero Danés, is a loop that should take you about one and a half hours to bird. The second, Sendero Sueco, leads off the far end of Sendero Danés and proceeds over a set of hills to a beautiful, secluded beach, where you will find picnic tables, showers, and potable water. Be forewarned, this is an eight-kilometer roundtrip hike along a steep trail with difficult footing. At the beach you will see Magnificent Frigatebird and Brown Pelican. Also, scan out to Cabo Blanco Island, which is an important nesting colony for Brown Boobies.

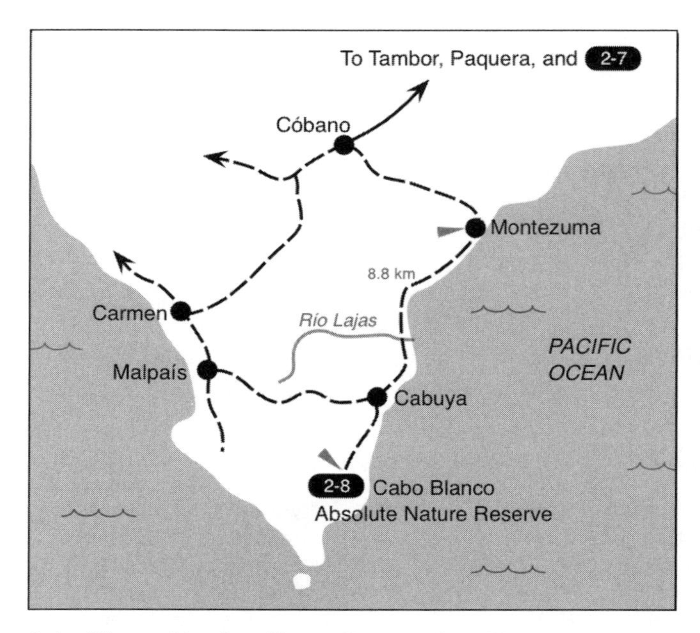

Cabo Blanco Absolute Nature Reserve Area Map

The habitat along these trails is almost exclusively second-growth forest. The more common species in the forest include Rufous-and-white Wren, Barred Antshrike, Rufous-capped Warbler, and Long-tailed Manakin. Other birds to keep an eye out for are Slate-headed Tody-Flycatcher, Thicket Tinamou, Black-headed Trogon, and Olivaceous Woodcreeper. Army ants are regular along the trails, and with them you might find Red-crowned Ant-Tanager, Gray-headed Tanager, and Ruddy Woodcreeper. Finally, Ivory-billed Woodcreeper and Stub-tailed Spadebill are two North Pacific Slope specialties that are difficult to find in most locations but are reliably found here.

Outside the Reserve

One quick stop is at the mouth of the Río Lajas, 4.8 km south of Montezuma and 1.6 km north of Cabuya. Park near the bridge and walk down along the bank of the river to get a better look at the trees along its edge. Here Boat-billed Herons make their nests, and roost during the day. Often you will see Yellow-crowned Night-Heron in the trees too. Other species to look for along the river include Great Blue Heron, Bare-throated Tiger-Heron, Green Kingfisher, and Amazon Kingfisher.

The dirt road that connects Cabuya and Malpaís offers excellent birding, and you should definitely spend some time exploring this area. The road is seven kilometers long and not in the best condition. There are some steep places and a few stream crossings. A 4×4 is recommended, though probably not necessary. From Cabuya, the road proceeds through cattle pastures, overgrown fields, and new-growth forest. The weedy fields often produce Blue Grosbeak, Gray-crowned Yellowthroat, Olive Sparrow, and Stripe-headed Sparrow. The scrubby trees may hide Striped Cuckoo, White-throated Magpie-Jay, Barred Antshrike, and Lineated Woodpecker. Along the road you will come to a few stream crossings. The forest near the water is generally lusher and is a good place to look for Long-tailed Manakin, Rose-throated Becard, Violaceous Trogon, and Blue-throated Goldentail.

As you approach Malpaís, the habitat around the road becomes secondary forest, and a stream runs parallel to the road. This wetter and darker habitat is a good place to find Red-crowned Ant-Tanager, Scaly-breasted Hummingbird, Ivory-billed Woodcreeper, and Little Tinamou.

Species to Expect

Little Tinamou (3) 12:17
Thicket Tinamou (3) 12:18
Brown Booby (15) 1:1
Brown Pelican (17) 4:1
Neotropic Cormorant (17) 4:4
Magnificent Frigatebird (19) 1:6
Bare-throated Tiger-Heron (21) 5:16
Great Blue Heron (23) 5:6
Little Blue Heron (23) 5:9
Cattle Egret (23) 5:13
Green Heron (25) 6:2
Yellow-crowned Night-Heron (27) 5:3
Boat-billed Heron (27) 5:2
White Ibis (29) 4:8
Black Vulture (31) 13:4
Turkey Vulture (31) 13:3
Double-toothed Kite (35) 16:1
Common Black-Hawk (45) 13:6
Roadside Hawk (43) 16:12
Gray Hawk (43) 16:14
Crested Caracara (53) 14:12
Laughing Falcon (53) 15:8
Gray-necked Wood-Rail (55) 6:13
Semipalmated Plover (65) 10:5
Spotted Sandpiper (73) 11:8
Whimbrel (71) 9:14
Red-billed Pigeon (87) 18:2
Inca Dove (89) 18:11
Common Ground-Dove (89) 18:8
White-tipped Dove (91) 18:14
Orange-fronted Parakeet (95) 19:13
Orange-chinned Parakeet (97) 19:14
White-fronted Parrot (99) 19:6
Yellow-naped Parrot (99) 19:5
Squirrel Cuckoo (103) 21:7
Striped Cuckoo (101) 21:5
Groove-billed Ani (103) 21:9
Common Pauraque (111) 21:18
White-collared Swift (117) 22:1
Scaly-breasted Hummingbird (125) 23:10
Green-breasted Mango (129) 23:13
Blue-throated Goldentail (129) 24:9
Steely-vented Hummingbird (127) 24:15
Rufous-tailed Hummingbird (129) 24:10
Cinnamon Hummingbird (129) 24:11
Black-headed Trogon (141) 26:10
Violaceous Trogon (141) 26:8
Blue-crowned Motmot (147) 27:8
Turquoise-browed Motmot (147) 27:11
Amazon Kingfisher (149) 27:4
Green Kingfisher (149) 27:5
***Hoffmann's Woodpecker (157) 28:16**
Lineated Woodpecker (161) 27:13
Pale-billed Woodpecker (161) 27:14
Plain Xenops (169) 29:3
Ruddy Woodcreeper (171) 29:13
Olivaceous Woodcreeper (171) 29:7
Northern Barred-Woodcreeper (169) 29:19
Ivory-billed Woodcreeper (173) 29:21
Streak-headed Woodcreeper (173) 29:8

Barred Antshrike (175) 31:3
Northern Beardless-Tyrannulet (191) 37:19
Ochre-bellied Flycatcher (199) 36:25
Slate-headed Tody-Flycatcher (197) 37:9
Common Tody-Flycatcher (197) 37:7
Yellow-olive Flycatcher (195) 37:16
Stub-tailed Spadebill (189) 37:2
Dusky-capped Flycatcher (209) 35:21
Brown-crested Flycatcher (209) 35:18
Great Kiskadee (211) 35:13
Boat-billed Flycatcher (211) 35:12
Social Flycatcher (211) 35:14
Streaked Flycatcher (213) 35:11
Tropical Kingbird (213) 35:1
Rose-throated Becard (217) 33:12
Masked Tityra (217) 34:1
Long-tailed Manakin (223) 33:4
Yellow-throated Vireo (225) 40:6
Lesser Greenlet (229) 40:7
White-throated Magpie-Jay (231) 39:18
Rufous-naped Wren (239) 38:2
Rufous-and-white Wren (241) 38:9
Banded Wren (241) 38:6
Plain Wren (241) 38:17
House Wren (243) 38:18
Long-billed Gnatwren (237) 32:15
Tropical Gnatcatcher (237) 41:2
Clay-colored Thrush (251) 39:8
Yellow Warbler (259) 42:2
Gray-crowned Yellowthroat (273) 42:13
Rufous-capped Warbler (275) 40:16
Gray-headed Tanager (283) 45:17
Red-crowned Ant-Tanager (277) 47:10
Red-legged Honeycreeper (293) 46:2
Blue-black Grassquit (297) 49:7
White-collared Seedeater (295) 49:2
Olive Sparrow (303) 50:15
Stripe-headed Sparrow (303) 50:7
Blue Grosbeak (313) 48:11
Painted Bunting (313) 48:13
Melodious Blackbird (315) 52:6
Great-tailed Grackle (317) 44:16
Baltimore Oriole (321) 44:7

Region 3
The South Pacific Slope

The **Fiery-billed Aracari** is a endemic toucan, restricted to the South Pacific region of Costa Rica and a small section of western Panama. It closely resembles the Collared Aracari of the Caribbean Slope, except that its bill and bellyband are red. The Fiery-billed Aracari is regularly encountered throughout the South Pacific region, though Las Cruces Biological Station (3B-3) is an especially reliable place to find it.

Regional Map: The South Pacific Slope

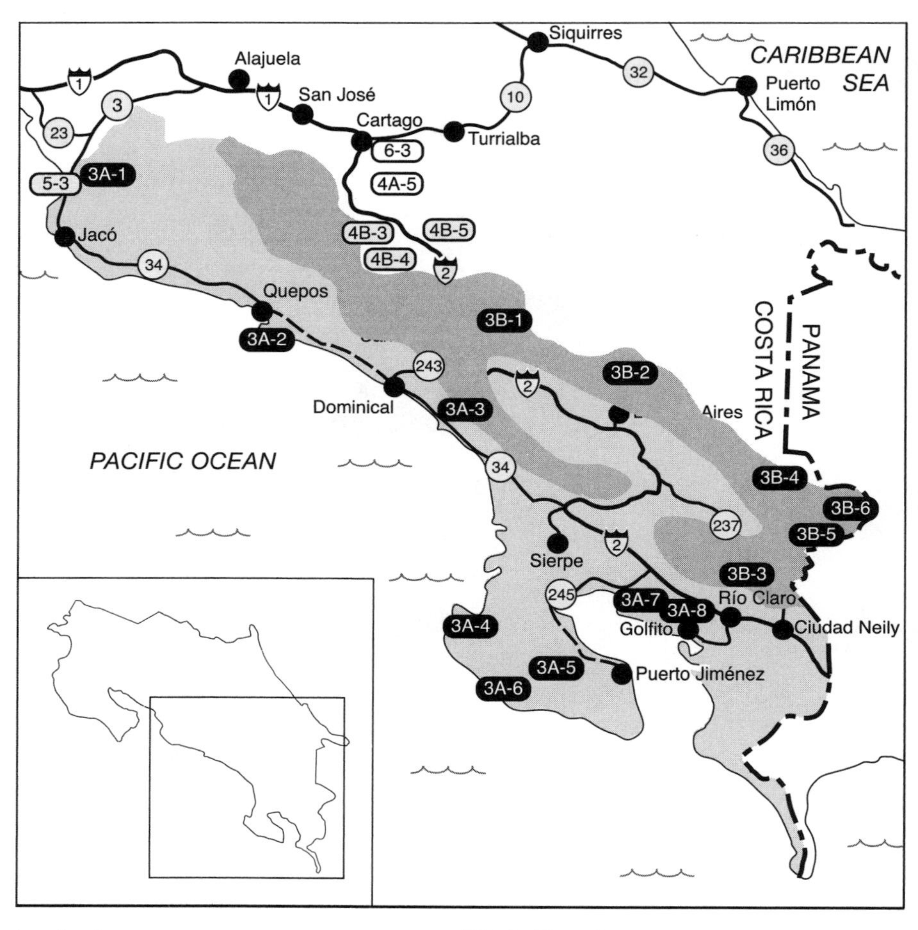

Lowlands
3A-1: Carara National Park
3A-2: Manuel Antonio National Park
3A-3: Oro Verde Biological Reserve
3A-4: Marenco Beach and Rainforest Lodge
3A-5: Bosque del Río Tigre
3A-6: Corcovado National Park
3A-7: Esquinas Rainforest Lodge
3A-8: Golfito Wildlife Refuge

Middle Elevations
3B-1: Talari Mountain Lodge
3B-2: Dúrika
3B-3: Las Cruces Biological Station
3B-4: Las Alturas
3B-5: Río Negro
3B-6: La Amistad Lodge

The South Pacific Slope starts roughly at the Río Tárcoles and runs south to the Panama border. This region is generally quite wet, unlike the dry-forest habitat found north of the Río Tárcoles. The eastern boundary of the South Pacific Slope abuts the Mountains region, which extends through the high elevations of the Cordillera de Talamanca. Like the Caribbean Slope, the South Pacific Slope has been split into two subregions, the Lowlands and the Middle-Elevations. The Lowlands runs from sea level to about 500 meters, while the Middle-Elevations subregion extends up to about 1,400 meters. The upper range of the Middle-Elevations is slightly higher on the South Pacific Slope than on the Caribbean Slope. This is due to differences in local climate, which allows similar ecosystems and avifauna to exist at a slightly higher elevation on the South Pacific side of the country.

Two river valleys, the Valle del General and the Valle de Coto Brus, are important geographic features in the region. The eastern wall of the valleys is created by the Talamanca Mountains, and the western boundary is formed by the Coastal Range, a small range of hills that parallels the Pacific Coast. The Río General, which runs southeast, meets the Río Coto Brus, which runs northwest to form the Río Grande de Térraba, which then cuts through the Coastal Range and meets the Pacific Ocean just north of the Osa Peninsula.

These two valleys have been heavily deforested for agriculture. Pineapple is a very important crop in the region, and you are sure to pass large fields of these tropical fruits if you drive through the Valle del General. In fact, Costa Rica is the number one exporter of pineapple in the world. Coffee is another important crop in the area, especially in the Coto Brus Valley, although much of the original coffee fields were converted to cow pastures after the coffee market crashed during the 1980s.

The unfortunate result of all this agriculture within the two river valleys is that very little middle-elevation forest has survived. The South Pacific Middle-Elevation sites in this guide generally contain forest fragments and small patches of second-growth forest. Only along the slopes of the Talamanca Mountains does true wilderness still exist, and it is accessible at sites 3B-4 and 3B-6.

Thanks to Costa Rica's excellent national park system, the South Pacific Lowlands subregion still retains a few large tracts of protected land, most notably the magnificent Osa Peninsula. The forest that covers most of this untamed peninsula is one of the largest lowland primary forests remaining in the entire neotropics. Some of Costa Rica's finest eco-lodges are located on the Osa, and although only two are described in this book, many others offer productive birding.

The South Pacific Slope is of special interest to visiting birders because of its many endemic species. Eighteen are confined to this region (plus a small section of western Panama), the high Talamanca Mountains helping to isolate these birds from the rest of the country. As a result, a number of the endemic species have closely related sister species living just on the other side of the mountains. Fiery-billed Aracari and Collared Aracari, Turquoise Cotinga and Lovely Cotinga, and Orange-collared Manakin and White-collared Manakin are examples of this phenomenon.

Most of the South Pacific Slope is logistically difficult to fit into many itineraries. In-country flights are one great way to cut down on travel time to this inaccessible section of the country. If you plan on driving, and are going to sites near the Panama border, I suggest allotting at least a week to explore the area. Indeed, it is possible to have a very successful Costa Rican birding trip focusing solely on the South Pacific Slope.

Of the 14 sites described in the region, Manuel Antonio National Park (3A-2) is by far the most popular general ecotourist destination. However, I find the birding there to be mediocre and recommend other locations if birds are your primary interest. One of my favorite sites in the lowlands is Carara National Park (3A-1). This site is a convenient day trip from San José, and because it sits on the transition between wet and dry forest, it is home to a huge diversity of species. Bosque del Río Tigre (3A-5) is another excellent destination, offering a comfortable birder's lodge and a wide range of unusual species tucked into the wilderness of the Osa Peninsula. Of the sites in the middle elevations, Las Cruces Biological Station (3B-3) is my favorite, providing good birding and comfortable accommodations. I also recommend the nearby Las Alturas (3B-4).

Regional Specialties

Marbled Wood-Quail (13) 12:9
Harpy Eagle (47) 17:9
Wattled Jacana (61) 6:17
Brown-throated Parakeet (95)
Smooth-billed Ani (103) 21:8
*Costa Rican Swift (119) 22:11
*White-crested Coquette (137) 25:5
*Garden Emerald (127) 24:13
*Charming Hummingbird (125) 24:18
*Snowy-bellied Hummingbird (133) 24:3
*White-tailed Emerald (133) 24:19
*Baird's Trogon (143) 26:6
*Fiery-billed Aracari (155) 27:16
Olivaceous Piculet (159) 29:1
*Golden-naped Woodpecker (157) 28:18
Red-crowned Woodpecker (157) 28:17
Red-rumped Woodpecker (159) 28:12
Pale-breasted Spinetail (163) 32:11
Ruddy Foliage-gleaner (167) 30:5
Tawny-winged Woodcreeper (171) 29:12
*Black-hooded Antshrike (177) 31:6
Yellow-bellied Tyrannulet (189) 37:14
Southern Beardless-Tyrannulet (191) 37:20
Mouse-colored Tyrannulet (191)
Yellow-crowned Tyrannulet (191) 37:15
Lesser Elaenia (193) 37:25
Black-tailed Flycatcher (199) 36:22
Bran-colored Flycatcher (199) 36:17
Rusty-margined Flycatcher (329)
Fork-tailed Flycatcher (207) 35:4
*Turquoise Cotinga (219) 34:5
*Yellow-billed Cotinga (219) 34:3
*Orange-collared Manakin (221) 33:2
Lance-tailed Manakin (223) 33:5
Blue-crowned Manakin (223) 33:7
Scrub Greenlet (229) 40:8
*Black-bellied Wren (243) 38:11
*Riverside Wren (241) 38:10
Rufous-breasted Wren (239) 38:8
Masked Yellowthroat (273) 42:12
Rosy Thrush-Tanager (279) 47:13

*Black-cheeked Ant-Tanager (277) 47:11
*Cherrie's Tanager (287) 47:4
Yellow-bellied Seedeater (295) 49:4
Ruddy-breasted Seedeater (297) 49:1
Wedge-tailed Grass-Finch (299) 50:10
Stripe-headed Brush-Finch (301) 48:14
Streaked Saltator (309) 48:4
Crested Oropendola (323)
Thick-billed Euphonia (327) 45:8
*Spot-crowned Euphonia (327) 45:2

Subregion 3A
The South Pacific Lowlands

The **Black-cheeked Ant-Tanager** is an extremely local endemic, found only in lowland forests around the Osa Peninsula. Noisy groups of these tanagers can be encountered moving through the understory of old growth forest, though it is most reliably found at Bosque del Río Tigre (3A-5), where it visits fruit feeders next to the lodge.

Common Birds to Know

Black Vulture (31) 13:4
Turkey Vulture (31) 13:3
Ruddy Ground-Dove (89) 18:7
White-tipped Dove (91) 18:14
Rufous-tailed Hummingbird (129) 24:10
Chestnut-mandibled Toucan (155) 27:19
Red-crowned Woodpecker (157) 28:17
*Black-hooded Antshrike (177) 31:6
Chestnut-backed Antbird (179) 31:11
Great Kiskadee (211) 35:13
Boat-billed Flycatcher (211) 35:12
Social Flycatcher (211) 35:14
Tropical Kingbird (213) 35:1
Lesser Greenlet (229) 40:7
*Riverside Wren (241) 38:10
Clay-colored Thrush (251) 39:8
Chestnut-sided Warbler (259) 43:4
Bananaquit (277) 40:24
*Cherrie's Tanager (287) 47:4
Blue-gray Tanager (291) 46:15
Palm Tanager (291) 45:19
Golden-hooded Tanager (289) 46:13
Variable Seedeater (295) 49:3
Buff-throated Saltator (309) 48:2

Site 3A-1: Carara National Park

Birding Time: 1–2 days
Elevation: Sea level
Trail Difficulty: 1
Reserve Hours: 7 a.m. to 4 p.m.
Entrance Fee: $10 (US 2008)
Trail Map Available on Site

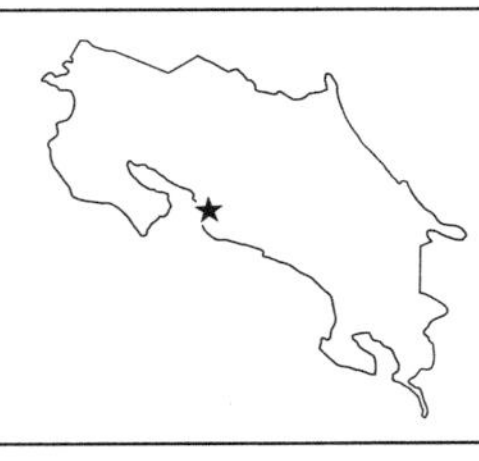

Carara National Park is one of Costa Rica's most famous birding destinations. It is the northernmost site of the South Pacific Slope, sitting on the southern bank of the Río Tárcoles, the river that creates the boundary between the North and South Pacific regions. Since it lies in a transition zone between tropical dry forest and tropical wet forest, Carara offers a unique and extremely diverse ecosystem. Here you will find species from both regions living side by side. Long-tailed Manakin, Rufous-and-white Wren, and Orange-fronted Parakeet are just a few examples of North Pacific species that can be found alongside South Pacific specialties such as Baird's Trogon, Black-bellied Wren, and Fiery-billed Aracari. Both foreign and local birders agree that Carara is a special place with an avifauna unlike anything else found in Costa Rica.

Carara National Park protects 11,600 acres of land and offers two well-maintained trail systems. While the trails can be hiked relatively quickly, the birding is generally so productive it is hard to explore the entire area. Frequently you will find yourself only a few hundred meters down a trail after hours of birding. The variety of habitats accessible in the immediate vicinity of Carara is another reason why the area is so productive. Not only do you get access to both dry- and wet-forest species, but also many water-associated birds are found nearby. The Tárcoles River Mouth (site 5-3), just 10 minutes away, is an important place to visit if you are birding here, as it attracts a huge variety of herons, shorebirds, terns, and mangrove specialists.

One of the best attributes of Carara is its convenience. While most South Pacific Slope sites are isolated and difficult to incorporate into an itinerary, Carara is less than two hours from San José and just 40 minutes from Rt. 1. Even if you cannot fit any other South Pacific Slope sites into your travel plans, you should not pass up Carara.

Target Birds

Great Tinamou (3) 12:6
Boat-billed Heron (27) 5:2
King Vulture (31) 13:5
Crane Hawk (41) 14:4
Collared Forest-Falcon (55) 16:7
Ruddy Quail-Dove (93) 18:17
Scarlet Macaw (99) 19:1
Striped Owl (105) 20:2
Purple-crowned Fairy (125) 23:14
*Baird's Trogon (143) 26:6
American Pygmy Kingfisher (149) 27:6
White-whiskered Puffbird (151) 28:6
*Fiery-billed Aracari (155) 27:16
*Golden-naped Woodpecker (157) 28:18
Scaly-throated Leaftosser (167) 30:9
Tawny-winged Woodcreeper (171) 29:12
Long-tailed Woodcreeper (171) 29:10
*Black-hooded Antshrike (177) 31:6
Dot-winged Antwren (183) 32:1
Streak-chested Antpitta (185) 30:18
Greenish Elaenia (195) 37:21
Northern Bentbill (197) 37:6
Slate-headed Tody-Flycatcher (197) 37:9
Stub-tailed Spadebill (189) 37:2
Royal Flycatcher (201) 35:23
Thrush-like Schiffornis (215) 33:10
Rufous Piha (215) 34:10
*Yellow-billed Cotinga (219) 34:3
*Orange-collared Manakin (221) 33:2
Blue-crowned Manakin (223) 33:7
Scrub Greenlet (229) 40:8
Green Shrike-Vireo (229) 40:1
*Black-bellied Wren (243) 38:11
*Riverside Wren (241) 38:10
White-shouldered Tanager (281) 46:18

Access

From the intersection of Rt. 23 and Rt. 34: Take Rt. 34 south toward Jacó and Quepos. In 11.2 km you will come to a bridge over the Río Tárcoles, and the Carara National Park headquarters will be on your left 2.4 km beyond that. (Driving time from Rt. 23: 10 min)

From Jacó: Take Rt. 34 north toward Rt. 23 and Orotina. In 23.8 km the Carara National Park headquarters will be on your right. (Driving time from Jacó: 20 min)

Logistics: A variety of options are available to you when staying in the Carara area. Jacó, 20 minutes south of the park, is a popular beach destination and offers a full range of accommodations. Closer to the park, two hotels are worthy of mention. The Tárcol Lodge is run by the same management as Rancho Naturalista (site 1B-7) and sits along the bank of the Río Tárcoles. It caters to birders, offering a package stay complete with a bird guide, although the facilities are a bit rundown. Hotel Villa Lapas, located along the waterfall road, offers more comfortable, upscale accommodations, though it lacks the personal attention to birders.

Tárcol Lodge
Tel: (506) 2554-8101
E-mail: info@ranchonaturalista.net

Hotel Villa Lapas
Tel: (506) 2637-0232
Website: www.villalapas.com
E-mail: info@villalapas.com

Birding Sites

Headquarters Trails

The trail system found at the headquarters, a series of three loops, starts at the right side of the parking lot and leads into old-growth forest. This forest is older and darker than that found along the River Trail (described below), and while the birding may not be quite as productive, it is still well worth your time. Look for Great Tinamou, Sulphur-rumped Flycatcher, Blue-crowned Manakin, Buff-rumped Warbler, and White-throated Shrike-Tanager. This is one of the best places in Costa Rica to find Streak-chested Antpitta, whose presence is most easily detected by its song, a series of repeated whistles. If you are lucky you may also find an Orange-collared Manakin lek, with 10 or more dancing males.

River Trail

The River Trail, located 1.4 km north of the park headquarters, is arguably the single most productive birding trail in all of Costa Rica. Here you will find seasonally flooded forest that harbors a host of interesting species. This area can be quite wet, especially during the rainy season, and I highly recommend wearing tall rubber boots. (If you do not have your own pair, you can rent them at the park headquarters.) Car theft is also a serious concern and you should be sure to take the proper precautions. Before entering the trail, you must pay your entrance fee at the park headquarters. As you do so, make sure there is a guard on duty at the River Trail parking lot. The guard will watch your car while you are birding, and you should give him a nice tip when you leave.

About 200 meters into the trail is a small stream crossing, where Royal Flycatcher is often found. As you continue along the River Trail, you will pass through patches of forest and clearings. The forest habitat is a great place to find mixed-species flocks. These can be extremely diverse, although Black-hooded Antshrike, Dot-winged Antwren, Long-billed Gnatwren, and Barred Antshrike are regularly seen. Other interesting species often found in the forest include Slate-headed Tody-Flycatcher, Northern Bentbill, Gray-headed Tanager, Baird's Trogon, and White-whiskered Puffbird. The clearings hold a different set of species. Here you may find Cherrie's Tanager, Dusky Antbird, Fiery-billed Aracari, and Black-bellied Wren.

About two kilometers down the trail you will come to a small path on the left that leads a very short distance to an overlook of the Meándrica Lagoon. Caiman are easily seen floating in the water or sunning on the shore, and many water

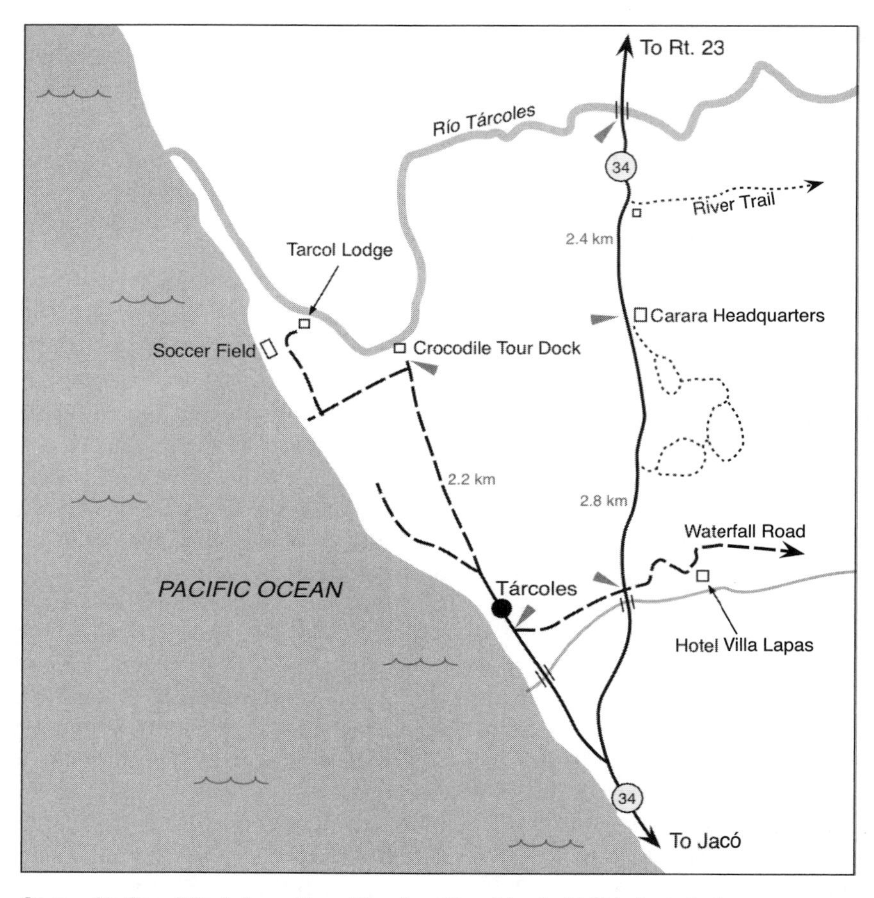

Carara National Park Area Map. Tárcoles River Mouth (5-3) is included.

birds inhabit the area. Look for Purple Gallinule, Northern Jacana, Green Heron, and Great Egret feeding around the floating vegetation and marsh grasses. Ringed Kingfisher, American Pygmy Kingfisher, and Anhinga are often seen perching in the trees around the lagoon, while Mangrove Swallows swoop low over the water. The main attraction of the lagoon is the Boat-billed Herons, which roost in the surrounding trees. The exact roost locations vary, but any tree that overhangs the water is worth a thorough check.

Tárcoles River Bridge

A quick stop at the bridge over the Tárcoles River is always worthwhile. The principle attraction here is not birds, but American Crocodiles, which like to sun on the exposed mud below the bridge. During the dry season it is not uncommon to find 20 or more large crocodiles basking at the water's edge.

The area can also produce some noteworthy birds, such as Wood Stork, Tricolored Heron, and Roseate Spoonbill, as well as other more common waders. It is a great place to see raptors, including Yellow-headed Caracara, Zone-tailed Hawk, and Gray Hawk, soaring by. Scarlet Macaws are frequently observed flying over the river in pairs.

You can park your car on either side of the bridge and then walk out along the sidewalk. Be sure to use caution, as the traffic over the bridge is dangerous.

Waterfall Road

The waterfall road is located 2.8 km south of the park headquarters and is a public dirt road that leads inland. Though the traffic on the road can occasionally become annoying, the area is appealing because the habitat is slightly drier than that found inside the park, so many of the dry-forest species are more easily found here.

At the beginning of the road there is a small field on your left. Check here for Cherrie's Tanager, Steely-vented Hummingbird, Scaly-breasted Hummingbird, Striped Cuckoo, and Rose-throated Becard. The small river on your right often holds Ringed Kingfisher. The forest found over the next two and a half kilometers is normally quite productive, and you will want to make frequent stops. Long-tailed Manakin is regularly seen, as well as Western Tanager, Rufous-capped Warbler, Blue-throated Goldentail, Turquoise-browed Motmot, and Black-headed Trogon. As you pass several small streams, you are likely to find Riverside Wren calling from the safety of a dense tangle. Keep an eye on the sky, as Scarlet Macaws and King Vultures are often seen flying overhead. At night, Mottled Owl can be heard calling from the hillsides.

Species to Expect

Great Tinamou (3) 12:6
Black-bellied Whistling-Duck (5) 8:7
Anhinga (17) 4:3
Bare-throated Tiger-Heron (21) 5:16
Great Egret (23) 5:14
Little Blue Heron (23) 5:9
Green Heron (25) 6:2
Boat-billed Heron (27) 5:2
Black Vulture (31) 13:4
Turkey Vulture (31) 13:3
King Vulture (31) 13:5
Gray-headed Kite (33) 17:3
Roadside Hawk (43) 16:12
Broad-winged Hawk (43) 16:13
Yellow-headed Caracara (53) 15:9
Laughing Falcon (53) 15:8
Gray-necked Wood-Rail (55) 6:13
Purple Gallinule (59) 6:15
Northern Jacana (61) 6:18
Spotted Sandpiper (73) 11:8
White-winged Dove (89) 18:12
Inca Dove (89) 18:11
Ruddy Ground-Dove (89) 18:7
White-tipped Dove (91) 18:14
Gray-chested Dove (91) 18:16
Ruddy Quail-Dove (93) 18:17
Scarlet Macaw (99) 19:1
Orange-chinned Parakeet (97) 19:14
White-crowned Parrot (97) 19:7
Red-lored Parrot (99) 19:4
Squirrel Cuckoo (103) 21:7
Striped Cuckoo (101) 21:5
Groove-billed Ani (103) 21:9
Mottled Owl (107) 20:6

Common Pauraque (111) 21:18
White-collared Swift (117) 22:1
*Costa Rican Swift (119) 22:11
Long-billed Hermit (121) 23:2
Stripe-throated Hermit (121) 23:1
Scaly-breasted Hummingbird (125) 23:10
White-necked Jacobin (125) 23:17
Violet-crowned Woodnymph (127) 24:4
Blue-throated Goldentail (129) 24:9
*Charming Hummingbird (125) 24:18
Steely-vented Hummingbird (127) 24:15
Rufous-tailed Hummingbird (129) 24:10
Purple-crowned Fairy (125) 23:14
Plain-capped Starthroat (135) 23:19
Black-headed Trogon (141) 26:10
*Baird's Trogon (143) 26:6
Violaceous Trogon (141) 26:8
Black-throated Trogon (141) 26:9
Slaty-tailed Trogon (145) 26:2
Blue-crowned Motmot (147) 27:8
Turquoise-browed Motmot (147) 27:11
Green Kingfisher (149) 27:5
White-whiskered Puffbird (151) 28:6
Rufous-tailed Jacamar (153) 26:12
*Fiery-billed Aracari (155) 27:16
Chestnut-mandibled Toucan (155) 27:19
Lineated Woodpecker (161) 27:13
Pale-billed Woodpecker (161) 27:14
Buff-throated Foliage-gleaner (165) 30:3
Plain Xenops (169) 29:3
Wedge-billed Woodcreeper (171) 29:6
Northern Barred-Woodcreeper (169) 29:19
Cocoa Woodcreeper (173) 29:17
Black-striped Woodcreeper (173) 29:16
Streak-headed Woodcreeper (173) 29:8
Barred Antshrike (175) 31:3
***Black-hooded Antshrike (177) 31:6**
Russet Antshrike (177) 31:4
Dot-winged Antwren (183) 32:1
Dusky Antbird (177) 31:8
Chestnut-backed Antbird (179) 31:11
Streak-chested Antpitta (185) 30:18
Ochre-bellied Flycatcher (199) 36:25
Paltry Tyrannulet (191) 37:10
Northern Bentbill (197) 37:6
Slate-headed Tody-Flycatcher (197) 37:9
Common Tody-Flycatcher (197) 37:7
Yellow-olive Flycatcher (195) 37:16
Royal Flycatcher (201) 35:23
Ruddy-tailed Flycatcher (199) 36:23
Sulphur-rumped Flycatcher (199) 36:21
Eastern Wood-Pewee (203) 36:8
Bright-rumped Attila (201) 35:6
Dusky-capped Flycatcher (209) 35:21
Great Crested Flycatcher (209) 35:17
Brown-crested Flycatcher (209) 35:18
Great Kiskadee (211) 35:13
Boat-billed Flycatcher (211) 35:12
Social Flycatcher (211) 35:14
Gray-capped Flycatcher (211) 35:15
Streaked Flycatcher (213) 35:11
Sulphur-bellied Flycatcher (213) 35:10
Piratic Flycatcher (193) 35:8
Tropical Kingbird (213) 35:1
White-winged Becard (217) 33:13
Rose-throated Becard (217) 33:12
Black-crowned Tityra (217) 34:2
***Orange-collared Manakin (221) 33:2**
White-ruffed Manakin (223) 33:9
Long-tailed Manakin (223) 33:4
Blue-crowned Manakin (223) 33:7
Yellow-throated Vireo (225) 40:6
Red-eyed Vireo (227) 40:3
Tawny-crowned Greenlet (229) 32:4
Lesser Greenlet (229) 40:7
Green Shrike-Vireo (229) 40:1
Brown Jay (231) 39:19
Mangrove Swallow (233) 22:23
Barn Swallow (237) 22:12
Rufous-naped Wren (239) 38:2
*Black-bellied Wren (243) 38:11
***Riverside Wren (241) 38:10**
Rufous-breasted Wren (239) 38:8
Rufous-and-white Wren (241) 38:9
House Wren (243) 38:18
Long-billed Gnatwren (237) 32:15
Tropical Gnatcatcher (237) 41:2
Clay-colored Thrush (251) 39:8
Golden-winged Warbler (257) 41:5
Tennessee Warbler (255) 40:22
Tropical Parula (257) 41:3
Yellow Warbler (259) 42:2
Chestnut-sided Warbler (259) 43:4
Black-and-white Warbler (267) 41:13

American Redstart (267) 41:10
Northern Waterthrush (269) 43:14
Canada Warbler (269) 42:8
Rufous-capped Warbler (275) 40:16
Buff-rumped Warbler (271) 40:23
Gray-headed Tanager (283) 45:17
*White-throated Shrike-Tanager (283) 47:1
White-shouldered Tanager (281) 46:18
Summer Tanager (285) 47:5
Western Tanager (285) 47:2
*Cherrie's Tanager (287) 47:4
Blue-gray Tanager (291) 46:15
Palm Tanager (291) 45:19
Bay-headed Tanager (289) 46:9
Golden-hooded Tanager (289) 46:13
Green Honeycreeper (293) 46:7
Red-legged Honeycreeper (293) 46:2
Blue-black Grassquit (297) 49:7
Variable Seedeater (295) 49:3
Orange-billed Sparrow (303) 48:17
Stripe-headed Sparrow (303) 50:7
Buff-throated Saltator (309) 48:2
Blue-black Grosbeak (311) 48:10
Great-tailed Grackle (317) 44:16
Baltimore Oriole (321) 44:7
*Yellow-crowned Euphonia (327) 45:1
White-vented Euphonia (327) 45:7

Note: The Carara area is a transition zone between Hoffmann's Woodpecker living to the north and the Red-crowned Woodpecker living to the south. Most individuals found at Carara are hybrids of the two species and show orange on their napes and bellies instead of yellow or red.

Site 3A-2: Manuel Antonio National Park

Birding Time: 1 day
Elevation: Sea level
Trail Difficulty: 1
Reserve Hours: 7 a.m. to 4 p.m.
Entrance Fee: $10 (US 2008)
Trail Map Available on Site

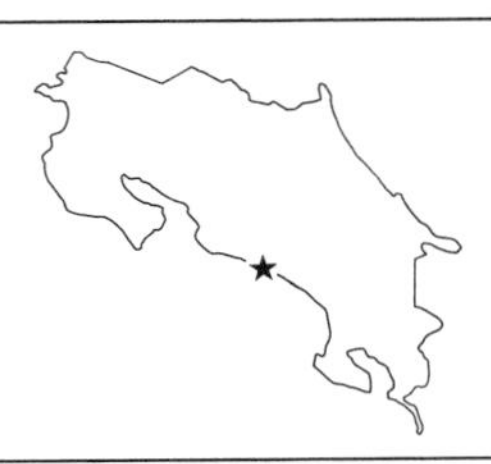

Manuel Antonio is one of Costa Rica's most popular tourist destinations and for obvious reasons. The area boasts stunning beaches, bordered by lush tropical forest overlooking the warm waters of the Pacific. Here it is easy to see many species of mammals, including Three-toed Sloth, Red-backed Squirrel Monkey, Mantled Howler Monkey, White-faced Capuchin, and White-nosed Coati. Finally, the entire Quepos/Manuel Antonio area is booming with businesses that cater to tourists, and visitors can find a wealth of hotels, restaurants, and activities to suit their needs.

For birding, Manuel Antonio is not a standout. Its inclusion in this book is based primarily on its popularity as a tourist destination. If your primary objective is birding, I recommend visiting other sites. However, if you do find yourself in the area, there are certainly worthwhile things to see, including many South Pacific endemics and a Brown Booby colony.

Target Birds

Brown Booby (15) 1:1	*Costa Rican Swift (119) 22:11
Boat-billed Heron (27) 5:2	*Fiery-billed Aracari (155) 27:16
Gray-necked Wood-Rail (55) 6:13	*Golden-naped Woodpecker (157) 28:18
Wandering Tattler (73) 10:8	*Black-hooded Antshrike (177) 31:6
Bridled Tern (83) 2:2	Yellow-crowned Tyrannulet (191) 37:15
Striped Owl (105) 20:2	*Black-bellied Wren (243) 38:11
Common Potoo (115) 20:4	*Riverside Wren (241) 38:10
Spot-fronted Swift (117) 22:5	*Cherrie's Tanager (287) 47:4

Access

From Quepos: As you arrive in Quepos, follow signs for Manuel Antonio, which should be very well marked. The road leaves from the southeast side of town and takes you 6.0 km up and over a hillside to the small town of Manuel Antonio. (Driving time from Quepos: 10 min)

Logistics: The Manuel Antonio/Quepos area has a full range of accommodations, from budget to luxury. You will have no problems finding lodging, food, or just about anything else during the low season. In high season, be sure to make reservations in advance.

Birding Sites

A nice variety of species can be found outside the park in the small town of Manuel Antonio. There are only two principal roads in town, the main road, which parallels the coast, and a secondary parallel road a few hundred meters inland. These two roads meet near the beginning of town. Near this intersection, behind some buildings, is a small wetland where herons, including Boat-billed Heron and Yellow-crowned Night-Heron, often roost in a dense clump of mangroves. The first access point from which to see the marsh is off the right side of the secondary road, about 50 meters beyond the intersection. Alternatively, there is a small footpath that runs along the opposite side of the wetland. White-throated Crakes live in the wet grasses, and though they are difficult to see, you can often hear their long, descending, reedy rattle. You may also find Gray-necked Wood-Rail and Scaly-breasted Hummingbird.

From the wetland, I suggest walking down the secondary road, as this area usually holds the most bird activity. You will cross a small stream, where you should look for Green Kingfisher and possibly Bare-throated Tiger-Heron or White Ibis. Continuing down the road, the trees and hedgerows on either side often yield Cherrie's Tanager, Bananaquit, Shining Honeycreeper, and Blue Dacnis. A number of euphonia species are possibilities here, although the Yellow-crowned Euphonia is the most prevalent. Other species to look for are Pale-vented Pigeon, Blue-throated Goldentail, Masked Tityra, Red-crowned Wood-

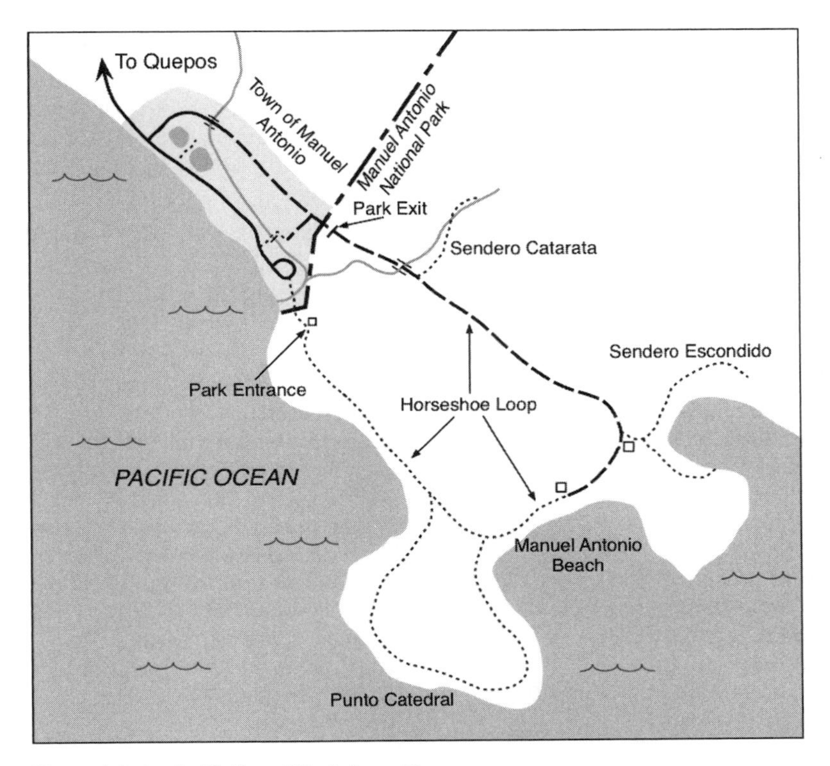

Manuel Antonio National Park Area Map

pecker, and Melodious Blackbird. The road ends at the national park exit, although there is a small cutover on your right that will bring you back to the main road, completing a loop.

Inside Manuel Antonio National Park you will encounter a different variety of species. To get to the entrance to the park, drive to the end of the main road and park. From here you must cross a small section of beach with a stream running through it to get to the ticketing booth. At high tide this can be flooded, but there are always some men with small boats who will ferry you across for a small tip. Check the mangroves at the edge of the beach for Yellow-crowned Night-Heron, Green Heron, and Spotted Sandpiper.

When you enter the park, be sure to pick up a trail map at the ticketing booth. The Horseshoe Loop leads out to Manuel Antonio Beach, then back to the park exit. The second half of this loop is a wide road, providing access for ranger vehicles, and is a nice place to bird, with good visibility. Along the trail you will encounter many Black-hooded Antshrike and Riverside Wren, two South Pacific endemics that are abundant within the park. You may also find Chestnut-backed Antbird, Olivaceous Piculet, Cocoa Woodcreeper, Purple-crowned Fairy, and

Long-billed Gnatwren. Some of the smaller trails such as Sendero Catarata and Sendero Escondido are also worth walking.

Hiking the loop out to Punto Catedral will provide you with beautiful views of the Pacific from atop high bluffs. This is the closest vantage point to Olocuita Island, where there is a colony of Brown Boobies. White Ibis and Snowy Egret often take refuge on the island and Bridled Tern can also be found. Spot-fronted Swift is sometimes reported from these high lookouts because the light conditions are often good enough to make this tough identification. Be aware that the trail here is a bit more challenging than in other parts of the park.

Species to Expect

Gray-headed Chachalaca (11) 12:1
Brown Booby (15) 1:1
Brown Pelican (17) 4:1
Magnificent Frigatebird (19) 1:6
Bare-throated Tiger-Heron (21) 5:16
Great Egret (23) 5:14
Snowy Egret (23) 5:10
Little Blue Heron (23) 5:9
Cattle Egret (23) 5:13
Green Heron (25) 6:2
Yellow-crowned Night-Heron (27) 5:3
Boat-billed Heron (27) 5:2
White Ibis (29) 4:8
Black Vulture (31) 13:4
Turkey Vulture (31) 13:3
Common Black-Hawk (45) 13:6
Roadside Hawk (43) 16:12
Short-tailed Hawk (43) 16:11
Yellow-headed Caracara (53) 15:9
Laughing Falcon (53) 15:8
White-throated Crake (57) 6:9
Gray-necked Wood-Rail (55) 6:13
Purple Gallinule (59) 6:15
Spotted Sandpiper (73) 11:8
Willet (71) 9:6
Laughing Gull (81) 3:8
Pale-vented Pigeon (87) 18:3
Inca Dove (89) 18:11
Ruddy Ground-Dove (89) 18:7
White-tipped Dove (91) 18:14
Ruddy Quail-Dove (93) 18:17
Orange-chinned Parakeet (97) 19:14
Brown-hooded Parrot (97) 19:9
White-crowned Parrot (97) 19:7
Squirrel Cuckoo (103) 21:7
Groove-billed Ani (103) 21:9
Long-billed Hermit (121) 23:2
Stripe-throated Hermit (121) 23:1
Scaly-breasted Hummingbird (125) 23:10
Violet-crowned Woodnymph (127) 24:4
Blue-throated Goldentail (129) 24:9
Rufous-tailed Hummingbird (129) 24:10
Purple-crowned Fairy (125) 23:14
Slaty-tailed Trogon (145) 26:2
Blue-crowned Motmot (147) 27:8
Ringed Kingfisher (149) 27:1
Green Kingfisher (149) 27:5
*Fiery-billed Aracari (155) 27:16
Chestnut-mandibled Toucan (155) 27:19
Olivaceous Piculet (159) 29:1
*Golden-naped Woodpecker (157) 28:18
Red-crowned Woodpecker (157) 28:17
Lineated Woodpecker (161) 27:13
Pale-billed Woodpecker (161) 27:14
Plain Xenops (169) 29:3
Tawny-winged Woodcreeper (171) 29:12
Wedge-billed Woodcreeper (171) 29:6
Cocoa Woodcreeper (173) 29:17
Streak-headed Woodcreeper (173) 29:8
***Black-hooded Antshrike (177) 31:6**
Chestnut-backed Antbird (179) 31:11
Ochre-bellied Flycatcher (199) 36:25
Paltry Tyrannulet (191) 37:10
Yellow-olive Flycatcher (195) 37:16
Bright-rumped Attila (201) 35:6
Rufous Mourner (201) 34:9
Dusky-capped Flycatcher (209) 35:21
Great Crested Flycatcher (209) 35:17
Great Kiskadee (211) 35:13
Boat-billed Flycatcher (211) 35:12
Social Flycatcher (211) 35:14

Gray-capped Flycatcher (211) 35:15
Streaked Flycatcher (213) 35:11
Tropical Kingbird (213) 35:1
Masked Tityra (217) 34:1
Black-crowned Tityra (217) 34:2
Red-capped Manakin (223) 33:6
Yellow-throated Vireo (225) 40:6
Philadelphia Vireo (227) 40:15
Red-eyed Vireo (227) 40:3
Lesser Greenlet (229) 40:7
Brown Jay (231) 39:19
Gray-breasted Martin (235) 22:15
Mangrove Swallow (233) 22:23
***Riverside Wren (241) 38:10**
House Wren (243) 38:18
Long-billed Gnatwren (237) 32:15
Tropical Gnatcatcher (237) 41:2
Swainson's Thrush (249) 39:2
Clay-colored Thrush (251) 39:8
Tennessee Warbler (255) 40:22
Yellow Warbler (259) 42:2
Chestnut-sided Warbler (259) 43:4
Black-and-white Warbler (267) 41:13
Northern Waterthrush (269) 43:14
Kentucky Warbler (271) 42:15
Bananaquit (277) 40:24
Summer Tanager (285) 47:5
***Cherrie's Tanager (287) 47:4**
Blue-gray Tanager (291) 46:15
Palm Tanager (291) 45:19
Golden-hooded Tanager (289) 46:13
Blue Dacnis (291) 46:3
Shining Honeycreeper (293) 46:1
Red-legged Honeycreeper (293) 46:2
Blue-black Grassquit (297) 49:7
Variable Seedeater (295) 49:3
Orange-billed Sparrow (303) 48:17
Black-striped Sparrow (303) 50:14
Melodious Blackbird (315) 52:6
Great-tailed Grackle (317) 44:16
Baltimore Oriole (321) 44:7
***Yellow-crowned Euphonia (327) 45:1**
Thick-billed Euphonia (327) 45:8
*Spot-crowned Euphonia (327) 45:2
White-vented Euphonia (327) 45:7

Nearby Birding Opportunities

The Parrita River Mouth

The Parrita River, which meets the sea about halfway between Jacó and Quepos, provides a good habitat for coastal species such as sandpipers, terns, and gulls, as well as mangrove specialists. To get to this birding area start in the town of Parrita (24 km north of Quepos) and head south on Rt. 34, almost immediately crossing a bridge over the Río Parrita. Continue 0.4 km beyond the bridge and then make a right onto a small paved road, which will end at the ocean in 3.8 km.

The Río Parrita runs quite close to the road 2.2 km after turning off Rt. 34, and this is a good place to scan for Short-billed Dowitcher, Black-necked Stilt, Lesser Yellowlegs, and Least Sandpiper. Mangrove Swallows and Gray-breasted Martins often feed over the river, and Wood Stork, Roseate Spoonbill, and many herons can be seen along the riverbank. Raptors such as Osprey, Yellow-headed Caracara, and Common Black-Hawk are also good possibilities.

When you arrive at the ocean you will come to a T-intersection. Turn right and proceed 0.4 km to the end of the road. From here it is a short walk along the beach to the mouth of the Río Parrita. The mudflats along the banks of the river provide excellent habitat for many shorebirds such as Western Sandpiper, Whimbrel, Greater Yellowlegs, and Sanderling. Terns and gulls often roost at the mouth of the river, and a small patch of mangroves, located a short way upstream, may

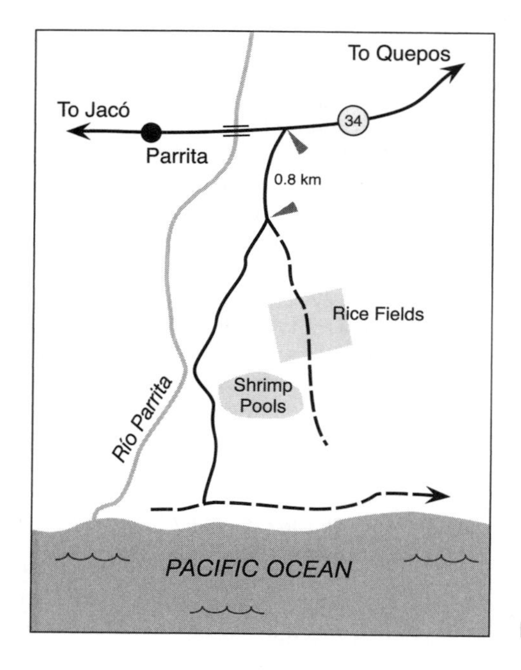

Parrita River Mouth Area Map

produce Yellow (Mangrove) Warbler, Prothonotary Warbler, Panama Flycatcher, and Mangrove Hummingbird.

Another place to explore is a dirt road that forks left 0.8 km from Rt. 34. This leads back through some rice fields, where you should look for open-habitat species such as Gray-crowned Yellowthroat, White-collared Seedeater, and Dickcissel. The road terminates near the back of a shrimp farm in a large mangrove, where many of the mangrove species can be found.

Site 3A-3: Oro Verde Biological Reserve

Birding Time: 1 day
Elevation: 250 meters
Trail Difficulty: 2
Reserve Hours: 6 a.m. to 5 p.m.
Entrance Fee: $30 (US 2007) with guided bird walk and meal
4×4 Recommended

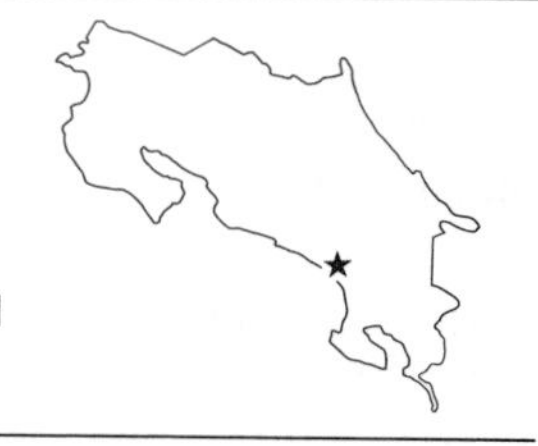

The Oro Verde Biological Reserve is a little known, locally owned reserve that is home to many exciting birds. Six brothers started the 370-acre reserve during the early 1990s. The family has lived on the land for 50 years, and all still reside in the vicinity. They are warm and welcoming to visitors, and you will feel their kind hospitality when you visit their project. The entrance fee includes a three-hour guided bird walk through the reserve, as well as a home-cooked meal.

The area is located about 250 meters above sea level, along the lower slopes of the Filas Costeñas (Costeña Hills). These hills are a part of the small coastal range of hills that help to create the General River Valley to the east. While the trail system is not extensive, if you take your time and bird the area thoroughly, you are sure to compile an impressive list. This site is a great place to find many difficult South Pacific endemics, such as Golden-naped Woodpecker and White-crested Coquette.

Oro Verde is located close to Marine Ballena National Park and the beach town of Dominical, both of which have been growing in popularity among tourists. In addition to picturesque beaches, the area offers two nearby reserves, Hacienda Barú National Wildlife Refuge and Rancho La Merced National Wildlife Refuge. Both, described in "Nearby Birding Opportunities" below, provide birders with other areas to explore if they want to spend more time in the vicinity.

Target Birds

King Vulture (31) 13:5
White Hawk (41) 17:2
Black Hawk-Eagle (49) 13:9
Scaled Pigeon (87) 18:4
Crested Owl (105) 20:3
*White-crested Coquette (137) 25:5
*Charming Hummingbird (125) 24:18
*Baird's Trogon (143) 26:6
*Fiery-billed Aracari (155) 27:16
Olivaceous Piculet (159) 29:1
*Golden-naped Woodpecker (157) 28:18
Tawny-winged Woodcreeper (171) 29:12
*Black-hooded Antshrike (177) 31:6
Southern Beardless-Tyrannulet (191) 37:20
Yellow-crowned Tyrannulet (191) 37:15
*Turquoise Cotinga (219) 34:5
*Orange-collared Manakin (221) 33:2
Blue-crowned Manakin (223) 33:7
Red-capped Manakin (223) 33:6
*Riverside Wren (241) 38:10
*White-throated Shrike-Tanager (283) 47:1
Thick-billed Euphonia (327) 45:8

Access

From Dominical: Head south on Rt. 34 for 15.2 km and turn left onto a dirt road heading inland. Follow this for 3.4 km to the Oro Verde Biological Reserve, which will be on your left. (Driving time from Dominical: 25 min)

From the intersection of Rt. 2 and Rt. 34: Take Rt. 34 north toward Dominical for 45.0 km, passing through the town of Uvita, and then turn right onto a dirt road heading inland. Follow this road for 3.4 km to the Oro Verde Biological Reserve, which will be on your left. (Driving time from Rt. 2: 50 min)

Logistics: You must make a reservation to visit the reserve at least a day or two in advance. You cannot stay at Oro Verde itself, but there are many hotels in the nearby beach towns of Uvita and Dominical.

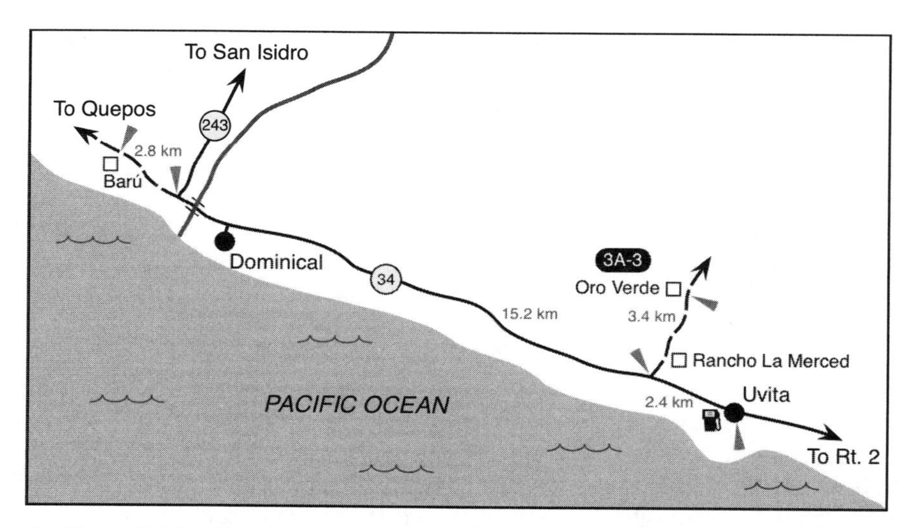

Oro Verde Driving Map. Rancho la Merced and Hacienda Barú national wildlife refuges are shown.

Oro Verde Biological Reserve

Tel: (506) 2743-8072 or 8843-8833
Website: www.costarica-birding-oroverde.com
E-mail: oro.reserva@gmail.com

Birding Sites

Upon arriving at Oro Verde you will find a small reception building and office on the left side of the road. In front of this building is a grassy yard surrounded by trees and shrubs. This is a good place to find a wide range of species such as Shining Honeycreeper, White-shouldered Tanager, Cherrie's Tanager, White-winged Becard, Yellow-crowned Tyrannulet, and Southern-beardless Tyrannulet. You are also likely to encounter both euphonia species that are restricted to the South Pacific Slope, Thick-billed Euphonia and Spot-crowned Euphonia. Similar habitat is accessible off a small dirt road that leads around behind the buildings.

Scanning the forested ridge behind Oro Verde is a good way to pick up some interesting raptors. Among the numerous Turkey Vultures and Black Vultures you may spot a King Vulture, White Hawk, Gray-headed Kite, or Black Hawk-Eagle. Also look for White-collared Swift and Costa Rican Swift cruising overhead.

One of the best birding areas at Oro Verde is the Manakin Trail, which starts a few hundred meters beyond the reception building by two metal posts. This trail leads through new-growth forest and, as its name implies, is a good place to find manakins. Red-capped Manakin, Blue-crowned Manakin, and Orange-collared Manakin are the most likely species to be encountered. The trail quickly comes to a small sugar mill and then to an observation platform. This area is good for Black-hooded Antshrike, White-necked Jacobin, Scaly-breasted Hummingbird, Baird's Trogon, and Dot-winged Antwren. This is also a great area to look for the endemic White-crested Coquette. Continuing up the trail through similar habitat, look for Golden-naped Woodpecker, Olivaceous Piculet, Black-bellied Wren, and Ruddy-tailed Flycatcher.

Higher up on the hillside, other trails lead through much older forest. While the birding here is generally slower, you may be able to find some different species, including White-tipped Sicklebill, Long-tailed Woodcreeper, Tawny-winged Woodcreeper, and Scaly-breasted Wren.

Species to Expect

Great Tinamou (3) 12:6
Little Tinamou (3) 12:17
Black Vulture (31) 13:4
Turkey Vulture (31) 13:3
King Vulture (31) 13:5
Gray-headed Kite (33) 17:3
Double-toothed Kite (35) 16:1
White Hawk (41) 17:2
Roadside Hawk (43) 16:12
Broad-winged Hawk (43) 16:13
Short-tailed Hawk (43) 16:11
Black Hawk-Eagle (49) 13:9
Laughing Falcon (53) 15:8
Scaled Pigeon (87) 18:4
Short-billed Pigeon (87) 18:5
Ruddy Ground-Dove (89) 18:7
White-tipped Dove (91) 18:14
Ruddy Quail-Dove (93) 18:17
Orange-fronted Parakeet (95) 19:13
White-crowned Parrot (97) 19:7
Red-lored Parrot (99) 19:4
Squirrel Cuckoo (103) 21:7
Crested Owl (105) 20:3
White-collared Swift (117) 22:1
***Costa Rican Swift (119) 22:11**
Long-billed Hermit (121) 23:2
Stripe-throated Hermit (121) 23:1
Scaly-breasted Hummingbird (125) 23:10
White-necked Jacobin (125) 23:17
Violet-headed Hummingbird (137) 25:11
*White-crested Coquette (137) 25:5
Violet-crowned Woodnymph (127) 24:4
Blue-throated Goldentail (129) 24:9
***Charming Hummingbird (125) 24:18**
Rufous-tailed Hummingbird (129) 24:10
Purple-crowned Fairy (125) 23:14
Long-billed Starthroat (135) 23:18
*Baird's Trogon (143) 26:6
Violaceous Trogon (141) 26:8
Black-throated Trogon (141) 26:9
Slaty-tailed Trogon (145) 26:2
Blue-crowned Motmot (147) 27:8
*Fiery-billed Aracari (155) 27:16
Chestnut-mandibled Toucan (155) 27:19
*Golden-naped Woodpecker (157) 28:18
Red-crowned Woodpecker (157) 28:17
Pale-billed Woodpecker (161) 27:14
Plain Xenops (169) 29:3
Tawny-winged Woodcreeper (171) 29:12
Wedge-billed Woodcreeper (171) 29:6
Cocoa Woodcreeper (173) 29:17
Black-striped Woodcreeper (173) 29:16
Streak-headed Woodcreeper (173) 29:8
Barred Antshrike (175) 31:3
***Black-hooded Antshrike (177) 31:6**
Dot-winged Antwren (183) 32:1
Chestnut-backed Antbird (179) 31:11
Southern Beardless-Tyrannulet (191) 37:20
Yellow-crowned Tyrannulet (191) 37:15
Yellow-bellied Elaenia (193) 37:26
Ochre-bellied Flycatcher (199) 36:25
Paltry Tyrannulet (191) 37:10
Northern Bentbill (197) 37:6
Common Tody-Flycatcher (197) 37:7
Eye-ringed Flatbill (201) 37:23
Yellow-olive Flycatcher (195) 37:16

Ruddy-tailed Flycatcher (199) 36:23
Tropical Pewee (203) 36:9
Bright-rumped Attila (201) 35:6
Great Crested Flycatcher (209) 35:17
Great Kiskadee (211) 35:13
Boat-billed Flycatcher (211) 35:12
Social Flycatcher (211) 35:14
Gray-capped Flycatcher (211) 35:15
Streaked Flycatcher (213) 35:11
Sulphur-bellied Flycatcher (213) 35:10
Piratic Flycatcher (193) 35:8
Tropical Kingbird (213) 35:1
White-winged Becard (217) 33:13
Masked Tityra (217) 34:1
Black-crowned Tityra (217) 34:2
***Orange-collared Manakin (221) 33:2**
White-ruffed Manakin (223) 33:9
Blue-crowned Manakin (223) 33:7
Yellow-throated Vireo (225) 40:6
Yellow-green Vireo (227) 40:5
Tawny-crowned Greenlet (229) 32:4
Lesser Greenlet (229) 40:7
Brown Jay (231) 39:19
Southern Rough-winged Swallow (235) 22:19
***Black-bellied Wren (243) 38:11**
***Riverside Wren (241) 38:10**
Plain Wren (241) 38:17
House Wren (243) 38:18
Scaly-breasted Wren (245) 38:20
Long-billed Gnatwren (237) 32:15
Tropical Gnatcatcher (237) 41:2
Clay-colored Thrush (251) 39:8
Tennessee Warbler (255) 40:22
Yellow Warbler (259) 42:2
Chestnut-sided Warbler (259) 43:4
Black-and-white Warbler (267) 41:13
Buff-rumped Warbler (271) 40:23
Bananaquit (277) 40:24
White-shouldered Tanager (281) 46:18
***Cherrie's Tanager (287) 47:4**
Blue-gray Tanager (291) 46:15
Palm Tanager (291) 45:19
Bay-headed Tanager (289) 46:9
Golden-hooded Tanager (289) 46:13
Blue Dacnis (291) 46:3
Green Honeycreeper (293) 46:7
Shining Honeycreeper (293) 46:1
Variable Seedeater (295) 49:3
Orange-billed Sparrow (303) 48:17
Black-striped Sparrow (303) 50:14
Buff-throated Saltator (309) 48:2
Blue-black Grosbeak (311) 48:10
***Yellow-crowned Euphonia (327) 45:1**
Thick-billed Euphonia (327) 45:8
*Spot-crowned Euphonia (327) 45:2

Nearby Birding Opportunities

Rancho La Merced National Wildlife Refuge

Just a few hundred meters up the dirt road leading to Oro Verde Reserve you will find Rancho La Merced National Wildlife Refuge. Although the birding here is generally not as productive as it is upslope at Oro Verde, Rancho La Merced does offer an extensive trail system through secondary forest and is well worth exploring. This is a good place to find Golden-crowned Spadebill, White-throated Shrike-Tanager, Olivaceous Piculet, and Blue-crowned Manakin. Even Turquoise Cotinga is seen from time to time. The hummingbird feeders at the entrance and porterweed flowers along the road attract Charming Hummingbird and Violet-crowned Woodnymph.

With prior reservation, Rancho La Merced offers a half-day bird tour starting at 6 a.m. ($35 US 2008), although you can walk the trails on your own for a $6 entrance fee. The refuge also offers horseback riding and a hands-on tour of the working cattle ranch, where you can try throwing a lasso and other cowboy skills.

Tel: (506) 8861-5147
Website: www.rancholamerced.com
E-mail: rancholamerced@ice.co.cr

Hacienda Barú National Wildlife Refuge

Hacienda Barú is a small reserve located just north of Dominical. The facility offers comfortable cabins, a small restaurant, and a well-maintained trail system. Aside from being a great place to find White-faced Capuchin Monkeys, it is also home to many birds. Along the trails on the west side of the road look for Streaked Flycatcher, Charming Hummingbird, Crested Caracara, and Gray-necked Wood-Rail. Lookout Trail climbs steeply up the hills on the east side of the road. Here the forest is older and home to Baird's Trogon, Ruddy Quail-Dove, and Gray-headed Tanager.

Hacienda Barú is located 2.8 km north of Dominical along the dirt road that leads to Quepos. It is open 7 a.m. to 5 p.m., and the entrance fee is $6 (US 2008).

Tel: (506) 2787-0003
Website: www.haciendabaru.com
E-mail: info@haciendabaru.com

Site 3A-4: Marenco Beach and Rainforest Lodge

Birding Time: 2 days
Elevation: Sea level
Trail Difficulty: 2
Reserve Hours: N/A
Entrance Fee: Included with overnight
Trail Map Available on Site

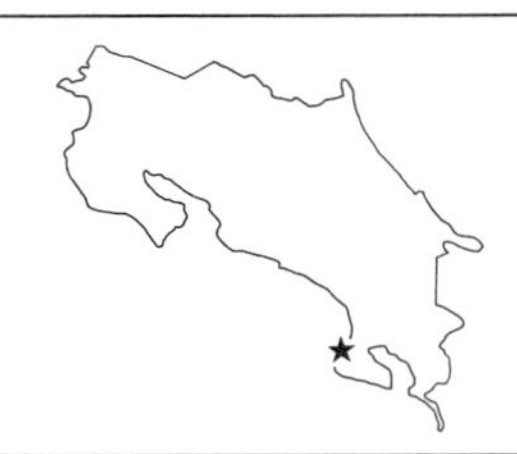

The Osa Peninsula is renowned as one of Costa Rica's wildest and most remote areas, and Marenco Beach and Rainforest Lodge is a very comfortable way to experience its untamed magnificence. The lodge, constructed in the early 1980s on a 1,230-acre private reserve, was one of the first eco-lodges on the peninsula. Accessible only via boat, it sits atop a forested ridge dramatically overlooking the Pacific Ocean. From this high vantage point, you can see monkeys and sloth in the trees, and whales and dolphins in the ocean below.

The facility can accommodate about 50 people, and there is a restaurant that serves three meals daily. Most of the lodging is in private bungalows, which are simple, clean, and comfortable. While there is no hot water, there is limited electricity produced by a small generator during the evenings. You can arrange tours of the property with on-site natural history guides, who can also take you to the nearby Corcovado National Park and Isla del Caño Biological Reserve.

The birding at Marenco is productive, and there are plenty of trails to explore. It is a great place to look for forest species such as Streak-chested Antpitta and Golden-crowned Spadebill. The true appeal of Marenco Lodge, however, is its isolation, set in one Central America's largest and most beautiful lowland rainforests.

Target Birds

Brown Booby (15) 1:1
Crested Eagle (47) 17:10
Harpy Eagle (47) 17:9
Wandering Tattler (73) 10:8
Scarlet Macaw (99) 19:1
*Costa Rican Swift (119) 22:11
*White-crested Coquette (137) 25:5
*Charming Hummingbird (125) 24:18
*Baird's Trogon (143) 26:6
*Fiery-billed Aracari (155) 27:16
*Golden-naped Woodpecker (157) 28:18
Long-tailed Woodcreeper (171) 29:10
Streak-chested Antpitta (185) 30:18
Black-tailed Flycatcher (199) 36:22
*Turquoise Cotinga (219) 34:5
*Yellow-billed Cotinga (219) 34:3
Blue-crowned Manakin (223) 33:7
*White-throated Shrike-Tanager (283) 47:1
*Black-cheeked Ant-Tanager (277) 47:11

Access

Marenco's remote location along the western shoreline of the Osa Peninsula makes access more complicated than at most other sites. There are three ways to get there. The easiest method is to drive to the town of Sierpe and take an hour and a half boat ride, provided by the lodge, down the Río Sierpe and along the Pacific coast to Marenco. (The driving directions below describe how to get to the docks at Sierpe.) Alternatively, you can fly on NatureAir or Sansa into the small town of Drake Bay, where someone from the lodge will pick you up for a 10-minute boat ride to Marenco. Drake Bay is most easily reached by boat or small airplane, but if you must arrive by car, there is a long dirt road that connects Drake Bay to Rincón (refer to site 3A-5). This route is discouraged, even for 4×4 vehicles, owing to some treacherous river crossings that could be impassible if the water level is high.

From the intersection of Rt. 34 and Rt. 2 (Palmar Norte): Take Rt. 2 south, immediately crossing a large bridge over the Río Grande de Térraba. Turn right immediately after the bridge onto a paved road, and proceed 1.2 km to a T-intersection. Turn right here and follow the paved road for 14.0 km to the town of Sierpe. (There are many sharp turns along this section, so be sure you stay on the paved road.) Go through the town of Sierpe, one block beyond the town park, to where the pavement ends. Turn left here at the Oleaje Sereno Hotel and Restaurant. Marenco maintains a guarded parking lot across the street, and the docks will be right in front of you. (Driving time from Rt. 34: 20 min)

Logistics: Once you are at Marenco, the lodge will take care of all logistical concerns. Just be sure to bring in any personal items that you might need.

Marenco Beach and Rainforest Lodge
Tel: (506) 2259-1919
Website: www.marencolodge.com
E-mail: info@marencolodge.com

Birding Sites

If you enter via the boat from Sierpe, you will pass through extensive mangroves on your way to the ocean. Here you may be able to see Mangrove Hummingbird, Panama Flycatcher, and the rare Yellow-billed Cotinga. Shorebirds such as Whimbrel and Willet can also be expected.

Birding the grounds of the lodge is often productive, especially first thing in the morning when it is still dark inside the forest. Look for fruiting trees, which attract Fiery-billed Aracari, Spot-crowned Euphonia, Blue Dacnis, and Shining Honeycreeper. Southern Beardless-Tyrannulet, Golden-naped Woodpecker, and Charming Hummingbird are also likely. Scarlet Macaws can often be seen flying by, and Scaly-breasted Wren calls from inside the forest edge.

The trails leading through the forest around the lodge are extensive and home to a large variety of birds. (If you choose to go out without a guide, be sure to pick up a trail map from reception before venturing into the forest.) For the physically strong, I suggest walking a long loop that starts on La Fila Trail and then turns right onto Public Trail and proceeds down to where the Río Claro meets the ocean. From here you can walk the beach back to the lodge. This loop, through mostly primary forest, will take about five hours to bird. If this is too long, you can shorten the loop considerably by taking the trail that leads past the Giant Tree.

Mixed flocks in the forest normally include Dot-winged Antwren, Tawny-crowned Greenlet, and Black-hooded Antshrike, although you should also look for White-throated Shrike-Tanager, Tawny-winged Woodcreeper, and Buff-throated Foliage-gleaner. Other birds of interest that are often found along the trails are Golden-crowned Spadebill, Bicolored Antbird, Streak-chested Antpitta, Thrush-like Schiffornis, and Black-cheeked Ant-Tanager. Both Rufous Piha and Rufous Mourner are common around Marenco, and the distinction between the two is one of the hardest visual identifications in the country. (Be advised, however, that vocally they are very different.) The first stretch of Public Trail offers access to new-growth and scrubby-field habitat. Here you might find Gray-crowned Yellowthroat, Yellow Tyrannulet, Cherrie's Tanager, and White-shouldered Tanager. The section of trail along the beach is a good place to look for Common Black-Hawk. The trails at Marenco are one of the few places in Costa Rica where Harpy Eagle can be found, although even here this is an extremely rare species.

Marenco also offers day trips to both the San Pedrillo Station of Corcovado National Park as well as to Isla del Caño Biological Reserve. The forest and trails at San Pedrillo are quite similar to those found at Marenco, and you should expect to see the same range of species. Isla del Caño is a renowned snorkeling and scuba diving destination, and joining one of those trips will yield some interesting birds. Red-footed Booby is occasionally seen with the Brown Boobies on the island. Red-necked Phalarope, Brown Noddy, Wedge-rumped Storm-Petrel, and Audubon's Shearwater are other seabirds that can be encountered on a trip out to the island.

Species to Expect

Brown Booby (15) 1:1
Brown Pelican (17) 4:1
Magnificent Frigatebird (19) 1:6
Black Vulture (31) 13:4
Turkey Vulture (31) 13:3
White Hawk (41) 17:2
Common Black-Hawk (45) 13:6
Broad-winged Hawk (43) 16:13
Collared Forest-Falcon (55) 16:7
Yellow-headed Caracara (53) 15:9
Spotted Sandpiper (73) 11:8
Royal Tern (81) 3:2
Short-billed Pigeon (87) 18:5
White-tipped Dove (91) 18:14
Scarlet Macaw (99) 19:1
Red-lored Parrot (99) 19:4
Mealy Parrot (99) 19:3
Common Pauraque (111) 21:18
Long-billed Hermit (121) 23:2
Stripe-throated Hermit (121) 23:1
Scaly-breasted Hummingbird (125) 23:10
*Charming Hummingbird (125) 24:18
Rufous-tailed Hummingbird (129) 24:10
Purple-crowned Fairy (125) 23:14
Violaceous Trogon (141) 26:8
Slaty-tailed Trogon (145) 26:2
White-whiskered Puffbird (151) 28:6
***Fiery-billed Aracari (155) 27:16**
Chestnut-mandibled Toucan (155) 27:19
*Golden-naped Woodpecker (157) 28:18
*Rufous-winged Woodpecker (159) 28:10
Pale-billed Woodpecker (161) 27:14
Buff-throated Foliage-gleaner (165) 30:3
Plain Xenops (169) 29:3
Tawny-winged Woodcreeper (171) 29:12
Northern Barred-Woodcreeper (169) 29:19
Cocoa Woodcreeper (173) 29:17
Black-striped Woodcreeper (173) 29:16
***Black-hooded Antshrike (177) 31:6**
Russet Antshrike (177) 31:4
Slaty Antwren (183) 32:3
Dot-winged Antwren (183) 32:1
Chestnut-backed Antbird (179) 31:11
Bicolored Antbird (179) 31:9
Black-faced Antthrush (187) 30:15
Streak-chested Antpitta (185) 30:18
Southern Beardless-Tyrannulet (191) 37:20
Yellow Tyrannulet (189) 37:11
Paltry Tyrannulet (191) 37:10
Scale-crested Pygmy-Tyrant (197) 37:4
Northern Bentbill (197) 37:6
Common Tody-Flycatcher (197) 37:7
Golden-crowned Spadebill (189) 37:3
Sulphur-rumped Flycatcher (199) 36:21
Bright-rumped Attila (201) 35:6
Rufous Mourner (201) 34:9
Dusky-capped Flycatcher (209) 35:21
Great Crested Flycatcher (209) 35:17
Social Flycatcher (211) 35:14
Piratic Flycatcher (193) 35:8
Tropical Kingbird (213) 35:1
Thrush-like Schiffornis (215) 33:10
Rufous Piha (215) 34:10
Masked Tityra (217) 34:1
*Turquoise Cotinga (219) 34:5
Blue-crowned Manakin (223) 33:7
Red-capped Manakin (223) 33:6
Yellow-throated Vireo (225) 40:6
Red-eyed Vireo (227) 40:3

Tawny-crowned Greenlet (229) 32:4
Lesser Greenlet (229) 40:7
***Riverside Wren (241) 38:10**
Scaly-breasted Wren (245) 38:20
Long-billed Gnatwren (237) 32:15
Tropical Gnatcatcher (237) 41:2
Swainson's Thrush (249) 39:2
White-throated Thrush (251) 39:9
Tennessee Warbler (255) 40:22
Yellow Warbler (259) 42:2
Chestnut-sided Warbler (259) 43:4
Northern Waterthrush (269) 43:14
Gray-crowned Yellowthroat (273) 42:13
Bananaquit (277) 40:24
Gray-headed Tanager (283) 45:17
*White-throated Shrike-Tanager (283) 47:1
White-shouldered Tanager (281) 46:18
***Black-cheeked Ant-Tanager (277) 47:11**
Summer Tanager (285) 47:5
***Cherrie's Tanager (287) 47:4**
Blue-gray Tanager (291) 46:15
Palm Tanager (291) 45:19
Bay-headed Tanager (289) 46:9
Golden-hooded Tanager (289) 46:13
Scarlet-thighed Dacnis (291) 46:4
Blue Dacnis (291) 46:3
Green Honeycreeper (293) 46:7
Shining Honeycreeper (293) 46:1
Blue-black Grassquit (297) 49:7
Variable Seedeater (295) 49:3
Orange-billed Sparrow (303) 48:17
Black-striped Sparrow (303) 50:14
Buff-throated Saltator (309) 48:2
Blue-black Grosbeak (311) 48:10
Baltimore Oriole (321) 44:7
Scarlet-rumped Cacique (321) 44:11
***Spot-crowned Euphonia (327) 45:2**
White-vented Euphonia (327) 45:7

Site 3A-5: Bosque del Río Tigre

Birding Time: 2–3 days
Elevation: 50 meters
Trail Difficulty: 2
Reserve Hours: N/A
Entrance Fee: Included with overnight
4×4 Recommended
Bird Guides Available

Bosque del Río Tigre is a birder's paradise. The owners of the small lodge, Liz Jones and Abraham Gallo, are avid birders and have an extensive knowledge of the diverse birdlife in the area. Of the 18 endemic species confined to Costa Rica's South Pacific Slope, 15 are regularly found around Bosque del Río Tigre, along with many other unusual birds. Expert bird guides, provided by the lodge, know exactly where to look for these key species, and you should expect to fill in many blanks on your target list.

It is the varied habitat around the lodge that makes for such exciting birding. The primary forest around Bosque del Río Tigre is part of the Golfo Dulce Forest Reserve, which runs continuously into Corcovado National Park. This enormous protected area makes for a diverse community of forest species. Better yet, the area also includes easily accessible secondary forest, new growth, lagoons, cattle pastures, and even nearby mangroves. It is not hard to find well over 100 species in a day of birding, and an extended stay should yield close to 200.

The lodge itself, which is tucked within the forest on the banks of a rushing river, has a feeling of total seclusion and manages to combine rustic flavor with luxury and comfort in a way that no other facility in Costa Rica can match. The small two-story building is almost completely open air, and even the four bedrooms upstairs have railings instead of walls on the sides that face into the forest. This creates the feeling of sleeping in a tree house in the middle of the rainforest, like something out of *Swiss Family Robinson*. While this may sound intimidating and uncomfortable, mosquito netting around the beds eliminates any insect problem. The facility is immaculately clean, the food is outstanding, and the personal service is first rate. Bosque del Río Tigre is a fun place to visit that will leave you with a long-lasting impression.

Target Birds

Marbled Wood-Quail (13) 12:9
Fasciated Tiger-Heron (21) 5:15
Boat-billed Heron (27) 5:2
King Vulture (31) 13:5
Gray-headed Kite (33) 17:3
Tiny Hawk (39) 16:2
Bat Falcon (51) 15:14
Uniform Crake (57) 6:8
Scarlet Macaw (99) 19:1
Striped Cuckoo (101) 21:5
Band-tailed Barbthroat (121) 23:5
White-tipped Sicklebill (123) 23:8
*White-crested Coquette (137) 25:5
*Charming Hummingbird (125) 24:18
*Baird's Trogon (143) 26:6
*Fiery-billed Aracari (155) 27:16
*Golden-naped Woodpecker (157) 28:18
Red-rumped Woodpecker (159) 28:12
Scaly-throated Leaftosser (167) 30:9
Tawny-winged Woodcreeper (171) 29:12
Long-tailed Woodcreeper (171) 29:10
Great Antshrike (177) 31:2
Black-faced Antthrush (187) 30:15
Yellow-bellied Tyrannulet (189) 37:14
Southern Beardless-Tyrannulet (191) 37:20
Slate-headed Tody-Flycatcher (197) 37:9
Golden-crowned Spadebill (189) 37:3
Black-tailed Flycatcher (199) 36:22
*Turquoise Cotinga (219) 34:5
*Orange-collared Manakin (221) 33:2
Scrub Greenlet (229) 40:8
Green Shrike-Vireo (229) 40:1
*Black-bellied Wren (243) 38:11
*White-throated Shrike-Tanager (283) 47:1
*Black-cheeked Ant-Tanager (277) 47:11
Thick-billed Euphonia (327) 45:8
*Spot-crowned Euphonia (327) 45:2

Access

Access is the largest drawback to this area, as it is isolated on the Osa Peninsula. Driving directions are provided below, although there are other transportation options that are worth considering. In-country flights offered by NatureAir and Sansa access the nearby town of Puerto Jiménez, where someone from the lodge will pick you up. It is also possible to rent a car in Puerto Jiménez and return it elsewhere (San José, for example), making the option of flying in and driving out quite feasible.

From the intersection of Rt. 2 and Rt. 245 (Chacarita): Take Rt. 245 toward Puerto Jiménez for 70.8 km. Turn right onto a small dirt road, following signs for

Dos Brazos. (If you cross the Río Tigre, you've gone 0.4 km too far.) Continue 2.8 km to the small town of Gallardo, where you should take your second left after the soccer field. Follow this road 5.4 km to Dos Brazos. As you enter town you will come to a T-intersection. Turn left here. After 0.4 km you will come to a small *soda* (restaurant) at the edge of town and a river crossing just beyond it. If the water level is high, you should contact Bosque del Río Tigre via radio at the soda, and someone will come to pick you up. If the water level is low enough and you have a 4×4, you can continue the final 0.4 km to the lodge. (Driving time from Rt. 2: 120 min)

Logistics: While there is one small hostel in Dos Brazos, staying at Bosque del Río Tigre is by far your best option, gaining you access to all the best trails, as well as the local bird guides.

Bosque del Río Tigre
Website: www.osaadventures.com
E-mail: info@osaadventures.com

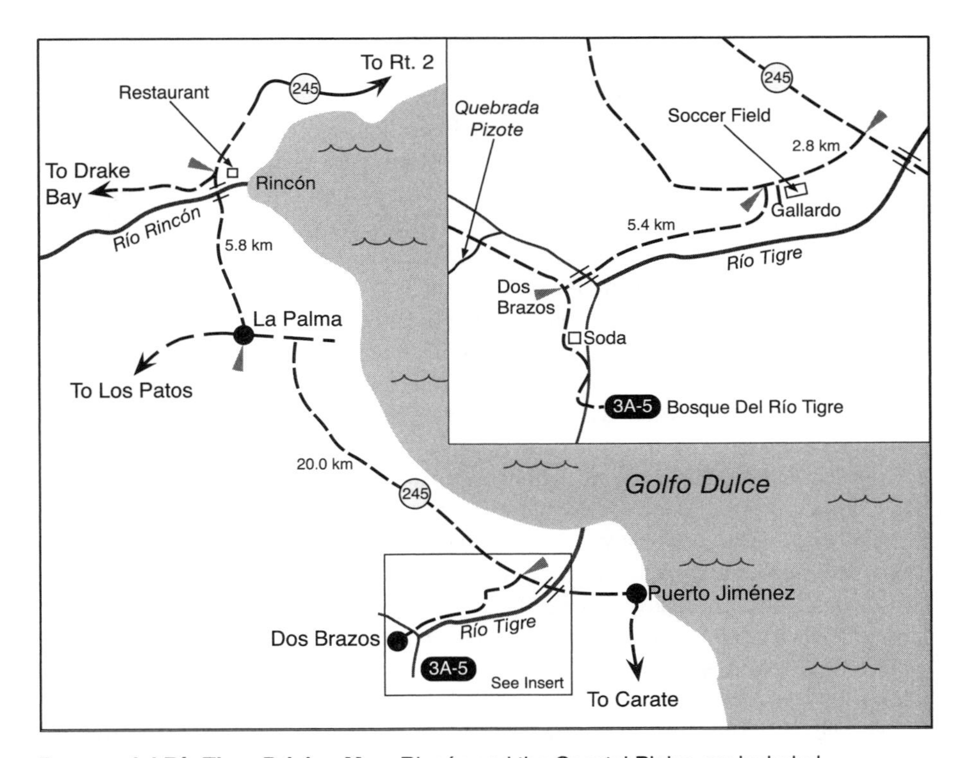

Bosque del Río Tigre Driving Map. Rincón and the Coastal Plains are included.

Birding Sites

Around the lodge are hummingbird and fruit feeders, which attract Charming Hummingbird, Golden-hooded Tanager, Cherrie's Tanager, and Spot-crowned Euphonia. Even the endemic Black-cheeked Ant-Tanager is a regular at the fruit feeders. Rice is also thrown on the ground and brings in Blue Ground-Dove and Little Tinamou, normally a very difficult bird to see. The forest edge around the lodge is a good place to look for Yellow-bellied Tyrannulet, Southern Beardless-Tyrannulet, and Green Shrike-Vireo.

The Ocelot, Gallinazo, and Bananal trails wind through the primary forest on the hillside behind the lodge. At first light Black-faced Antthrush is regularly seen walking along the trails, and Scaly-throated Leaftosser often searches for food in the leaf litter on the trail's edge. There is great diversity of species found along these trails, and all four trogons of the South Pacific Lowlands are common. Other birds to look for include Golden-crowned Spadebill, Black-striped Woodcreeper, Rufous Piha, Bicolored Antbird, and Red-capped Manakin.

These trails eventually lead to a larger cart path that proceeds along the top of the ridge. There is a lek of Orange-collared Manakins right off the path, and when they are displaying, you will certainly get good looks at these cute little orange birds hopping from branch to branch. The cart path is also a good place to find mixed-species flocks, which might include Long-tailed Woodcreeper, Gray-headed Tanager, and White-winged Becard.

Following the cart path uphill, you will soon enter pasture and have an expansive look at the forest below, somewhat like being on a canopy tower. This is a good place to spot raptors such as Bat Falcon, Gray-headed Kite, and Tiny Hawk. King Vulture is quite common in the area. Looking down at the treetops, you can also find Turquoise Cotinga as well as many species of tanagers, parrots, and flycatchers. The pasture itself often yields Yellow Tyrannulet, Thick-billed Seed-Finch, and Black-bellied Wren. Walking downhill on the cart path will bring you through second-growth habitat, where you might find Great Antshrike, Olivaceous Piculet, Black-bellied Wren, and Red-rumped Woodpecker.

The Boat-billed Heron Trail, which starts at the lodge, is another productive trail. This is a good place to encounter Bronzy Hermit, Band-tailed Barbthroat, as well as Ruddy Quail-Dove. A few hundred meters down the trail you will come to a wet area thick with heliconias. This is a reliable place to find the rare Uniform Crake, especially during the dry season, when there is only a small amount of suitable wet habitat remaining. The trail continues to a small pond, where Boat-billed Herons make their nests, and you are also likely to find Least Grebe.

Another trail, which follows the river upstream, is a pretty and often productive walk that yields a different array of species. To walk this trail I suggest having tall rubber boots or water shoes, because the trail crosses the river a number of times. Most of the habitat in this area is new growth, where you might find Black-bellied Wren, Buff-rumped Warbler, Dusky Antbird, and Mourning Warbler. This is also one of the best areas to look for the Black-tailed Flycatcher and Great Antshrike. Flowering trees often attract White-necked Jacobin and Long-billed Starthroat, and the endemic White-crested Coquette is a regular here. If you walk the river early in the morning, you have a good chance of finding Fasciated Tiger-Heron feeding along the rocks.

Walking through the town of Dos Brazos can be quite productive. A small wetland across from the soda sometimes holds American Pygmy Kingfisher. Look for Double-toothed Kite and White-necked Puffbird in the treetops on the hillsides around town, and listen for the rough squawks of Scarlet Macaws as the birds pass overhead. Proceeding through town, keep an eye open for Smooth-billed Ani, Scaly-breasted Hummingbird, Blue-throated Goldentail, Scrub Greenlet, and Red-rumped Woodpecker. The road heading back toward Gallardo often yields Striped Cuckoo.

At the far end of town you will find the Quebrada Pizote. If you walk up this small stream, you can often find some interesting species such as Striped Woodhaunter, Russet Antshrike, and Tawny-crowned Greenlet. However, the real attraction of the area is White-tipped Sicklebill, which nests along the stream. The guides at the lodge always know where to find this exciting species.

Species to Expect

Little Tinamou (3) 12:17
Gray-headed Chachalaca (11) 12:1
Marbled Wood-Quail (13) 12:9
Least Grebe (9) 7:4
Neotropic Cormorant (17) 4:4
Snowy Egret (23) 5:10
Little Blue Heron (23) 5:9
Cattle Egret (23) 5:13
Green Heron (25) 6:2
Boat-billed Heron (27) 5:2
White Ibis (29) 4:8
Black Vulture (31) 13:4
Turkey Vulture (31) 13:3
King Vulture (31) 13:5
Gray-headed Kite (33) 17:3
Swallow-tailed Kite (37) 15:2
Double-toothed Kite (35) 16:1
Roadside Hawk (43) 16:12
Short-tailed Hawk (43) 16:11
Crested Caracara (53) 14:12
Yellow-headed Caracara (53) 15:9
Laughing Falcon (53) 15:8
Bat Falcon (51) 15:14
White-throated Crake (57) 6:9
Gray-necked Wood-Rail (55) 6:13
Uniform Crake (57) 6:8
Purple Gallinule (59) 6:15
Northern Jacana (61) 6:18
Spotted Sandpiper (73) 11:8
Pale-vented Pigeon (87) 18:3
Short-billed Pigeon (87) 18:5
Ruddy Ground-Dove (89) 18:7
Blue Ground-Dove (91) 18:6
White-tipped Dove (91) 18:14
Gray-chested Dove (91) 18:16
Ruddy Quail-Dove (93) 18:17
*Crimson-fronted Parakeet (95) 19:10
Scarlet Macaw (99) 19:1
Orange-chinned Parakeet (97) 19:14
Brown-hooded Parrot (97) 19:9
White-crowned Parrot (97) 19:7
Red-lored Parrot (99) 19:4
Mealy Parrot (99) 19:3
Squirrel Cuckoo (103) 21:7
Striped Cuckoo (101) 21:5
Smooth-billed Ani (103) 21:8
White-collared Swift (117) 22:1
***Costa Rican Swift (119) 22:11**

Band-tailed Barbthroat (121) 23:5
Long-billed Hermit (121) 23:2
Stripe-throated Hermit (121) 23:1
White-tipped Sicklebill (123) 23:8
Scaly-breasted Hummingbird (125) 23:10
White-necked Jacobin (125) 23:17
*White-crested Coquette (137) 25:5
Blue-throated Goldentail (129) 24:9
***Charming Hummingbird (125) 24:18**
Rufous-tailed Hummingbird (129) 24:10
Purple-crowned Fairy (125) 23:14
Long-billed Starthroat (135) 23:18
***Baird's Trogon (143) 26:6**
Violaceous Trogon (141) 26:8
Black-throated Trogon (141) 26:9
Slaty-tailed Trogon (145) 26:2
Blue-crowned Motmot (147) 27:8
Ringed Kingfisher (149) 27:1
Amazon Kingfisher (149) 27:4
Green Kingfisher (149) 27:5
White-necked Puffbird (151) 28:3
Rufous-tailed Jacamar (153) 26:12
*Fiery-billed Aracari (155) 27:16
Chestnut-mandibled Toucan (155) 27:19
Olivaceous Piculet (159) 29:1
*Golden-naped Woodpecker (157) 28:18
Red-crowned Woodpecker (157) 28:17
Red-rumped Woodpecker (159) 28:12
Lineated Woodpecker (161) 27:13
Pale-billed Woodpecker (161) 27:14
Slaty Spinetail (163) 32:9
Buff-throated Foliage-gleaner (165) 30:3
Plain Xenops (169) 29:3
Scaly-throated Leaftosser (167) 30:9
Long-tailed Woodcreeper (171) 29:10
Wedge-billed Woodcreeper (171) 29:6
Northern Barred-Woodcreeper (169) 29:19
Cocoa Woodcreeper (173) 29:17
Black-striped Woodcreeper (173) 29:16
Great Antshrike (177) 31:2
***Black-hooded Antshrike (177) 31:6**
Russet Antshrike (177) 31:4
Dot-winged Antwren (183) 32:1
Dusky Antbird (177) 31:8
Chestnut-backed Antbird (179) 31:11
Bicolored Antbird (179) 31:9
Black-faced Antthrush (187) 30:15
Southern Beardless-Tyrannulet (191) 37:20
Yellow Tyrannulet (189) 37:11
Yellow-bellied Elaenia (193) 37:26
Ochre-bellied Flycatcher (199) 36:25
Paltry Tyrannulet (191) 37:10
Northern Bentbill (197) 37:6
Common Tody-Flycatcher (197) 37:7
Eye-ringed Flatbill (201) 37:23
Yellow-olive Flycatcher (195) 37:16
Golden-crowned Spadebill (189) 37:3
Sulphur-rumped Flycatcher (199) 36:21
Black-tailed Flycatcher (199) 36:22
Tropical Pewee (203) 36:9
Yellow-bellied Flycatcher (205) 36:20
Bright-rumped Attila (201) 35:6
Rufous Mourner (201) 34:9
Dusky-capped Flycatcher (209) 35:21
Great Crested Flycatcher (209) 35:17
Great Kiskadee (211) 35:13
Boat-billed Flycatcher (211) 35:12
Social Flycatcher (211) 35:14
Gray-capped Flycatcher (211) 35:15
Streaked Flycatcher (213) 35:11
Piratic Flycatcher (193) 35:8
Tropical Kingbird (213) 35:1
Rufous Piha (215) 34:10
White-winged Becard (217) 33:13
Rose-throated Becard (217) 33:12
Masked Tityra (217) 34:1
Black-crowned Tityra (217) 34:2
*Turquoise Cotinga (219) 34:5
***Orange-collared Manakin (221) 33:2**
Blue-crowned Manakin (223) 33:7
Red-capped Manakin (223) 33:6
Yellow-throated Vireo (225) 40:6
Philadelphia Vireo (227) 40:15
Red-eyed Vireo (227) 40:3
Yellow-green Vireo (227) 40:5
Tawny-crowned Greenlet (229) 32:4
Lesser Greenlet (229) 40:7
Green Shrike-Vireo (229) 40:1
Southern Rough-winged Swallow (235) 22:19
*Black-bellied Wren (243) 38:11
***Riverside Wren (241) 38:10**
House Wren (243) 38:18
Scaly-breasted Wren (245) 38:20

Long-billed Gnatwren (237) 32:15
Tropical Gnatcatcher (237) 41:2
Clay-colored Thrush (251) 39:8
White-throated Thrush (251) 39:9
Tennessee Warbler (255) 40:22
Yellow Warbler (259) 42:2
Chestnut-sided Warbler (259) 43:4
American Redstart (267) 41:10
Northern Waterthrush (269) 43:14
Mourning Warbler (271) 42:14
Buff-rumped Warbler (271) 40:23
Bananaquit (277) 40:24
Gray-headed Tanager (283) 45:17
*White-throated Shrike-Tanager (283) 47:1
White-shouldered Tanager (281) 46:18
***Black-cheeked Ant-Tanager (277) 47:11**
Summer Tanager (285) 47:5
***Cherrie's Tanager (287) 47:4**
Blue-gray Tanager (291) 46:15
Palm Tanager (291) 45:19
Bay-headed Tanager (289) 46:9
Golden-hooded Tanager (289) 46:13
Scarlet-thighed Dacnis (291) 46:4
Blue Dacnis (291) 46:3
Green Honeycreeper (293) 46:7
Shining Honeycreeper (293) 46:1
Red-legged Honeycreeper (293) 46:2
Blue-black Grassquit (297) 49:7
Variable Seedeater (295) 49:3
White-collared Seedeater (295) 49:2
Thick-billed Seed-Finch (295) 49:10
Orange-billed Sparrow (303) 48:17
Black-striped Sparrow (303) 50:14
Buff-throated Saltator (309) 48:2
Blue-black Grosbeak (311) 48:10
Great-tailed Grackle (317) 44:16
Bronzed Cowbird (317) 44:15
Baltimore Oriole (321) 44:7
Yellow-billed Cacique (321) 44:12
Scarlet-rumped Cacique (321) 44:11
*Yellow-crowned Euphonia (327) 45:1
Thick-billed Euphonia (327) 45:8
***Spot-crowned Euphonia (327) 45:2**

Nearby Birding Opportunities

Rincón and the Coastal Plains

The town of Rincón is located 5.8 km north of La Palma along Rt. 245. This is a special place to bird because it has the most accessible lek site of the endemic Yellow-billed Cotinga. Between December and April, these birds gather in the area to display and mate. Park your car near the restaurant on the northern bank of the Río Rincón. Mangrove Swallow, shorebirds, and herons can often be seen from the bridge. This is a good place to scan for Yellow-billed Cotinga as well as Turquoise Cotinga. These birds can also be found in the trees on the steep hillside in front of the restaurant.

The first few hundred meters of the road that leads to Drake Bay can also yield some interesting species, including Black-throated Trogon, Golden-naped Woodpecker, Blue Dacnis, and Bright-rumped Attila. You should also take the time to walk north along Rt. 245 in the direction of Rt. 2, where you will soon find mangroves on your right. If you have not yet seen the two Cotinga species, this is another good place to check. It also often yields Common Black-Hawk, Panama Flycatcher, and the endemic Mangrove Hummingbird.

Between Rincón and Bosque del Río Tigre, Rt. 245 proceeds through an area known as the Coastal Plains, which are mostly flat cattle pastures and fields. This area offers excellent birding. Some of the more exciting species that you can find here are Ruddy-breasted Seedeater, Yellow-bellied Seedeater, Pale-breasted Spinetail, Red-breasted Blackbird, and Red-rumped Woodpecker. Scarlet Macaws and Red-lored Parrots are regularly found feeding in the trees, and the tiny Pearl Kite is often seen surveying the surroundings from a high snag. The exact location of these birds changes from year to year, and the guides at Bosque del Río Tigre will have up-to-date knowledge of their current locations.

Site 3A-6: Corcovado National Park

Birding Time: 3–8 days
Elevation: Sea level
Trail Difficulty: 2–3
Reserve Hours: N/A
Entrance/Camping Fee: $14 (US 2008)
4×4 Recommended
Trail Map Available on Site

Named by National Geographic as the second most biodiverse place on earth, Corcovado National Park is the jewel in the crown of Costa Rica's extensive national park system. Corcovado protects one of the largest expanses of primary lowland rainforest remaining in the Americas. All four of Costa Rica's monkey species live within the reserve, as well as larger mammals such as White-lipped Peccary, Baird's Tapir, puma, and jaguar. Birds, of course, are also plentiful. Over 350 species can be found in the park, including many South Pacific endemics.

Established in 1975, the 100,000-acre park sits on the western edge of the Osa Peninsula. During its early years Corcovado National Park suffered serious threats from gold miners, squatters, and international lumber companies. The Costa Rican government, however, acted quickly and decisively to protect the park. Today, Corcovado is a great success story in conservation, though some illegal activities still occur along its borders.

The Corcovado experience is one of hiking and camping. Visitors can stay at four ranger stations distributed around the reserve, each equipped with a campsite. These stations are connected by a series of long trails, and hiking between stations generally takes about a day. Sirena Station, the largest of the four, also has an airstrip, making it accessible via small aircraft. Unlike Costa Rica's other national parks, advance reservation is required in order to visit Corcovado, so be sure to make reservations with the contact information provided below.

With its incredible biodiversity and numerous endemics, Corcovado might sound like a birding paradise. In reality, however, most birders find Corcovado

National Park to be a logistical hassle, physical nightmare, and a huge drain on time that could be better spent visiting a variety of locations. The two other sites described on the Osa Peninsula, Marenco Beach and Rainforest Lodge and Bosque del Río Tigre, offer similar birding with much easier accessibility. For hikers and backpackers, however, Corcovado is an extremely popular destination. If rugged outdoor adventure is as important to you as birding, then a trip to Corcovado could be an ideal vacation.

Target Birds

Great Curassow (11) 12:3
Brown Booby (15) 1:1
Red-footed Booby (15) 1:2
King Vulture (31) 13:5
Crested Eagle (47) 17:10
Harpy Eagle (47) 17:9
Black-and-white Hawk-Eagle (49) 17:13
Red-throated Caracara (49) 14:11
Scarlet Macaw (99) 19:1
Pheasant Cuckoo (101) 21:6
*Baird's Trogon (143) 26:6
Red-rumped Woodpecker (159) 28:12
Streak-chested Antpitta (185) 30:18
Golden-crowned Spadebill (189) 37:3
Black-tailed Flycatcher (199) 36:22
Thrush-like Schiffornis (215) 33:10
*Turquoise Cotinga (219) 34:5
*Yellow-billed Cotinga (219) 34:3
Gray-headed Tanager (283) 45:17
*White-throated Shrike-Tanager (283) 47:1
*Black-cheeked Ant-Tanager (277) 47:11

Access

Access to Corcovado National Park is not straightforward. First, you *must* have prior reservations with the park office. Be sure to make these well in advance of your trip, especially if you are visiting during the high season. Second, since there are a number of different ranger stations where you can stay, as well as three primary entrances and exits, your visit will require careful planning.

The easiest way to enter Corcovado National Park is to charter a small plane to fly you directly to the Sirena Station. Located in the middle of the park along the shoreline, Sirena is the largest of the Corcovado ranger stations, serving as the central node for the long trails that lead between stations. Sirena offers an extensive network of local trails, a large campsite, and bunkhouse and even provides food service with prior reservation. The only company currently flying to Sirena is Alfa Romeo Air Charters.

If you do not fly, you must hike in from one of three towns that provide access points to the park. The first is a small town called Carate, located about 43 kilometers southwest of Puerto Jiménez along a very bumpy road (4×4 required). Once at Carate, it is a relatively short, 3-kilometer (2-mile) hike along the beach to La Leona Station. This station offers some trails and a campsite, but its facilities are limited in comparison to the Sirena Station. A 16-kilometer (10-mile) trail connects La Leona and Sirena. This path proceeds mostly along the beach, making the walk hot and difficult. It is best to hike around low tide, as high tide

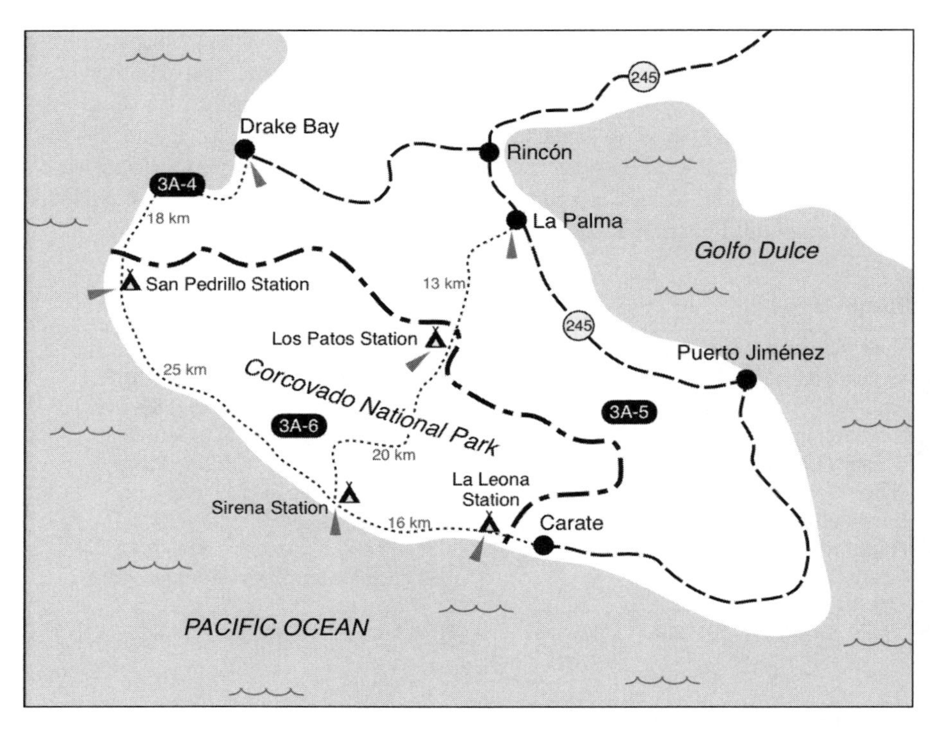

Corcovado National Park Area Map. Marenco Beach and Rainforest Lodge (3A-4) and Bosque del Río Tigre (3A-5) are also shown.

covers the trail in certain locations and makes for some treacherous river crossings owing to crocodiles and sharks.

Alternatively, you can start in the town of La Palma, 24 kilometers northwest of Puerto Jiménez along Rt. 245. From here it is a 13-kilometer (8-mile) hike to Los Patos Station, a small ranger station that offers a campsite and a few bunk beds. Los Patos connects to the Sirena Station via a beautiful 20-kilometer (12.5-mile) trail that leads through inland primary forest.

The third main entrance to Corcovado National Park is from the town of Drake Bay, located on the western side of the Osa Peninsula. Drake Bay is most easily accessible via water taxi from the town of Sierpe. (The road between Rincón and Drake Bay, if even passable, requires a 4×4 vehicle.) From Drake Bay, an 18-kilometer (11-mile) trail leads to the San Pedrillo Station, which offers a bunkroom and camping facility. From San Pedrillo you can access the Sirena Station with a 25-kilometer (16-mile) hike along the coast.

Logistics: Whichever way you enter Corcovado, having a rental car while inside the park is useless. From the town of Puerto Jiménez you can take a taxi or public bus to Carate and La Palma or fly directly into Sirena. While limited bunk-rooms are available at Sirena, San Pedrillo, and Los Patos, most visitors camp, and I recommend bringing all your camping supplies with you. Alternatively, there are a few companies in Puerto Jiménez that rent camping equipment. You must also carry in all the food that you will need while in the park, although there is a food service available at Sirena with prior reservation. Potable water is available at all ranger stations, but having a small water filter while on long hikes can be useful.

Corcovado National Park
Tel: (506) 2735-5580
E-mail: pncorcovado@gmail.com

Alfa Romeo Air Charters
Website: www.alfaromeoair.com

Birding Sites

Interesting birding can be found throughout the park, although the best places to focus your time are around the Sirena and Los Patos stations. The Sirena Station offers the most extensive local trail system, which provides access to all the habitats in the area. Walking the trails that lead inland into the forest could yield Black-cheeked Ant-Tanager, White-throated Shrike-Tanager, Baird's Trogon, and Black-striped Woodcreeper. The tiny Golden-crowned Spadebill inhabits the understory, along with Streak-chested Antpitta, although both are more easily heard than seen. The large forest edge around the station itself also holds potential for interesting birds such as Southern Beardless-Tyrannulet, Yellow-bellied Tyrannulet, Fiery-billed Aracari, and Spot-crowned Euphonia. Fruiting trees in the area could produce Turquoise Cotinga along with a variety of tanager species.

If you walk the trails along the beach, you are sure to see numerous Scarlet Macaws flying overhead and feeding in the trees. Yellow-billed Cotinga is another possibility along the shoreline, especially around mangroves. Be sure to check any rocky outcroppings for Surfbird and Wandering Tattler.

Los Patos Station is the only station not located near the coast. The birding in this interior forest is excellent and well worth exploring. In particular, both this area and the trail that connects Los Patos to Sirena offer chances to find Red-throated Caracara, now a very rare bird in Costa Rica.

Harpy Eagle is Corcovado's most sought-after species, and it is believed that at least a few individuals permanently reside within the park because both juveniles and adults have been reported from various locations. This exciting species is a possibility anywhere within the park, but actually finding it is a real long shot.

Site 3A-7: Esquinas Rainforest Lodge

Birding Time: 1–2 days
Elevation: 100 meters
Trail Difficulty: 1–3
Reserve Hours: N/A
Entrance Fee: Included with overnight
4×4 Recommended
Trail Map Available on Site

Esquinas Rainforest Lodge abuts the little-known Piedras Blancas National Park and is a shining example of how ecotourism can benefit local communities. The lodge was started in 1993 by residents of the nearby town of La Gamba in an effort to derive economic benefit from the national park. At the time, these locals did not have the funds to start the project on their own, and received help from a European-based non-profit organization called Rainforest of the Austrians. The project was a success, offering employment to many local people. Furthermore, in 2005, when the lodge was sold to a private owner, $210,000 was donated to the town of La Gamba to be used for the public good.

The hotel sits along a small stream and is surrounded on three sides by forested hills. The main lodge is a large open-air building with thatched roof, which holds the restaurant, bar, reception, and a nice lounge and gathering area. Guests stay in clean and comfortable cabins that are scattered around the grounds. Other amenities include laundry as well as a small swimming pool.

The birding at Esquinas Lodge is exciting and diverse. The lodge offers a nice trail system through the forest, where one can find many interior-forest species. In addition, new-growth, second-growth, and edge habitats are easily accessible. Finally, one of the most enticing areas for birders at Esquinas is not actually at the lodge, but around the nearby town of La Gamba, where rice patties and other fields attract an interesting array of species.

Target Birds

Marbled Wood-Quail (13) 12:9
Black Hawk-Eagle (49) 13:9
Uniform Crake (57) 6:8
Wattled Jacana (61) 6:17
Ruddy Quail-Dove (93) 18:17
Brown-throated Parakeet (95)
Blue-headed Parrot (97) 19:8
Barn Owl (105) 20:9
Bronzy Hermit (121) 23:4
Band-tailed Barbthroat (121) 23:5
*White-crested Coquette (137) 25:5
*Charming Hummingbird (125) 24:18
*Baird's Trogon (143) 26:6
*Golden-naped Woodpecker (157) 28:18
Red-rumped Woodpecker (159) 28:12
*Rufous-winged Woodpecker (159) 28:10
Pale-breasted Spinetail (163) 32:11
Striped Woodhaunter (165) 30:4
Long-tailed Woodcreeper (171) 29:10
Yellow-crowned Tyrannulet (191) 37:15
Lesser Elaenia (193) 37:25
Black-tailed Flycatcher (199) 36:22

Rusty-margined Flycatcher (329)
Fork-tailed Flycatcher (207) 35:4
Scrub Greenlet (229) 40:8
Green Shrike-Vireo (229) 40:1
*Black-bellied Wren (243) 38:11
*White-throated Shrike-Tanager (283) 47:1
*Black-cheeked Ant-Tanager (277) 47:11
Yellow-bellied Seedeater (295) 49:4
Ruddy-breasted Seedeater (297) 49:1
Red-breasted Blackbird (315) 44:13
*Spot-crowned Euphonia (327) 45:2

Access

From the intersection of Rt. 2 and Rt. 245: Take Rt. 2 south for 14.2 km and turn right onto a dirt road heading to La Gamba. (**From Río Claro:** Take Rt. 2 north for 13.0 km and then turn left onto a dirt road heading to La Gamba.) Go 3.6 km, through La Gamba center, and turn right. The road immediately crosses a bridge and comes to a T-intersection, where you should turn left. In 0.2 km turn right, and this road will end at Esquinas Rainforest Lodge in 1.6 km. (Driving time from Rt. 245: 30 min; from Río Claro: 30 min)

Logistics: To bird this area it is best to stay at Esquinas Rainforest Lodge, which serves three meals daily.

Esquinas Rainforest Lodge
Tel: (506) 2741-8001
Website: www.esquinaslodge.com
E-mail: esquinas@racsa.co.cr

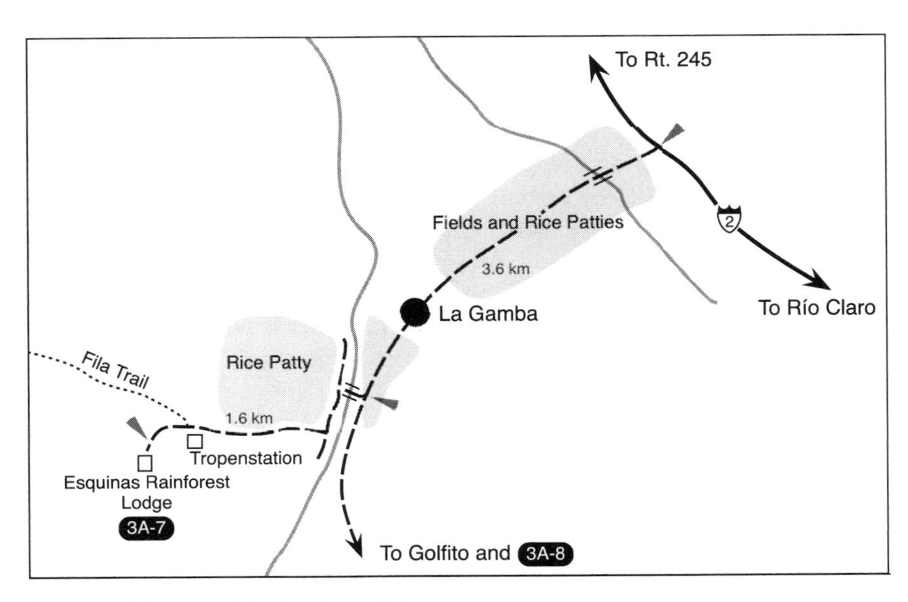

Esquinas Rainforest Lodge Area Map

Birding Sites

Esquinas Rainforest Lodge

Esquinas Lodge maintains six trails that lead through the nearby forest. The short Bird Trail, which leaves from behind the main lodge, generally lives up to its name and is a relatively easy walk over flat terrain. The first half leads through young forest, where you should be able to find Gray-chested Dove, Rufous-tailed Jacamar, Orange-billed Sparrow, Riverside Wren, and Dot-winged Antwren moving through the understory. Higher up in the trees, you might find Purple-crowned Fairy, White-shouldered Tanager, Green Honeycreeper, and Scarlet-rumped Cacique. The second half of the trail winds through a former cocoa plantation and is a good place to see Buff-rumped Warbler and Blue-crowned Motmot.

The most consistently productive trail to walk at Esquinas is the first section of Fila Trail. Start by walking out the driveway, almost to the Tropenstation (Austrian biological station), to where the Fila Trail begins on your left. The first few hundred meters lead through new growth, where you can often find Black-bellied Wren. Soon after this, the trail proceeds up into old secondary forest and along a ridgeline. Mixed flocks composed of Black-hooded Antshrike, Dot-winged Antwren, and Tawny-crowned Greenlet are common, and you should not be surprised to find Rufous Mourner, Rufous Piha, Striped Woodhaunter, Eye-ringed Flatbill, and Sulphur-rumped Flycatcher feeding with them. Other regulars along the trail include White-throated Shrike-Tanager, Yellow-billed Cacique, Black-cheeked Ant-Tanager, and Baird's Trogon. You can loop back to the lodge by taking the Ocelot Trail, or if you want a longer walk, continue to La Trocha Trail, which will also bring you back to the lodge.

Riverbed Trail is located below Fila Trail and proceeds along a stream at the base of a small valley. While this trail offers a shorter and flatter walk than the Fila Trail, most of the vegetation is above your head, so viewing birds is not quite as easy. However, the birding here can be excellent. In particular, the beginning of this trail, which passes by a small pond, is a good place to look for Uniform Crake.

As you walk the driveway out of Esquinas Lodge, you will pass gardened grounds on your right. Scaly-breasted Hummingbird and Cherrie's Tanager are abundant in the area, although you should also keep an eye out for Spot-crowned Euphonia, Thick-billed Euphonia, Charming Hummingbird, Fiery-billed Aracari, Yellow-bellied Tyrannulet, and Yellow-crowned Tyrannulet. Farther down the road, near the Tropenstation, is a good place to find Orange-collared Manakin and Gray-headed Chachalaca.

La Gamba

The fields and rice patties around the small town of La Gamba offer some of the most interesting birding in the area, providing a good opportunity to find a few species with very limited ranges in Costa Rica. As you leave Esquinas Rainforest Lodge you will come to a large rice patty on your left in about 0.8 km. Blue Ground-Dove is abundant here, along with Ruddy Ground-Dove and White-tipped Dove. The trees around the fields often hold Scrub Greenlet, and the patties themselves can yield a variety of sparrow species, including the uncommon Ruddy-breasted Seedeater and Yellow-bellied Seedeater. Wattled Jacana is also occasionally found in the area along with the more numerous Northern Jacana. At first light, it is not uncommon to see Barn Owl hunting over the rice.

Other rice patties and fields found along the road between La Gamba and Rt. 2 yield many of the same species and are worth birding thoroughly. These fields also produce Fork-tailed Flycatcher, Lesser Elaenia, Pale-breasted Spinetail, and Red-breasted Blackbird. This is the best area to see Brown-throated Parakeet, a recent invasive from Panama with a stable population at La Gamba.

Another Panamanian invasive that can be found nearby is Rusty-margined Flycatcher. To get to the area where this bird has been seen, start at Rt. 2 and head toward La Gamba for 0.4 km to a small bridge. Continue 0.2 km farther to a palm plantation on your right. The birds are usually seen around here. It is currently unclear whether this species will continue to expand or will retreat back to Panama.

Species to Expect

Great Tinamou (3) 12:6
Little Tinamou (3) 12:17
Black-bellied Whistling-Duck (5) 8:7
Gray-headed Chachalaca (11) 12:1
Crested Guan (11) 12:4
Least Grebe (9) 7:4
Great Egret (23) 5:14
Little Blue Heron (23) 5:9
Cattle Egret (23) 5:13
Green Heron (25) 6:2
White Ibis (29) 4:8
Black Vulture (31) 13:4
Turkey Vulture (31) 13:3
King Vulture (31) 13:5
Swallow-tailed Kite (37) 15:2
White-tailed Kite (37) 15:1
Roadside Hawk (43) 16:12
Broad-winged Hawk (43) 16:13
Black Hawk-Eagle (49) 13:9
Yellow-headed Caracara (53) 15:9
Laughing Falcon (53) 15:8
Peregrine Falcon (51) 15:5
White-throated Crake (57) 6:9
Gray-necked Wood-Rail (55) 6:13
Uniform Crake (57) 6:8
Purple Gallinule (59) 6:15
Northern Jacana (61) 6:18
Pale-vented Pigeon (87) 18:3
Short-billed Pigeon (87) 18:5
Ruddy Ground-Dove (89) 18:7
Blue Ground-Dove (91) 18:6
White-tipped Dove (91) 18:14
Gray-chested Dove (91) 18:16
Ruddy Quail-Dove (93) 18:17
Brown-throated Parakeet (95)
Orange-chinned Parakeet (97) 19:14
Blue-headed Parrot (97) 19:8
White-crowned Parrot (97) 19:7
Red-lored Parrot (99) 19:4
Squirrel Cuckoo (103) 21:7
Smooth-billed Ani (103) 21:8
Barn Owl (105) 20:9
Common Pauraque (111) 21:18
White-collared Swift (117) 22:1
***Costa Rican Swift (119) 22:11**
Bronzy Hermit (121) 23:4
Band-tailed Barbthroat (121) 23:5
Long-billed Hermit (121) 23:2
Stripe-throated Hermit (121) 23:1

Scaly-breasted Hummingbird (125) 23:10
White-necked Jacobin (125) 23:17
Violet-crowned Woodnymph (127) 24:4
*Charming Hummingbird (125) 24:18
Rufous-tailed Hummingbird (129) 24:10
Purple-crowned Fairy (125) 23:14
***Baird's Trogon (143) 26:6**
Violaceous Trogon (141) 26:8
Black-throated Trogon (141) 26:9
Slaty-tailed Trogon (145) 26:2
Blue-crowned Motmot (147) 27:8
Ringed Kingfisher (149) 27:1
Amazon Kingfisher (149) 27:4
Rufous-tailed Jacamar (153) 26:12
*Fiery-billed Aracari (155) 27:16
Chestnut-mandibled Toucan (155) 27:19
*Golden-naped Woodpecker (157) 28:18
Red-crowned Woodpecker (157) 28:17
*Rufous-winged Woodpecker (159) 28:10
Lineated Woodpecker (161) 27:13
Pale-breasted Spinetail (163) 32:11
Striped Woodhaunter (165) 30:4
Buff-throated Foliage-gleaner (165) 30:3
Plain Xenops (169) 29:3
Wedge-billed Woodcreeper (171) 29:6
Cocoa Woodcreeper (173) 29:17
Black-striped Woodcreeper (173) 29:16
Streak-headed Woodcreeper (173) 29:8
***Black-hooded Antshrike (177) 31:6**
Dot-winged Antwren (183) 32:1
Chestnut-backed Antbird (179) 31:11
Black-faced Antthrush (187) 30:15
Yellow-bellied Tyrannulet (189) 37:14
Yellow Tyrannulet (189) 37:11
Yellow-crowned Tyrannulet (191) 37:15
Yellow-bellied Elaenia (193) 37:26
Lesser Elaenia (193) 37:25
Ochre-bellied Flycatcher (199) 36:25
Paltry Tyrannulet (191) 37:10
Scale-crested Pygmy-Tyrant (197) 37:4
Northern Bentbill (197) 37:6
Slate-headed Tody-Flycatcher (197) 37:9
Common Tody-Flycatcher (197) 37:7
Eye-ringed Flatbill (201) 37:23
Yellow-olive Flycatcher (195) 37:16
Sulphur-rumped Flycatcher (199) 36:21
Bright-rumped Attila (201) 35:6
Rufous Mourner (201) 34:9
Dusky-capped Flycatcher (209) 35:21
Great Crested Flycatcher (209) 35:17
Great Kiskadee (211) 35:13
Boat-billed Flycatcher (211) 35:12
Rusty-margined Flycatcher (329)
Social Flycatcher (211) 35:14
Gray-capped Flycatcher (211) 35:15
Sulphur-bellied Flycatcher (213) 35:10
Piratic Flycatcher (193) 35:8
Tropical Kingbird (213) 35:1
Fork-tailed Flycatcher (207) 35:4
Rufous Piha (215) 34:10
White-winged Becard (217) 33:13
Rose-throated Becard (217) 33:12
Masked Tityra (217) 34:1
Black-crowned Tityra (217) 34:2
***Orange-collared Manakin (221) 33:2**
Blue-crowned Manakin (223) 33:7
Red-capped Manakin (223) 33:6
Red-eyed Vireo (227) 40:3
Yellow-green Vireo (227) 40:5
Scrub Greenlet (229) 40:8
Tawny-crowned Greenlet (229) 32:4
Lesser Greenlet (229) 40:7
Green Shrike-Vireo (229) 40:1
Gray-breasted Martin (235) 22:15
Mangrove Swallow (233) 22:23
Southern Rough-winged Swallow (235) 22:19
*Black-bellied Wren (243) 38:11
***Riverside Wren (241) 38:10**
Plain Wren (241) 38:17
House Wren (243) 38:18
Scaly-breasted Wren (245) 38:20
Long-billed Gnatwren (237) 32:15
Tropical Gnatcatcher (237) 41:2
Clay-colored Thrush (251) 39:8
Tennessee Warbler (255) 40:22
Yellow Warbler (259) 42:2
Chestnut-sided Warbler (259) 43:4
Northern Waterthrush (269) 43:14
Mourning Warbler (271) 42:14
Buff-rumped Warbler (271) 40:23
Bananaquit (277) 40:24
Gray-headed Tanager (283) 45:17
*White-throated Shrike-Tanager (283) 47:1
White-shouldered Tanager (281) 46:18
*Black-cheeked Ant-Tanager (277) 47:11

Summer Tanager (285) 47:5
***Cherrie's Tanager (287) 47:4**
Blue-gray Tanager (291) 46:15
Palm Tanager (291) 45:19
Bay-headed Tanager (289) 46:9
Golden-hooded Tanager (289) 46:13
Blue Dacnis (291) 46:3
Green Honeycreeper (293) 46:7
Shining Honeycreeper (293) 46:1
Red-legged Honeycreeper (293) 46:2
Blue-black Grassquit (297) 49:7
Variable Seedeater (295) 49:3
White-collared Seedeater (295) 49:2
Yellow-bellied Seedeater (295) 49:4
Ruddy-breasted Seedeater (297) 49:1
Thick-billed Seed-Finch (295) 49:10
Orange-billed Sparrow (303) 48:17
Black-striped Sparrow (303) 50:14
Buff-throated Saltator (309) 48:2
Blue-black Grosbeak (311) 48:10
Red-breasted Blackbird (315) 44:13
Eastern Meadowlark (315) 50:16
Great-tailed Grackle (317) 44:16
Bronzed Cowbird (317) 44:15
Baltimore Oriole (321) 44:7
Yellow-billed Cacique (321) 44:12
Scarlet-rumped Cacique (321) 44:11
*Yellow-crowned Euphonia (327) 45:1
Thick-billed Euphonia (327) 45:8
***Spot-crowned Euphonia (327) 45:2**

Site 3A-8: Golfito Wildlife Refuge

Birding Time: 5 hours
Elevation: 400 meters
Trail Difficulty: 2 (or by car)
Reserve Hours: N/A
Entrance Fee: N/A

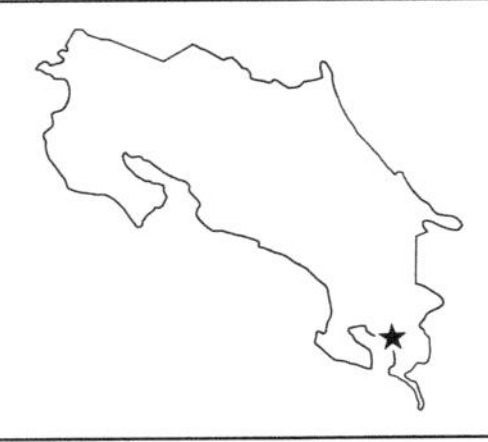

The coastal city of Golfito is one of the largest cities in the South Pacific region of Costa Rica. It sits along the calm waters of the Golfo Dulce and supports a large fishing community. Golfito is home to the only tax-free zone in Costa Rica, and Ticos travel here from around the country to buy imports duty-free, but the prices are nothing special for foreigners.

The Golfito Wildlife Refuge surrounds the city of Golfito and protects 7,400 acres of land. The reserve was created in 1985 with the principal purpose of protecting the steep mountainous slopes that abut the city. While there are not any public trails in the reserve, you can gain access to the area via a public road that leads up into the hills behind the city to a set of radio towers. The road takes you through a nice combination of forest and semi-cleared land where you can find many of the South Pacific endemic species. Even the Black-cheeked Ant-Tanager is often found along this roadside.

Access

From Río Claro: Turn off Rt. 2 onto Rt. 14 toward Golfito and proceed for 18.6 km. At this point, the road will begin to parallel the shoreline. Go 0.6 km farther, just past a small soccer field, and then turn right onto a dirt road. This road proceeds past a few houses and then quickly enters forest, leading all the way to the Golfito radio towers. (Driving time from Río Claro: 20 min)

Logistics: The city of Golfito offers many cheap cabins and a few mid-level hotels as well. See a travel guide for details.

Birding Sites

The road to the radio towers is about seven kilometers long and quite steep in places. You can explore the area either by parking at the bottom and hiking up the entire road, a good four- to five-hour roundtrip walk, or by parking periodically along the road to bird.

The first two kilometers offers some of the best birding as the road winds its way up the steep forested hillside. This forest is older than the second-growth forest found above, and offers a good opportunity to find Black-throated Trogon, Blue-crowned Motmot, Rufous-tailed Jacamar, Black-striped Woodcreeper, Black-faced Antthrush, and Scaly-breasted Wren. Because of the steep slope, the canopies of the trees on the downhill side are often nearly at eye level. This makes canopy flocks unusually easy to observe, and they often include White-shouldered Tanager, Golden-hooded Tanager, Blue Dacnis, Shining Honeycreeper, and Tropical Gnatcatcher.

As the road reaches the top of the ridgeline, it flattens out, and the influences of man on the surrounding habitat become obvious. There are a few houses surrounded by cleared land along the road, but you will find large patches of second-growth forest in between them. Here you should look for Golden-naped Woodpecker, Gray-headed Tanager, Black-cheeked Ant-Tanager, Northern Bentbill, and Black-hooded Antshrike. Small flocks of Chestnut-mandibled Toucan and Fiery-billed Aracari frequently move through the treetops, and it is not uncommon to spot a circling King Vulture or White Hawk.

Subregion 3B

The South Pacific Middle-Elevations

During the breeding season, male **Lance-tailed Manakins** form dancing pairs and perform intricate displays in order to impress potential female mates (as does its close relative, the Long-tailed Manakin). The Lance-tailed Manakin has a very limited range within Costa Rica, occupying middle-elevation forests at the upper end of the Coto Brus Valley. It should be looked for at Río Negro (3B-5) and La Amistad Lodge (3B-6).

Common Birds to Know

Black Vulture (31) 13:4
Turkey Vulture (31) 13:3
White-tipped Dove (91) 18:14
Rufous-tailed Hummingbird (129) 24:10
Red-crowned Woodpecker (157) 28:17
Streak-headed Woodcreeper (173) 29:8
Paltry Tyrannulet (191) 37:10
Scale-crested Pygmy-Tyrant (197) 37:4
Boat-billed Flycatcher (211) 35:12
Social Flycatcher (211) 35:14
Tropical Kingbird (213) 35:1
Yellow-throated Vireo (225) 40:6
Lesser Greenlet (229) 40:7
Rufous-breasted Wren (239) 38:8
House Wren (243) 38:18
Orange-billed Nightingale-Thrush (247) 38:26
Clay-colored Thrush (251) 39:8
Chestnut-sided Warbler (259) 43:4
*Cherrie's Tanager (287) 47:4
Blue-gray Tanager (291) 46:15
Golden-hooded Tanager (289) 46:13
Buff-throated Saltator (309) 48:2

3B-1: Talari Mountain Lodge

Birding Time: 1 day
Elevation: 800 meters
Trail Difficulty: 1
Reserve Hours: N/A
Entrance Fee: Included with overnight

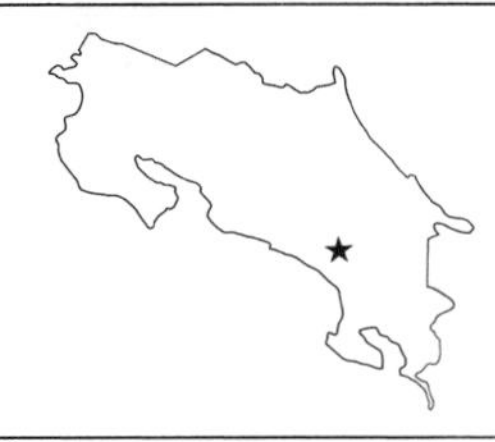

Talari Mountain Lodge, a small, 20-acre property located just 15 minutes from the city of San Isidro, is an unlikely birding hotspot. Yet it is extremely popular among birders, and if you choose to visit the area, you will be pleasantly surprised. Most of the habitat is gardens or young second-growth forest, but the property was planted with a wide variety of fruiting trees, which attract a wealth of birds. While other sites with older and larger forests have more potential species, the density of birds found around Talari, as well as the easy viewing conditions, make this lodge a fun place to go birding. It is an ideal destination for beginning and intermediate neotropical birders.

The lodge itself is not very big, with only 10 guest rooms, each with hot water and simple, yet comfortable furnishings. A small restaurant serves three meals a day (breakfast is included in your overnight), and a swimming pool is available. Talari's most important asset, however, is its location. Centrally positioned between San José and popular birding sites to the south, Talari is an ideal stopover point when traveling between these two areas.

Though the birding at Talari Mountain Lodge is excellent, it is still a small area. If you plan to spend more than five hours birding the vicinity, I suggest exploring some of the locations described in "Nearby Birding Opportunities," especially Los Cusingos. Not only does this location hold a number of interesting species not found at Talari, it was the home of the late Dr. Alexander Skutch,

Costa Rica's most famous ornithologist. His house is now a small museum where you can learn about the life of this interesting and influential man.

Target Birds

Pearl Kite (37) 15:12
Scaled Pigeon (87) 18:4
Tropical Screech-Owl (107) 20:15
Common Potoo (115) 20:4
*White-crested Coquette (137) 25:5
*Garden Emerald (127) 24:13
*Snowy-bellied Hummingbird (133) 24:3
Long-billed Starthroat (135) 23:18
*Fiery-billed Aracari (155) 27:16
Olivaceous Piculet (159) 29:1
Pale-breasted Spinetail (163) 32:11
Southern Beardless-Tyrannulet (191) 37:20
Yellow-crowned Tyrannulet (191) 37:15
Lesser Elaenia (193) 37:25
Bran-colored Flycatcher (199) 36:17
Fork-tailed Flycatcher (207) 35:4
*Turquoise Cotinga (219) 34:5
*Orange-collared Manakin (221) 33:2
Scrub Greenlet (229) 40:8
Speckled Tanager (289) 46:8
Stripe-headed Brush-Finch (301) 48:14
Streaked Saltator (309) 48:4

Access

From San Isidro: Turn your odometer to zero at the McDonald's along Rt. 2 in San Isidro Center. From this point proceed 1.2 km south on Rt. 2 and then fork left off the highway onto a road that goes up a small hill. At the top of this hill turn left, following the main road, and continue for 5.4 km. The driveway to Talari Mountain Lodge will be on your right. (Driving time from San Isidro: 15 min)

Logistics: Lodging on site is by far your most convenient option. Food is available on site or in the nearby towns.

Talari Mountain Lodge
Tel: (506) 2771-0341
Website: www.talari.co.cr
E-mail: talaricostarica@gmail.com

Birding Sites

It is best to start your birding at Talari by exploring the gardens and grounds around the restaurant, rooms, and swimming pool. This area has an abundance of fruiting trees and shrubs, and as a result, an abundance of birds. Look for Fiery-billed Aracari, Scaled Pigeon, Streaked Saltator, and White-winged Becard. Flowers attract Long-billed Starthroat, Snowy-bellied Hummingbird, and Garden Emerald among others. This is also one of the most reliable places in Costa Rica to find Turquoise Cotinga. The hotel puts out bananas and other fruit at a number of feeding stations throughout the grounds. These attract many tanager species, including Cherrie's Tanager, Blue Dacnis, and Speckled Tanager. Blue-crowned Motmot is also a regular at the feeders. Walking along the Río General, which passes behind the lodge, often yields Amazon Kingfisher, Ringed Kingfisher, Buff-rumped Warbler, and the aptly named Riverside Wren.

Scan the cow pastures on both sides of the Talari property. These scrubby fields often hold some interesting species such as Bran-colored Flycatcher, Fork-tailed Flycatcher, Indigo Bunting, and Pearl Kite.

The hillside in front of the Talari lodge is forested with young second-growth trees. A series of paths winding through this area offer a good chance to find species different from those seen in the gardens. The Stripe-headed Brush-Finch is an uncommon species that can sometimes be found feeding on the ground. Also, look for Ovenbird, Orange-billed Nightingale-Thrush, and Orange-billed Sparrow in the lower levels of the forest. Gray-headed Tanager, Orange-collared Manakin, and American Restart are other possibilities.

Species to Expect

Gray-headed Chachalaca (11) 12:1
Great Egret (23) 5:14
Little Blue Heron (23) 5:9
Cattle Egret (23) 5:13
Black Vulture (31) 13:4
Turkey Vulture (31) 13:3
Swallow-tailed Kite (37) 15:2
Roadside Hawk (43) 16:12
Broad-winged Hawk (43) 16:13
Short-tailed Hawk (43) 16:11
Zone-tailed Hawk (45) 14:2
Yellow-headed Caracara (53) 15:9
Laughing Falcon (53) 15:8
Bat Falcon (51) 15:14
White-throated Crake (57) 6:9
Gray-necked Wood-Rail (55) 6:13
Spotted Sandpiper (73) 11:8
Scaled Pigeon (87) 18:4
Ruddy Ground-Dove (89) 18:7
White-tipped Dove (91) 18:14
Orange-chinned Parakeet (97) 19:14
Brown-hooded Parrot (97) 19:9
White-crowned Parrot (97) 19:7
Squirrel Cuckoo (103) 21:7
Striped Cuckoo (101) 21:5
Smooth-billed Ani (103) 21:8
Common Pauraque (111) 21:18
Chestnut-collared Swift (117) 22:3
White-collared Swift (117) 22:1
***Costa Rican Swift (119) 22:11**
Stripe-throated Hermit (121) 23:1
Scaly-breasted Hummingbird (125) 23:10
*Garden Emerald (127) 24:13
Blue-throated Goldentail (129) 24:9
*Snowy-bellied Hummingbird (133) 24:3
Rufous-tailed Hummingbird (129) 24:10
Long-billed Starthroat (135) 23:18
Blue-crowned Motmot (147) 27:8
Amazon Kingfisher (149) 27:4
***Fiery-billed Aracari (155) 27:16**
Olivaceous Piculet (159) 29:1
Red-crowned Woodpecker (157) 28:17
Lineated Woodpecker (161) 27:13
Wedge-billed Woodcreeper (171) 29:6
Cocoa Woodcreeper (173) 29:17
Streak-headed Woodcreeper (173) 29:8
Yellow Tyrannulet (189) 37:11
Yellow-crowned Tyrannulet (191) 37:15
Yellow-bellied Elaenia (193) 37:26
Mountain Elaenia (193) 37:24
Ochre-bellied Flycatcher (199) 36:25
Paltry Tyrannulet (191) 37:10
Common Tody-Flycatcher (197) 37:7
Yellow-olive Flycatcher (195) 37:16
Tropical Pewee (203) 36:9

Yellow-bellied Flycatcher (205) 36:20
Dusky-capped Flycatcher (209) 35:21
Great Crested Flycatcher (209) 35:17
Great Kiskadee (211) 35:13
Boat-billed Flycatcher (211) 35:12
Social Flycatcher (211) 35:14
Gray-capped Flycatcher (211) 35:15
Streaked Flycatcher (213) 35:11
Piratic Flycatcher (193) 35:8
Tropical Kingbird (213) 35:1
Fork-tailed Flycatcher (207) 35:4
White-winged Becard (217) 33:13
Masked Tityra (217) 34:1
*Turquoise Cotinga (219) 34:5
***Orange-collared Manakin (221) 33:2**
White-ruffed Manakin (223) 33:9
Yellow-throated Vireo (225) 40:6
Philadelphia Vireo (227) 40:15
Yellow-green Vireo (227) 40:5
Lesser Greenlet (229) 40:7
Brown Jay (231) 39:19
Blue-and-white Swallow (233) 22:20
***Riverside Wren (241) 38:10**
Rufous-breasted Wren (239) 38:8
Plain Wren (241) 38:17
House Wren (243) 38:18
Tropical Gnatcatcher (237) 41:2
Orange-billed Nightingale-Thrush (247) 38:26
Swainson's Thrush (249) 39:2
Wood Thrush (249) 39:4
Clay-colored Thrush (251) 39:8
White-throated Thrush (251) 39:9
Golden-winged Warbler (257) 41:5
Tennessee Warbler (255) 40:22
Yellow Warbler (259) 42:2
Chestnut-sided Warbler (259) 43:4
Black-and-white Warbler (267) 41:13
American Redstart (267) 41:10
Ovenbird (267) 43:12
Mourning Warbler (271) 42:14
Rufous-capped Warbler (275) 40:16
Buff-rumped Warbler (271) 40:23
Bananaquit (277) 40:24
Gray-headed Tanager (283) 45:17
Summer Tanager (285) 47:5
***Cherrie's Tanager (287) 47:4**
Blue-gray Tanager (291) 46:15
Palm Tanager (291) 45:19
Silver-throated Tanager (291) 46:6
Speckled Tanager (289) 46:8
Bay-headed Tanager (289) 46:9
Golden-hooded Tanager (289) 46:13
Blue Dacnis (291) 46:3
Green Honeycreeper (293) 46:7
Red-legged Honeycreeper (293) 46:2
Variable Seedeater (295) 49:3
Yellow-faced Grassquit (297) 49:6
Stripe-headed Brush-Finch (301) 48:14
Orange-billed Sparrow (303) 48:17
Black-striped Sparrow (303) 50:14
Rufous-collared Sparrow (307) 50:13
Streaked Saltator (309) 48:4
Buff-throated Saltator (309) 48:2
Rose-breasted Grosbeak (311) 48:8
Indigo Bunting (313) 48:12
Great-tailed Grackle (317) 44:16
Baltimore Oriole (321) 44:7
***Yellow-crowned Euphonia (327) 45:1**

Nearby Birding Opportunities

Los Cusingos

Los Cusingos is the farm of the late Dr. Alexander Skutch, one of Costa Rica's most beloved naturalists. While *A Guide to the Birds of Costa Rica* (coauthored with Gary Stiles) is his most well-known work, Skutch authored nearly 35 books on birds during his life. Skutch passed away in 2004, and today his house is maintained as a small museum in his honor. Here you can see where he lived, browse his book collection, and explore the beautiful property that helped inspire his love of nature throughout much of his life.

One of the most productive areas to bird at Los Cusingos is around the gardened grounds and fruiting trees near Skutch's house. You might find Red-headed Barbet, Long-billed Starthroat, White-crested Coquette, and Yellow-bellied Tyrannulet. At the tops of the large surrounding trees you might be lucky enough to find a Turquoise Cotinga.

Two kilometers of trails leads through the forest behind the house. While this forest is not primary, it is significantly older and larger than what is found at Talari Mountain Lodge, and as a result is home to a larger diversity of birds. Even though it is at the same elevation as Talari, species such as Bicolored Antbird, Tawny-winged Woodcreeper, Rufous Piha, Golden-crowned Spadebill, and Red-capped Manakin are much easier to find here.

The reserve is open from 7 a.m. to 4 p.m. (7 a.m. to noon on Sunday) and costs $10 per person to enter. Expect to spend about four hours birding the area.

To get to Los Cusingos from Talari Lodge, turn right out of the driveway and go 2.2 km to the town of Rivas. As the paved road turns sharply left, you should turn right onto a dirt road. Follow this for 10.4 km (it will become paved midway and pass through the town of General Viejo) and then turn sharply left onto a dirt road, following signs for Quizarrá. In 2.2 km fork right, and 0.8 km beyond this you will arrive at Los Cusingos on your right. (Driving time from Talari: 30 min)

Quebradas Biological Reserve

Only 20 minutes from Talari Mountain Lodge is a small, locally run biological reserve with a nice trail system. The reserve sits at about 1,250 meters, so there are a number of higher-elevation species that you can find here that do not occur at Talari. The terrain in the area is fairly steep, but the trails do access some nice forest where you will have a good chance of finding Tawny-throated Leaftosser, Green Hermit, Violet Sabrewing, Silvery-fronted Tapaculo, Spotted Barbtail, and Silver-throated Tanager. One nice two-hour walk starts on Sendero Quebrada and then connects to Sendero Uña del Tigre, which loops back to the station.

The reserve is open 7 a.m. to 3 p.m. but is closed on Mondays. To get there from Talari Mountain Lodge, turn right out of the driveway, go 0.4 km and then turn left. Proceed 2.0 km to the community of Miravalles, where you should turn left onto a dirt road. Go 2.0 km to a T-intersection, turn right and then immediately left. In 0.6 km you will arrive in the town of Quebradas, where you should turn right onto a paved road that quickly turns to dirt. Follow this road 1.2 km, crossing over a bridge, and then turn right again. In 2.0 km the road will end at the Quebradas Biological Reserve.

San Isidro Sewage Ponds

If you have 15 extra minutes while in San Isidro, a quick stop at the sewage ponds could be productive. The area is best scanned from the roadside because

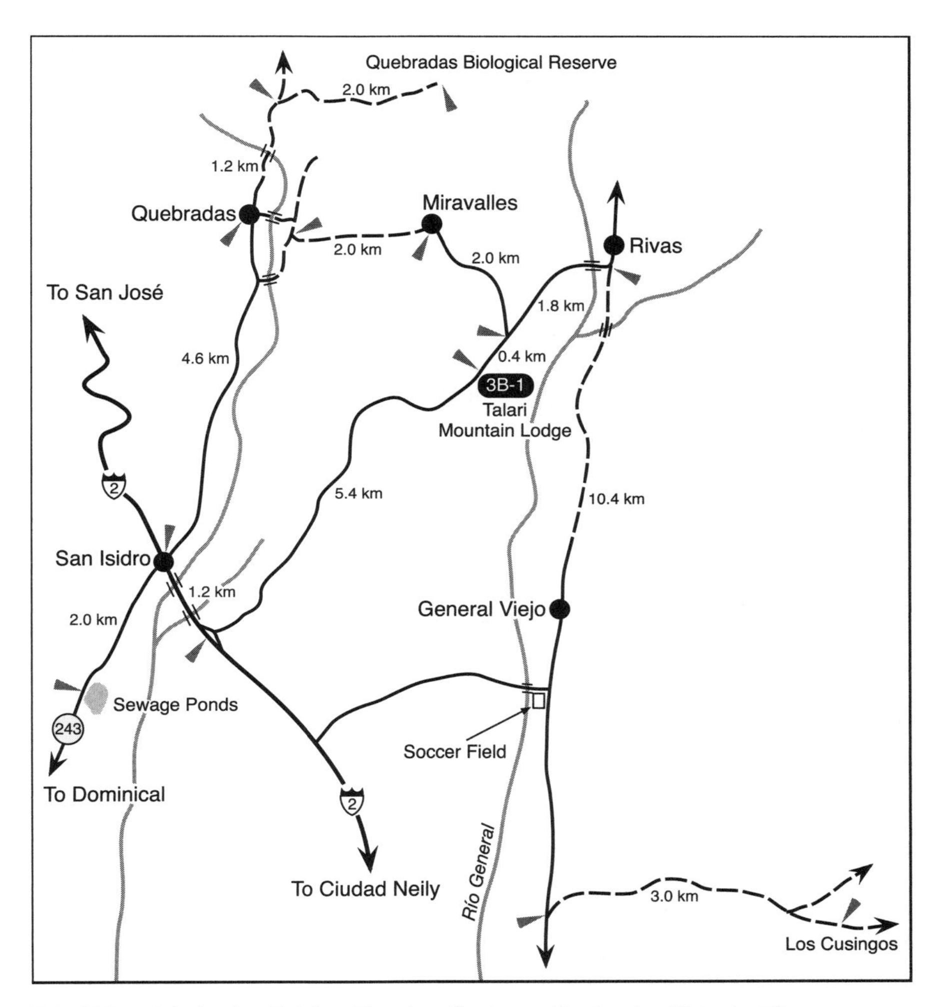

Talari Mountain Lodge Driving Map. Los Cusingos, Quebradas Biological Reserve, and the San Isidro sewage ponds are also included.

the gates are generally locked. The key bird for the area is Southern Lapwing, an invasive species moving into Costa Rica and still uncommon in most places. You will also probably find Black Phoebe, Northern Jacana, and Black-bellied Whistling-Duck.

To get to the sewage ponds, turn off Rt. 2 at the McDonald's in San Isidro and head toward Dominical. Continue for 2.0 km, passing the San Isidro central park and a soccer stadium on your left. The sewage ponds will be on your left.

Birding Time: 2.5 hours
Elevation: 1,000 meters
Trail Difficulty: 1
Reserve Hours: N/A
Entrance Fee: N/A
4×4 Recommended

Dúrika is a small site set in the picturesque foothills of the Valle del General. The principal attractions here are Rosy Thrush-Tanager and Wedge-tailed Grass-Finch, rarities that are very difficult to find elsewhere in the country. Generally, birding in the area is not spectacular, and unless you are specifically in search of these species, I do not suggest including this site in your itinerary. You should also be aware that the dirt road leading to Dúrika is in rough condition and will require a four-wheel-drive vehicle.

Target Birds

Bat Falcon (51) 15:14
Ocellated Crake (57) 6:7
Southern Lapwing (67)
White-tailed Nightjar (113) 21:20
Pale-breasted Spinetail (163) 32:11
Rosy Thrush-Tanager (279) 47:13
Wedge-tailed Grass-Finch (299) 50:10

Access

From Buenos Aires: Turn off Rt. 2 toward Buenos Aires center. Continue 3.8 km, through Buenos Aires, to a T-intersection. Turn right, go one block, and turn left onto a dirt road. After 1.2 km turn left. From here go 1.4 km to a fork and bear right. Follow this dirt road for 12.0 km until you arrive at a fork in the road just before the Dúrika Biological Reserve. (Driving time from Buenos Aires: 55 min)

Logistics: The logistics at this site are difficult. There are a few local options available in and around Buenos Aires. However, you may be more comfortable staying in San Isidro or San Vito, despite the additional driving time. Food is available in Buenos Aires, although I suggest bringing a snack and water with you to Dúrika.

Birding Sites

The first birding spot you will come to is a small marsh in Buenos Aires. The marsh will be on your right 0.6 km after turning off Rt. 2. It is best viewed from the road, although you will want to be careful of the traffic. This is a great place

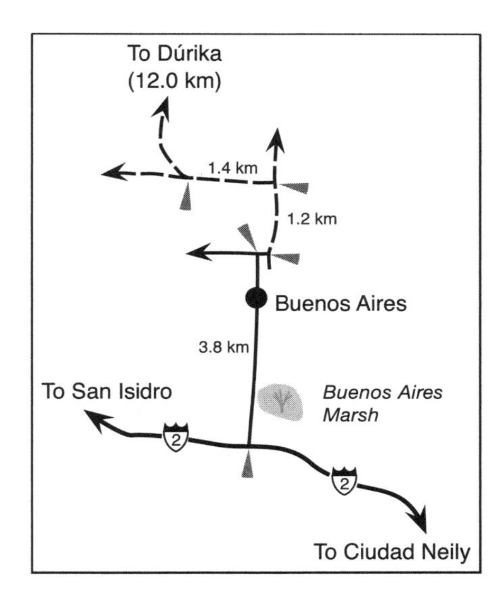

Dúrika Driving Map

to find Southern Lapwing, and you will also encounter Purple Gallinule, Snowy Egret, and Northern Jacana.

Continue according to the directions above, and after bearing right onto the long dirt road that leads to Dúrika, proceed 4.6 km. At this point the road begins to follow a high ridgeline covered in windswept grasses that extends for about 2.5 km. Aside from spectacular views, these grasslands consistently produce the rare Wedge-tailed Grass-Finch, but it can be difficult to spot in this dense habitat. It might also be possible to find Pale-breasted Spinetail and Bat Falcon, or even an Ocellated Crake, a true rarity. If you happen to be in the grasslands at night, White-tailed Nightjar is a regular.

After the grassland ends, proceed for 5.0 km (12.0 km after bearing right) until you come to a fork in the road. Bearing left leads to the Dúrika Biological Station while the right fork leads to Olán. Park your car at the fork and get out to explore. To your immediate left is dense new-growth forest on a hillside. This is the best place to find Rosy Thrush-Tanager, a beautiful bird that frequents the leaf litter under dense tangles. You will have the most luck locating this species by listening for its distinctive song: a rich whistled duet between male and female that calls to mind a Black-bellied Wren. Though you will encounter some other species such as Chestnut-backed Antbird and Rufous-breasted Wren, the birding here is otherwise not particularly noteworthy.

Site 3B-3: Las Cruces Biological Station

Birding Time: 1–2 days
Elevation: 1,150 meters
Trail Difficulty: 2
Reserve Hours: 7 a.m. to 5 p.m.
Entrance Fee: $18 (US 2008) or included with overnight
Bird Guides Available
Trail Map Available on Site

Located just outside the town of San Vito, Las Cruces Biological Station is one of three research stations run by the Organization for Tropical Studies (OTS) in Costa Rica. Las Cruces protects 625 acres of forest, much of which is primary. Most of the surrounding land in the San Vito area has been deforested for coffee production, and thus the Las Cruces forest is a quintessential example of a forest fragment. Much of the research that has gone on at Las Cruces has explored the effects of this fragmentation on plant and animal communities. Forest regeneration is another important area of study at the station. Las Cruces recently reforested an area known as Melissa's Meadow, and has plans to reforest a large corridor that will connect the station with a nearby indigenous reserve.

Attached to the Las Cruces Biological Station is the Wilson Botanical Garden. This garden, which was started in the early 1960s, is internationally recognized as one of the most important tropical botanical gardens in the world. It holds over 1,000 genera of plants, including a great variety of bromeliads, orchids, heliconias, and bamboos, as well as the second largest collection of palms in the world. Any nature lover will enjoy strolling through the beautiful grounds of the Wilson Botanical Garden.

Birders will certainly love spending a day or two exploring Las Cruces. The botanical garden, primary forest, and new-growth plots offer a wide range of accessible habitats. The birding is relatively easy here, as the trails are wide and the forest is not particularly dense. Las Cruces also provides the nicest visitor facilities of the three OTS stations. Feeling more like an eco-lodge than a biological station, Las Cruces offers comfortable and clean rooms as well as excellent food. I recommend spending a day or two birding here, although you may wish to stay at the station longer and use it as a base from which to bird other sites in the area. Las Alturas (3B-4) and Río Negro (3B-5) are both convenient day trips from Las Cruces.

Target Birds

Marbled Wood-Quail (13) 12:9
Barred Forest-Falcon (53) 16:5
Sunbittern (29) 6:16
Scaled Pigeon (87) 18:4
Blue-headed Parrot (97) 19:8
Mottled Owl (107) 20:6
Common Potoo (115) 20:4
*Garden Emerald (127) 24:13

*Charming Hummingbird (125) 24:18
*Snowy-bellied Hummingbird (133) 24:3
*White-tailed Emerald (133) 24:19
Long-billed Starthroat (135) 23:18
*Fiery-billed Aracari (155) 27:16
Pale-breasted Spinetail (163) 32:11
Ruddy Foliage-gleaner (167) 30:5
Brown-billed Scythebill (175) 29:14
Yellow-crowned Tyrannulet (191) 37:15
Greenish Elaenia (195) 37:21
Rough-legged Tyrannulet (191) 37:18
Bran-colored Flycatcher (199) 36:17
*Turquoise Cotinga (219) 34:5
Blue-crowned Manakin (223) 33:7
Masked Yellowthroat (273) 42:12
Speckled Tanager (289) 46:8
Stripe-headed Brush-Finch (301) 48:14
Streaked Saltator (309) 48:4
Crested Oropendola (323)
Thick-billed Euphonia (327) 45:8
Lesser Goldfinch (325) 50:2

Access

From San Vito: Take Rt. 237 toward Ciudad Neily for 5.8 km and Las Cruces Biological Station will be on your right. (Driving time from San Vito: 10 min)

Logistics: I highly recommend staying at the biological station when birding Las Cruces. While there are some other options around the San Vito area, the biological station is comfortable and convenient, with three meals served daily.

Las Cruces Biological Station
Tel: (506) 2524-0628
Website: www.ots.ac.cr
E-mail: nat-hist@ots.ac.cr

Birding Sites

The Wilson Botanical Garden is a nice place to begin birding at Las Cruces. The garden is relatively large, about 24 acres, and has many small trails and paths leading through it. As you explore, look for Long-billed Starthroat, Blue-headed Parrot, Rufous-breasted Wren, Streaked Saltator, and Fiery-billed Aracari. Keep an eye out for Crested Oropendola, a species that has recently invaded the South Pacific region of Costa Rica and can sometimes be seen here.

A few sections of the garden are worthy of special note. The hillside behind the guest cabins often yields Scaly-breasted Hummingbird, while the porterweed flowers in front of the cabins attract Garden Emerald, White-tailed Emerald, and Snowy-bellied Hummingbird. There are also some fruit feeders near the dining area that attract a variety of tanagers and other birds, including Speckled Tanager, Silver-throated Tanager, Cherrie's Tanager, Blue-crowned Motmot, and Red-crowned Woodpecker. You should also be sure to check along the main road in front of Las Cruces where some small coffee fields and new-growth habitat could produce interesting species such as Rose-breasted Grosbeak, Rufous-browed Peppershrike, and Lesser Goldfinch. At night, Vermiculated Screech-Owl, Tropical Screech-Owl, and Mottled Owl can often be heard around the garden.

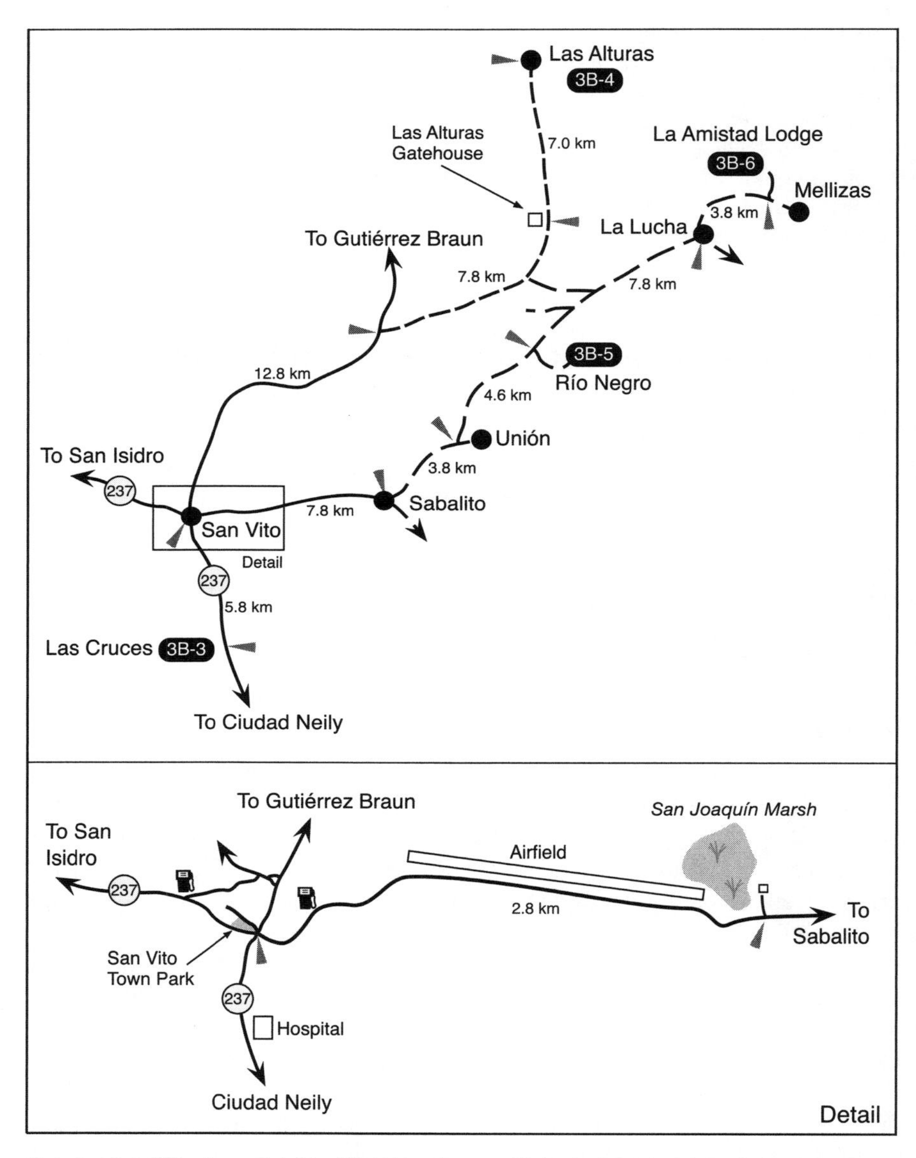

Greater San Vito Area Driving Map. Las Cruces Biological Station (3B-3), Las Alturas (3B-4), Río Negro (3B-5), and La Amistad Lodge (3B-6) are included. The insert shows details around San Vito, including the San Joaquín Marsh.

One area in the garden that is good for forest species is Tree Fern Hill, located near the trailhead for the Río Java Trail. Orange-billed Nightingale-Thrush and Orange-billed Sparrow are commonly seen foraging on the ground here. Also look for Gray-headed Tanager, Blue-crowned Manakin, Red-faced Spinetail, and Buff-throated Foliage-gleaner. (There are two trails in the garden known as the Jungle Trails. These are not especially productive, and your time is better spent elsewhere.)

The 625-acre Las Cruces Forest is accessed through the Río Java Trail, which is the flattest and widest of the six trails in the forest. The birding here can be excellent, and you may not feel the need to wander elsewhere. Mixed flocks usually hold Plain Antvireo, Slaty Antwren, Red-crowned Ant-Tanager, Tawny-crowned Greenlet, and Spotted Woodcreeper. Also, keep an eye open for Ruddy Foliage-gleaner, Brown-billed Scythebill, Sulphur-rumped Flycatcher, and Rose-throated Becard. If you are lucky you could see Marbled Wood-Quail, Black-faced Antthrush, or Scaly-breasted Wren skulking through the understory. At first light, you may hear Barred Forest-Falcon and Collared Forest-Falcon calling from within the forest.

Melissa's Meadow Trail leads off Río Java Trail just before it reaches the Java River, and will take you into new-growth forest that was recently cattle pasture. This area will yield some very different species. Look for Masked Yellowthroat, Pale-breasted Spinetail, Bran-colored Flycatcher, Orange-collared Manakin, and possibly Lesser Elaenia. The other trails that loop through the forest could produce Chiriqui Quail-Dove or Scaled Pigeon, but these steep trails may be more effort than they are worth.

Species to Expect

Great Tinamou (3) 12:6
Little Tinamou (3) 12:17
Crested Guan (11) 12:4
Marbled Wood-Quail (13) 12:9
Cattle Egret (23) 5:13
Black Vulture (31) 13:4
Turkey Vulture (31) 13:3
Swallow-tailed Kite (37) 15:2
Roadside Hawk (43) 16:12
Broad-winged Hawk (43) 16:13
Barred Forest-Falcon (53) 16:5
Yellow-headed Caracara (53) 15:9
Laughing Falcon (53) 15:8
Gray-necked Wood-Rail (55) 6:13
Scaled Pigeon (87) 18:4
Ruddy Pigeon (87) 18:5
Short-billed Pigeon (87) 18:5
Ruddy Ground-Dove (89) 18:7
White-tipped Dove (91) 18:14
Gray-chested Dove (91) 18:16
*Chiriqui Quail-Dove (93) 18:18
***Crimson-fronted Parakeet (95) 19:10**
Orange-chinned Parakeet (97) 19:14
Brown-hooded Parrot (97) 19:9
Blue-headed Parrot (97) 19:8
White-crowned Parrot (97) 19:7
Squirrel Cuckoo (103) 21:7
Smooth-billed Ani (103) 21:8
Mottled Owl (107) 20:6
White-collared Swift (117) 22:1
***Costa Rican Swift (119) 22:11**
Green Hermit (121) 23:3
Stripe-throated Hermit (121) 23:1
Scaly-breasted Hummingbird (125) 23:10
Violet Sabrewing (123) 23:9

*Garden Emerald (127) 24:13
Violet-crowned Woodnymph (127) 24:4
Blue-throated Goldentail (129) 24:9
*Charming Hummingbird (125) 24:18
Rufous-tailed Hummingbird (129) 24:10
*White-tailed Emerald (133) 24:19
Green-crowned Brilliant (123) 23:15
Violaceous Trogon (141) 26:8
Collared Trogon (143) 26:5
Blue-crowned Motmot (147) 27:8
Rufous-tailed Jacamar (153) 26:12
Emerald Toucanet (153) 27:17
***Fiery-billed Aracari (155) 27:16**
Chestnut-mandibled Toucan (155) 27:19
Red-crowned Woodpecker (157) 28:17
Smoky-brown Woodpecker (159) 28:13
Golden-olive Woodpecker (159) 28:11
Lineated Woodpecker (161) 27:13
Pale-billed Woodpecker (161) 27:14
Red-faced Spinetail (163) 30:12
Lineated Foliage-gleaner (165) 30:8
Buff-throated Foliage-gleaner (165) 30:3
Ruddy Foliage-gleaner (167) 30:5
Plain Xenops (169) 29:3
Tawny-winged Woodcreeper (171) 29:12
Olivaceous Woodcreeper (171) 29:7
Wedge-billed Woodcreeper (171) 29:6
Spotted Woodcreeper (173) 29:20
Streak-headed Woodcreeper (173) 29:8
Brown-billed Scythebill (175) 29:14
***Black-hooded Antshrike (177) 31:6**
Plain Antvireo (181) 32:6
Slaty Antwren (183) 32:3
Black-faced Antthrush (187) 30:15
Ochre-breasted Antpitta (185) 30:13
Yellow Tyrannulet (189) 37:11
Yellow-bellied Elaenia (193) 37:26
Mountain Elaenia (193) 37:24
Ochre-bellied Flycatcher (199) 36:25
Slaty-capped Flycatcher (195) 36:26
Paltry Tyrannulet (191) 37:10
Scale-crested Pygmy-Tyrant (197) 37:4
Common Tody-Flycatcher (197) 37:7
Yellow-olive Flycatcher (195) 37:16
White-throated Spadebill (189) 37:1
Sulphur-rumped Flycatcher (199) 36:21
Yellow-bellied Flycatcher (205) 36:20
Bright-rumped Attila (201) 35:6
Dusky-capped Flycatcher (209) 35:21
Great Crested Flycatcher (209) 35:17
Boat-billed Flycatcher (211) 35:12
Social Flycatcher (211) 35:14
Gray-capped Flycatcher (211) 35:15
Piratic Flycatcher (193) 35:8
Tropical Kingbird (213) 35:1
White-winged Becard (217) 33:13
Rose-throated Becard (217) 33:12
Masked Tityra (217) 34:1
*Orange-collared Manakin (221) 33:2
White-ruffed Manakin (223) 33:9
Blue-crowned Manakin (223) 33:7
Yellow-throated Vireo (225) 40:6
Tawny-crowned Greenlet (229) 32:4
Lesser Greenlet (229) 40:7
Rufous-browed Peppershrike (229) 40:2
Blue-and-white Swallow (233) 22:20
***Riverside Wren (241) 38:10**
Rufous-breasted Wren (239) 38:8
Plain Wren (241) 38:17
House Wren (243) 38:18
White-breasted Wood-Wren (245) 38:15
Scaly-breasted Wren (245) 38:20
Long-billed Gnatwren (237) 32:15
Orange-billed Nightingale-Thrush (247) 38:26
Swainson's Thrush (249) 39:2
Clay-colored Thrush (251) 39:8
White-throated Thrush (251) 39:9
Golden-winged Warbler (257) 41:5
Tennessee Warbler (255) 40:22
Tropical Parula (257) 41:3
Chestnut-sided Warbler (259) 43:4
Blackburnian Warbler (261) 41:8
Black-and-white Warbler (267) 41:13
American Redstart (267) 41:10
Ovenbird (267) 43:12
Northern Waterthrush (269) 43:14
Mourning Warbler (271) 42:14
Wilson's Warbler (269) 42:4
Slate-throated Redstart (273) 42:7
Rufous-capped Warbler (275) 40:16
Buff-rumped Warbler (271) 40:23
Bananaquit (277) 40:24
Common Bush-Tanager (279) 45:13
Gray-headed Tanager (283) 45:17

Red-crowned Ant-Tanager (277) 47:10
Summer Tanager (285) 47:5
White-winged Tanager (285) 47:7
***Cherrie's Tanager (287) 47:4**
Blue-gray Tanager (291) 46:15
Palm Tanager (291) 45:19
Silver-throated Tanager (291) 46:6
Speckled Tanager (289) 46:8
Bay-headed Tanager (289) 46:9
Golden-hooded Tanager (289) 46:13
Scarlet-thighed Dacnis (291) 46:4
Green Honeycreeper (293) 46:7
Variable Seedeater (295) 49:3
Yellow-faced Grassquit (297) 49:6
Orange-billed Sparrow (303) 48:17
Black-striped Sparrow (303) 50:14
Rufous-collared Sparrow (307) 50:13
Streaked Saltator (309) 48:4
Buff-throated Saltator (309) 48:2
Rose-breasted Grosbeak (311) 48:8
Blue-black Grosbeak (311) 48:10
Great-tailed Grackle (317) 44:16
Baltimore Oriole (321) 44:7
Thick-billed Euphonia (327) 45:8
***Spot-crowned Euphonia (327) 45:2**
Lesser Goldfinch (325) 50:2

Nearby Birding Opportunities

San Joaquín Marsh

This small wetland, located between San Vito and Sabalito, can be very productive, especially during the dry season when there are lots of migrant ducks. To get to the San Joaquín Marsh from San Vito, take the main road toward Sabalito for 2.8 km. Soon after passing an airstrip on your left, you will be able to see a part of the marsh. Turn left into the very first driveway after the marsh and ask the owners of this small house for permission to access the wetland. The family is accustomed to birders and is happy to receive tips for their hospitality. (About $2 per person is appropriate.)

Regulars in the San Joaquín Marsh include Blue-winged Teal, Least Grebe, Northern Jacana, Purple Gallinule, and Common Moorhen. This may also be the best place in the country to find Masked Duck, and occasionally you will see Ring-necked Duck and Wattled Jacana. Many of the kingfishers and herons are frequently found here. The marsh grasses are a reliable place to find Masked Yellowthroat. Finally, the new-growth scrub and small farm fields along the edge of the marsh often produce interesting birds such as Bran-colored Flycatcher and Yellow-bellied Seedeater.

Centro Turístico Chocoacos

Centro Turístico Chocoacos is located along Rt. 237, 8.0 km south of the intersection of Rt. 237 and Rt. 2 (37.6 km north of San Vito) near the town of Las Vueltas. This location makes a nice rest stop when traveling to or from San Vito along Rt. 237. The main attraction for birders is a small pond that is home to some 30 pairs of Boat-billed Herons. Other water birds, such as Ringed Kingfisher and Bare-throated Tiger-Heron, are also found here. There is a small restaurant on site where you can get a cold drink and some food. A minimal entrance fee is required.

Site 3B-4: Las Alturas

Birding Time: 1 day
Elevation: 1,300 meters
Trail Difficulty: 1–2
Reserve Hours: 5 a.m. to 6 p.m.
Entrance Fee: N/A
4×4 Recommended

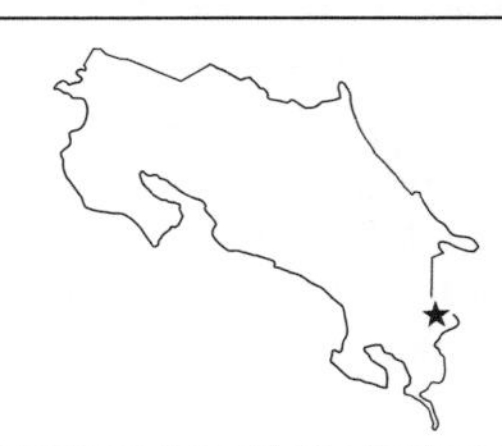

Las Alturas, a large, active cattle ranch, is not a typical birding destination in Costa Rica. About 80 families live and work here, and it is fun to stroll around the little town, admiring the simple ranching lifestyle. You will not find much more than a tiny general store, but the village is clean and proudly maintained. There is also a modest biological station on the ranch that is occasionally visited by students with the OTS, but it is usually closed.

While Las Alturas is a working ranch, the majority of its 24,700 acres has been protected, and the combination of open and forest habitats makes for excellent birding. Also, since Las Alturas borders the vast forest of La Amistad International Park, you have an excellent chance to see large and uncommon raptors, including hawk-eagles. Most of the birding is done from quiet dirt roads that wind through the property. It is efficient to use your car as a mobile base of operations, making this site especially appealing for birders who do not want to do too much walking.

Use caution, however, any time you step off the dirt roads. Chiggers are quite prolific in the grasses at Las Alturas. I highly recommend tucking your pants into your socks, using tall rubber boots, or applying some form of repellant to protect yourself from these annoying little pests.

Target Birds

Bicolored Hawk (39) 16:8
Solitary Eagle (45) 13:8
Black-and-white Hawk-Eagle (49) 17:13
Ornate Hawk-Eagle (49) 17:11
*Sulphur-winged Parakeet (95) 19:12
Pheasant Cuckoo (101) 21:6
*White-crested Coquette (137) 25:5
*Garden Emerald (127) 24:13
*Snowy-bellied Hummingbird (133) 24:3
*White-tailed Emerald (133) 24:19
Long-billed Starthroat (135) 23:18
Scaly-throated Foliage-gleaner (165) 30:2
Buff-fronted Foliage-gleaner (165) 30:6
Ruddy Foliage-gleaner (167) 30:5
Brown-billed Scythebill (175) 29:14
Ochre-breasted Antpitta (185) 30:13
Torrent Tyrannulet (201) 36:1
Bran-colored Flycatcher (199) 36:17
*Turquoise Cotinga (219) 34:5
*Three-wattled Bellbird (221) 34:12
Rufous-browed Peppershrike (229) 40:2
Masked Yellowthroat (273) 42:12
White-winged Tanager (285) 47:7
Blue Seedeater (297) 49:11
Stripe-headed Brush-Finch (301) 48:14
Streaked Saltator (309) 48:4
Lesser Goldfinch (325) 50:2

Access

From San Vito: (Refer to site 3B-3 for a map.) Take the road that leads north out of town toward Gutiérrez Braun for 12.8 km. Fork right onto a dirt road and follow this 7.8 km to the Las Alturas gatehouse. At the gatehouse you must present some form of identification, preferably a passport, and the gatekeeper will write you a pass. When you arrive in the small village of Las Alturas, 7.0 km farther down the road, you must present this pass at the office. Before you leave, you need to pick up your pass and return it to the gatekeeper on your way out. Keep in mind that the logistics of entering Las Alturas have been changing lately, and it may be a good idea to ask at Las Cruces Biological Station (3B-3) for the current procedure. (Driving time from San Vito to the gatehouse: 40 min)

Logistics: While it is possible to stay at Las Alturas Biological Station, the facilities are rundown, and it is difficult to make a reservation. I suggest staying in San Vito, at either Las Cruces Biological Station (3B-3) or another hotel in town, and entering Las Alturas first thing in the morning. While there is a small general store in the village, it is best to bring a boxed lunch.

Birding Sites

Immediately after the gatehouse, the habitat around the road changes to forest. Anywhere along this road can yield interesting species, and I suggest driving slowly on your way to the village and stopping anywhere you hear or see bird activity. Mixed flocks often produce species such as Smoky-brown Woodpecker, White-winged Tanager, Spotted Barbtail, and Slaty-capped Flycatcher. Keep an ear open for the whistles of the Rufous-browed Peppershrike.

You will arrive in the village of Las Alturas 7.0 km beyond the gatehouse. Here you should check the river that runs through town for Black Phoebe and Torrent Tyrannulet. You might also find Streaked Saltator in the vicinity. Just after crossing the river you will come to a series of forks. Bear left at the first fork to arrive at the second. If you bear right at the second fork, you will be on the road to the *tajo* (quarry), described below. Bearing left will put you on a road that loosely parallels the river for about 2.5 km. This is an excellent road to bird, and since it is flat, it offers comfortable walking. A wide range of species are possibilities here, including White-whiskered Puffbird, Olivaceous Woodcreeper, White-naped Brush-Finch, Brown Violetear, and White-crested Coquette. Fruiting trees in the area attract Fiery-billed Aracari as well as Turquoise Cotinga.

If you bear right at the second fork, you will be on the road to the tajo. The first part of this road leads through pastureland, which is a good place to find species such as Sulphur-winged Parakeet, Smooth-billed Ani, and Lesser Goldfinch. You are also likely to find Snowy-bellied Hummingbird and Long-billed Starthroat feeding in the flowering trees.

After following the road to the tajo for 1.6 km, you will find a trail on your right that leads up to the small biological station. The few hundred meters before the station offers expansive vistas of the surrounding countryside, a likely place to spot raptors. Broad-winged Hawk, Short-tailed Hawk, and Double-toothed Kite are a few of the more likely species in the area, and you have an excellent chance to spot something exciting, such as an Ornate Hawk-Eagle, Black-and-white Hawk-Eagle, or Solitary Eagle. You are also likely to encounter Pale-breasted Spinetail in the grassy fields. At the biological station there are some trails that lead into the forest behind. Use caution, however, as these trails are in poor condition and not well marked. Since this is the largest forest in Central America, it is not a place you want to get lost.

Back on the main road to the tajo, continue 0.6 km and then keep left at a small intersection. Soon after this, the road enters forest. This very productive area often yields Spotted Woodcreeper, Golden-crowned Warbler, Russet Antshrike, and Red-faced Spinetail. The very rare Pheasant Cuckoo is often heard singing near the road or by the tajo.

Species to Expect

Great Tinamou (3) 12:6
Little Tinamou (3) 12:17
Cattle Egret (23) 5:13
Black Vulture (31) 13:4
Turkey Vulture (31) 13:3
Swallow-tailed Kite (37) 15:2
Double-toothed Kite (35) 16:1
Roadside Hawk (43) 16:12
Broad-winged Hawk (43) 16:13
Short-tailed Hawk (43) 16:11
Zone-tailed Hawk (45) 14:2
Ornate Hawk-Eagle (49) 17:11
Crested Caracara (53) 14:12
Yellow-headed Caracara (53) 15:9
Laughing Falcon (53) 15:8
Gray-necked Wood-Rail (55) 6:13
Band-tailed Pigeon (87) 18:1
Ruddy Pigeon (87) 18:5
Short-billed Pigeon (87) 18:5
Ruddy Ground-Dove (89) 18:7
White-tipped Dove (91) 18:14
Gray-chested Dove (91) 18:16
***Sulphur-winged Parakeet (95) 19:12**
*Crimson-fronted Parakeet (95) 19:10
Orange-chinned Parakeet (97) 19:14
Brown-hooded Parrot (97) 19:9
Blue-headed Parrot (97) 19:8
White-crowned Parrot (97) 19:7
Squirrel Cuckoo (103) 21:7
Pheasant Cuckoo (101) 21:6
Smooth-billed Ani (103) 21:8
Chestnut-collared Swift (117) 22:3
White-collared Swift (117) 22:1
*Costa Rican Swift (119) 22:11
Green Hermit (121) 23:3
Stripe-throated Hermit (121) 23:1
Scaly-breasted Hummingbird (125) 23:10
Brown Violetear (131) 23:6
*White-crested Coquette (137) 25:5
*Garden Emerald (127) 24:13
*Snowy-bellied Hummingbird (133) 24:3
Rufous-tailed Hummingbird (129) 24:10
*White-tailed Emerald (133) 24:19
Purple-crowned Fairy (125) 23:14
Long-billed Starthroat (135) 23:18
Violaceous Trogon (141) 26:8
Collared Trogon (143) 26:5
Blue-crowned Motmot (147) 27:8
White-whiskered Puffbird (151) 28:6
Red-headed Barbet (153) 28:2
*Fiery-billed Aracari (155) 27:16
Chestnut-mandibled Toucan (155) 27:19
Acorn Woodpecker (157) 28:15
Red-crowned Woodpecker (157) 28:17

Smoky-brown Woodpecker (159) 28:13
Golden-olive Woodpecker (159) 28:11
Lineated Woodpecker (161) 27:13
Pale-billed Woodpecker (161) 27:14
Pale-breasted Spinetail (163) 32:11
Slaty Spinetail (163) 32:9
Red-faced Spinetail (163) 30:12
Spotted Barbtail (163) 29:5
Lineated Foliage-gleaner (165) 30:8
Buff-throated Foliage-gleaner (165) 30:3
Plain Xenops (169) 29:3
Ruddy Woodcreeper (171) 29:13
Olivaceous Woodcreeper (171) 29:7
Spotted Woodcreeper (173) 29:20
Streak-headed Woodcreeper (173) 29:8
Brown-billed Scythebill (175) 29:14
Russet Antshrike (177) 31:4
Slaty Antwren (183) 32:3
Bicolored Antbird (179) 31:9
Black-faced Antthrush (187) 30:15
Yellow Tyrannulet (189) 37:11
Yellow-bellied Elaenia (193) 37:26
Mountain Elaenia (193) 37:24
Torrent Tyrannulet (201) 36:1
Ochre-bellied Flycatcher (199) 36:25
Slaty-capped Flycatcher (195) 36:26
Paltry Tyrannulet (191) 37:10
Scale-crested Pygmy-Tyrant (197) 37:4
Common Tody-Flycatcher (197) 37:7
Eye-ringed Flatbill (201) 37:23
Yellow-olive Flycatcher (195) 37:16
Sulphur-rumped Flycatcher (199) 36:21
Olive-sided Flycatcher (203) 36:4
Western Wood-Pewee (203) 36:7
Yellow-bellied Flycatcher (205) 36:20
Black Phoebe (207) 36:5
Bright-rumped Attila (201) 35:6
Dusky-capped Flycatcher (209) 35:21
Great Kiskadee (211) 35:13
Boat-billed Flycatcher (211) 35:12
Social Flycatcher (211) 35:14
Streaked Flycatcher (213) 35:11
Piratic Flycatcher (193) 35:8
Tropical Kingbird (213) 35:1
Rose-throated Becard (217) 33:12
Masked Tityra (217) 34:1
*Turquoise Cotinga (219) 34:5
White-ruffed Manakin (223) 33:9
Blue-crowned Manakin (223) 33:7
Yellow-throated Vireo (225) 40:6
Philadelphia Vireo (227) 40:15
Tawny-crowned Greenlet (229) 32:4
Lesser Greenlet (229) 40:7
Rufous-browed Peppershrike (229) 40:2
Blue-and-white Swallow (233) 22:20
Rufous-breasted Wren (239) 38:8
Plain Wren (241) 38:17
House Wren (243) 38:18
White-breasted Wood-Wren (245) 38:15
Long-billed Gnatwren (237) 32:15
Tropical Gnatcatcher (237) 41:2
*Black-faced Solitaire (249) 39:13
Orange-billed Nightingale-Thrush (247) 38:26
Swainson's Thrush (249) 39:2
Mountain Thrush (251) 39:6
Clay-colored Thrush (251) 39:8
White-throated Thrush (251) 39:9
Golden-winged Warbler (257) 41:5
Tennessee Warbler (255) 40:22
Tropical Parula (257) 41:3
Chestnut-sided Warbler (259) 43:4
Black-throated Green Warbler (261) 43:9
Blackburnian Warbler (261) 41:8
Black-and-white Warbler (267) 41:13
American Redstart (267) 41:10
Mourning Warbler (271) 42:14
Wilson's Warbler (269) 42:4
Canada Warbler (269) 42:8
Slate-throated Redstart (273) 42:7
Golden-crowned Warbler (275) 40:18
Buff-rumped Warbler (271) 40:23
Bananaquit (277) 40:24
Common Bush-Tanager (279) 45:13
White-shouldered Tanager (281) 46:18
Red-crowned Ant-Tanager (277) 47:10
Summer Tanager (285) 47:5
White-winged Tanager (285) 47:7
***Cherrie's Tanager (287) 47:4**
Blue-gray Tanager (291) 46:15
Palm Tanager (291) 45:19
Silver-throated Tanager (291) 46:6
Speckled Tanager (289) 46:8
Bay-headed Tanager (289) 46:9
Golden-hooded Tanager (289) 46:13

Scarlet-thighed Dacnis (291) 46:4
Green Honeycreeper (293) 46:7
Variable Seedeater (295) 49:3
Yellow-faced Grassquit (297) 49:6
White-naped Brush-Finch (301) 48:15
Black-striped Sparrow (303) 50:14
Rufous-collared Sparrow (307) 50:13
Streaked Saltator (309) 48:4
Buff-throated Saltator (309) 48:2
Rose-breasted Grosbeak (311) 48:8
Blue-black Grosbeak (311) 48:10
Great-tailed Grackle (317) 44:16
Bronzed Cowbird (317) 44:15
Baltimore Oriole (321) 44:7
Thick-billed Euphonia (327) 45:8
Lesser Goldfinch (325) 50:2

Site 3B-5: Río Negro

Birding Time: 2.5 hours
Elevation: 1,050 meters
Trail Difficulty: 1
Reserve Hours: N/A
Entrance Fee: N/A
4×4 Recommended

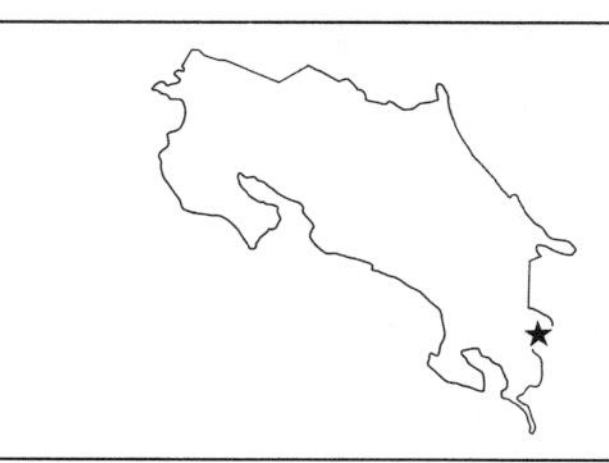

Río Negro is a nice birding area located 45 minutes from San Vito. What you will find here is a small patch of secondary forest set among coffee plantations and cattle pastures. This is not an official reserve; the property is maintained by a coffee company. It is completely open to the public, and there are no entrance fees or hours.

While a small patch of secondary forest might not sound that exciting, Río Negro has a lot to recommend it. The birding is relatively easy, thanks to a wide flat trail. Also, a number of interesting species can be found here, including many South Pacific Slope specialties. The most famous bird in the area is Lance-tailed Manakin, a bird that barely crosses north of the Panama border to make its home at Río Negro.

Target Birds

Scaled Pigeon (87) 18:4
Common Potoo (115) 20:4
*Snowy-bellied Hummingbird (133) 24:3
*White-tailed Emerald (133) 24:19
Sepia-capped Flycatcher (195) 36:27
Lance-tailed Manakin (223) 33:5
Black-chested Jay (231) 39:20
Stripe-headed Brush-Finch (301) 48:14
Crested Oropendola (323)

Access

From San Vito: (Refer to site 3B-3 for a map.) Take the main road to Sabalito for 7.8 km. At the gas station in Sabalito, fork left onto a dirt road. Go 3.8 km and then turn left, following signs for Mellizas. From this point continue straight for 4.6 km to a small ranch house on your right, which is currently painted purple.

Just after this house make a right, and follow the road to its end, in 1.2 km, where you will see a small house and stables. Park near the house and enter the trail on its left. (Driving time from San Vito: 45 min)

Logistics: The nearest major town in the area is San Vito, and I suggest looking for lodging there. Alternatively, both Las Cruces Biological Station (3B-3) and La Amistad Lodge (3B-6) offer accommodations and are good bases from which to explore Río Negro. I also suggest bringing water and a boxed lunch when birding Río Negro, as there is not much food in the area.

Birding Sites

The main road leading into the forest is a small flat cart path, which continues for about one kilometer before it reaches coffee fields. The habitat along the trail is quite homogeneous, though many interesting species can be found. Mixed flocks often include Plain Antvireo, Ruddy Woodcreeper, Plain Xenops, Sulphur-rumped Flycatcher, and White-ruffed Manakin. Also, keep an eye open for Ruddy Quail-Dove, Fiery-billed Aracari, Sepia-capped Flycatcher, and Olivaceous Piculet. Migrant warblers are quite prevalent during the proper time of year, and you may run into Worm-eating Warbler, Kentucky Warbler, and Blackburnian Warbler. The Lance-tailed Manakin is best found by listening for its song, which is very similar to that of the Long-tailed Manakin, whiny slurred phrases interspersed with rich whistles. Once you hear the birds singing, you may need to bushwhack a short way into the forest to find them displaying.

Toward its end, the cart path makes a sharp right turn and then quickly comes out into coffee fields. At the point where the road turns, a smaller footpath continues straight through more forest. While the walking here becomes slightly more difficult, it is one of the most reliable places to encounter the Lance-tailed Manakin if you have not found it already.

Species to Expect

Black Vulture (31) 13:4
Turkey Vulture (31) 13:3
Roadside Hawk (43) 16:12
Black Hawk-Eagle (49) 13:9
Barred Forest-Falcon (53) 16:5
Band-tailed Pigeon (87) 18:1
White-tipped Dove (91) 18:14
Ruddy Quail-Dove (93) 18:17
Orange-chinned Parakeet (97) 19:14
Squirrel Cuckoo (103) 21:7
Common Potoo (115) 20:4
Green Hermit (121) 23:3
Stripe-throated Hermit (121) 23:1
Violet-crowned Woodnymph (127) 24:4
*Snowy-bellied Hummingbird (133) 24:3
Rufous-tailed Hummingbird (129) 24:10
*White-tailed Emerald (133) 24:19
Violaceous Trogon (141) 26:8
Blue-crowned Motmot (147) 27:8
***Fiery-billed Aracari (155) 27:16**
Chestnut-mandibled Toucan (155) 27:19
Olivaceous Piculet (159) 29:1
Pale-billed Woodpecker (161) 27:14
Plain Xenops (169) 29:3
Ruddy Woodcreeper (171) 29:13
Olivaceous Woodcreeper (171) 29:7

Streak-headed Woodcreeper (173) 29:8
***Black-hooded Antshrike (177) 31:6**
Plain Antvireo (181) 32:6
Slaty Antwren (183) 32:3
Bicolored Antbird (179) 31:9
Yellow Tyrannulet (189) 37:11
Yellow-bellied Elaenia (193) 37:26
Sepia-capped Flycatcher (195) 36:27
Paltry Tyrannulet (191) 37:10
Scale-crested Pygmy-Tyrant (197) 37:4
Yellow-olive Flycatcher (195) 37:16
Sulphur-rumped Flycatcher (199) 36:21
Yellow-bellied Flycatcher (205) 36:20
Bright-rumped Attila (201) 35:6
Dusky-capped Flycatcher (209) 35:21
Great Crested Flycatcher (209) 35:17
Tropical Kingbird (213) 35:1
White-winged Becard (217) 33:13
White-ruffed Manakin (223) 33:9
Lance-tailed Manakin (223) 33:5
Blue-crowned Manakin (223) 33:7
Yellow-throated Vireo (225) 40:6
Philadelphia Vireo (227) 40:15
Tawny-crowned Greenlet (229) 32:4
Lesser Greenlet (229) 40:7
Rufous-browed Peppershrike (229) 40:2
Rufous-breasted Wren (239) 38:8
Plain Wren (241) 38:17
White-breasted Wood-Wren (245) 38:15
Long-billed Gnatwren (237) 32:15
Tropical Gnatcatcher (237) 41:2
Orange-billed Nightingale-Thrush (247) 38:26
Swainson's Thrush (249) 39:2
Clay-colored Thrush (251) 39:8
Tennessee Warbler (255) 40:22
Chestnut-sided Warbler (259) 43:4
Blackburnian Warbler (261) 41:8
Black-and-white Warbler (267) 41:13
Worm-eating Warbler (267) 40:21
Ovenbird (267) 43:12
Kentucky Warbler (271) 42:15
Mourning Warbler (271) 42:14
Wilson's Warbler (269) 42:4
Golden-crowned Warbler (275) 40:18
Rufous-capped Warbler (275) 40:16
Buff-rumped Warbler (271) 40:23
Gray-headed Tanager (283) 45:17
White-shouldered Tanager (281) 46:18
Red-crowned Ant-Tanager (277) 47:10
Summer Tanager (285) 47:5
White-winged Tanager (285) 47:7
*Cherrie's Tanager (287) 47:4
Blue-gray Tanager (291) 46:15
Bay-headed Tanager (289) 46:9
Golden-hooded Tanager (289) 46:13
Orange-billed Sparrow (303) 48:17
Black-striped Sparrow (303) 50:14
Buff-throated Saltator (309) 48:2

Site 3B-6: La Amistad Lodge

Birding Time: 1–4 days
Elevation: 1,200 meters
Trail Difficulty: 1–2
Reserve Hours: N/A
Entrance Fee: Included with overnight
4×4 Recommended

Hacienda La Amistad is a 23,500-acre private farm that strives to integrate socially responsible agriculture, conservation, and ecotourism as a means of living off the land and protecting it at the same time. The principal crop is shade-grown coffee, which is produced 100% organically. Most birders have heard of the benefits that shade-grown coffee has for birds in comparison with coffee grown in full sunlight. Visiting Hacienda La Amistad is a good way to experience these positive impacts firsthand. You will see that unlike in most agricultural fields, many resident and migrant species of birds utilize the fruits and flowers of the shade trees and make a home within the coffee fields.

While coffee production is the farm's primary economic interest, only 5% of the land is farmed. The remainder has been left for conservation and ecotourism, and most of this contains pristine primary forest. La Amistad Lodge serves as the center for the farm's ecotourism project. The main building, set among coffee fields, cow pastures, and forest, is quite elegant, with the feeling of a Swiss chalet. Guests enjoy hot showers, comfortable beds, cozy sofas, a large-screen TV, and three meals served daily. It is certainly not what one expects to find while making the drive into this remote region of the country.

For birders and nature lovers alike, the unique appeal of La Amistad Lodge is its four remote camps found deep within the reserve. These are permanent campsites where you sleep in rustic cabins, not a standard tent. Trails are maintained around and between the camps, offering visitors convenient access to the beautiful forest. La Amistad Lodge provides everything you need to visit these camps, including a sleeping bag and pad, food, and even a guide and cook. It is camping in style, although it still requires a sense of adventure.

Target Birds

Spotted Wood-Quail (13) 12:11
Bicolored Hawk (39) 16:8
Solitary Eagle (45) 13:8
Ornate Hawk-Eagle (49) 17:11
Barred Forest-Falcon (53) 16:5
*Sulphur-winged Parakeet (95) 19:12
Brown Violetear (131) 23:6
*Snowy-bellied Hummingbird (133) 24:3
*White-tailed Emerald (133) 24:19
*White-throated Mountain-gem (135) 24:6
Resplendent Quetzal (145) 26:1
Scaly-throated Foliage-gleaner (165) 30:2
Ruddy Foliage-gleaner (167) 30:5
Scaled Antpitta (185) 30:20
Ochre-breasted Antpitta (185) 30:13
Rough-legged Tyrannulet (191) 37:18
Bran-colored Flycatcher (199) 36:17
*Turquoise Cotinga (219) 34:5
*Three-wattled Bellbird (221) 34:12
Lance-tailed Manakin (223) 33:5
Masked Yellowthroat (273) 42:12
Flame-colored Tanager (287) 47:9
White-winged Tanager (285) 47:7
Yellow-bellied Seedeater (295) 49:4
Blue Seedeater (297) 49:11
Stripe-headed Brush-Finch (301) 48:14
Streaked Saltator (309) 48:4
Thick-billed Euphonia (327) 45:8

Access

From San Vito: (Refer to site 3B-3 for a map.) Take the road to Sabalito for 7.8 km. Fork left at the gas station and go 3.8 km on a dirt road. Turn left, following signs for Mellizas, and proceed 12.4 km to the town of La Lucha. Turn left at a T-intersection and go 3.8 km. The driveway to La Amistad Lodge will be on your left. Follow this road for 1.6 km to the lodge. (Driving time from San Vito: 90 min)

Logistics: Staying at La Amistad Lodge is your only option for birding the area. The facilities at the main lodge are very nice, while the campsites within the forest are rustic. Wherever you are staying, three meals are served daily.

Hacienda La Amistad
Tel: (506) 2289-7667
Website: www.laamistad.com
E-mail: amistad@racsa.co.cr

Birding Sites

The land around La Amistad Lodge is vast and wild and laced with an extensive network of trails and roads. As a result it is quite easy for visitors to get lost, and management prefers that you not walk alone. The lodge provides local guides to lead you through the property and help you spot birds, though they are not bird experts.

The best birding at La Amistad is generally found deep within the forest at the four permanent campsites. The Cottoncito Camp, located at 1,500 meters, is an especially productive place to bird. There are four trails leading into the forest around the camp. This is a good place to look for raptors such as Ornate Hawk-Eagle and Bicolored Hawk, as well as many middle-elevation forest species such as White-throated Spadebill, Collared Trogon, Ochre-breasted Antpitta, and White-winged Tanager. Other campsites, including the Coto Brus Camp and the Punto Mira Camp, reach nearly 2,000 meters and are good for many mountain species such as Resplendent Quetzal, Spotted Wood-Quail, and Black Guan.

The lands around the main lodge can also produce some interesting birding, and it is well worth spending a morning or a day exploring this area. Fruit feeders are maintained right in front of the lodge. As a result, the bushes in the area are usually filled with birds. Thick-billed Euphonia and Tennessee Warbler are abundant, and you should be able to find Streaked Saltator, Flame-colored Tanager, and Red-crowned Woodpecker.

The shade trees, planted throughout the coffee fields, are a variety favored by many birds. When these trees are flowering, they are a favorite of hummingbirds, including Snowy-bellied Hummingbird, Long-billed Starthroat, White-tailed Emerald, and Charming Hummingbird. When in fruit, they attract many species of tanagers. Look for Bay-headed Tanager, Golden-hooded Tanager, Flame-colored Tanager, and Cherrie's Tanager. Any wet grassy areas within the fields may hold Masked Yellowthroat.

A number of cattle pastures are found a short distance from the lodge, and these too can offer productive birding. Slaty Spinetail, White-naped Brush-Finch, and Yellow Tyrannulet can be found in the undergrowth. Yellow-headed Caracara, Mountain Elaenia, Brown-capped Vireo, Speckled Tanager, and Rufous-capped Warbler are other regulars in this habitat.

Forest habitat is also available near the lodge, though it is generally not as productive as the forest found around the campsites. One easy trail to walk starts at the back of the lodge and proceeds up into the forest behind. Here you might find Gray-headed Tanager, Russet Antshrike, and Sulphur-rumped Flycatcher. Lance-tailed Manakin, a bird confined to this tiny region of the country, is also a regular find in the forest patches near the lodge. Another interesting species at this site is the Three-wattled Bellbird, which is present between March and October. Its loud *BOK!* is easily heard, and with some patience you should be able to spot one, generally perched atop a tall tree.

Region 4
The Mountains

The beautiful **Long-tailed Silky-Flycatcher** can be seen perched atop tall trees in the Upper Mountains region. Loose flocks of this endemic species are often encountered feeding on small fruits and sallying for insects at Km 70 (4B-3) and San Gerardo de Dota (4B-4).

Regional Map: The Mountains

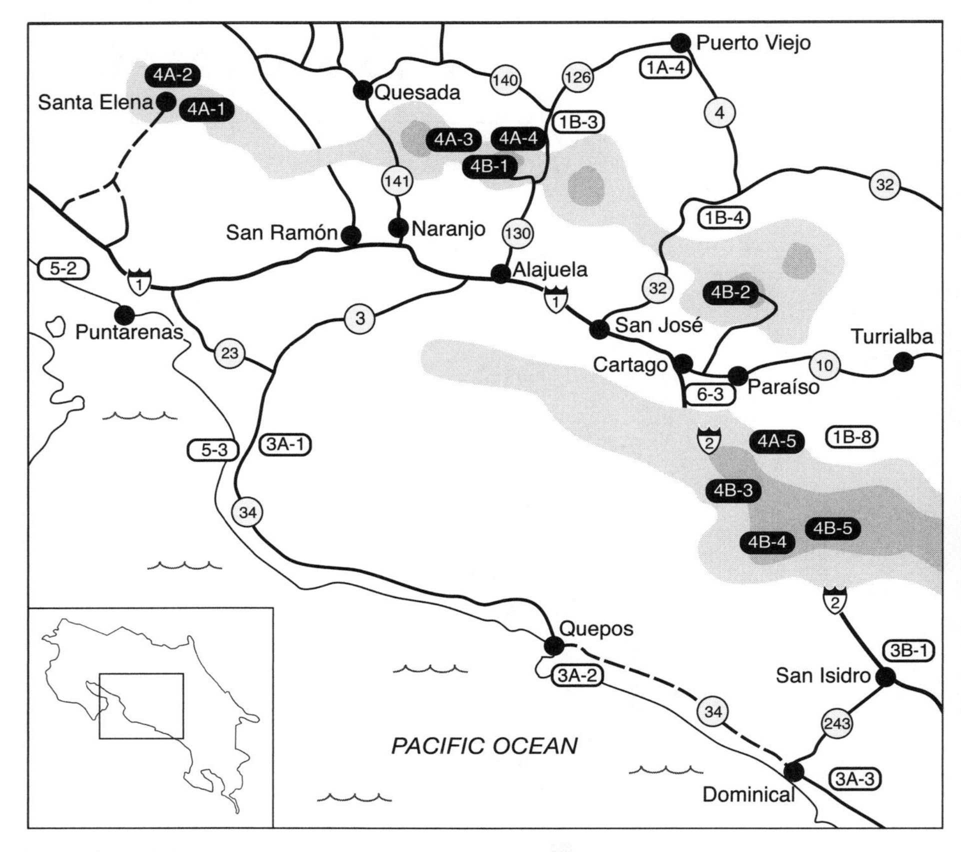

Lower Mountains
4A-1: Monteverde Cloud Forest Reserve
4A-2: Santa Elena Reserve
4A-3: Bosque de Paz Biological Reserve and Lodge
4A-4: La Paz Waterfall Gardens
4A-5: Tapantí National Park

Upper Mountains
4B-1: Poás Volcano National Park
4B-2: Irazú Volcano National Park
4B-3: Km 70
4B-4: San Gerardo de Dota
4B-5: Cerro de la Muerte

Introduction

For the purposes of this book, the Mountains region is loosely defined as elevated areas where the effects of slope, Caribbean or Pacific, on the avifauna are not readily noticeable. This means that sites covered here will harbor very few species that are confined to only one side of the mountains. (A few notable exceptions include Black-bellied Hummingbird, White-bellied Mountain-gem, Coppery-headed Emerald, Rufous-breasted Antthrush, and Tawny-capped Euphonia—all of which are Caribbean Slope specialties that occur at Lower Mountain sites.) The

Mountain "environment" tends to occur upward of 1,200 meters, though there is some variability caused by local conditions. For example, because of differences in climate between the Caribbean and Pacific slopes, similar habitats and species tend to reach a higher elevation on the Pacific side than they do on the Caribbean.

The distinction between the subregions, Lower Mountains and Upper Mountains, is similarly based more on avifauna and habitat than strictly on altitude. Upper Mountain sites, usually above 2,000 meters, are where you are likely to see only the truly high-elevation species. At Lower Mountain sites (1,200 to 2,000 meters) you will find greater diversity, and while many mountain specialties can be seen, you will also encounter crossover species from the middle-elevation sites.

One of the first things you will notice upon arriving in the mountains is the cooler temperature. While nowhere in Costa Rica is it cold enough to sustain permafrost, the average daily low atop Costa Rica's highest peak, Mt. Chirripó (3,820 meters), is only 2°C (36°F). Lower locations, such as the Monteverde Cloud Forest Reserve (1,400 meters), see an average daily low of about 13°C (56°F). While these are not exactly arctic conditions, it is not what most people expect in the tropics.

Interestingly, just as the temperature of the Mountains region may remind a visitor of the temperate zone, an unusually high percentage of species found in this region are temperate in origin. Thrushes, sparrows, and finches, all families with North American origins, are especially well represented in the mountains. The same pattern is seen in the plant kingdom, with oaks, thistles, alders, and blackberries occupying an important part of the flora found in the Costa Rican mountains.

Even though the mountains are an area of relatively low biodiversity, many visiting birders will be eager to spend time here because the area is a hotspot of endemism. Of Costa Rica's 89 endemic species, 39 are mountain specialties. To explain this phenomenon, it is helpful to step outside Costa Rica and look at the geography of the greater region. As the Cordillera de Guanacaste reaches the Nicaraguan border at the north end of Costa Rica, it yields to an expanse of lowlands that extends through most of Nicaragua. Similarly, the Cordillera de Talamanca succumbs to extensive lowlands that begin in central Panama and continue all the way into Colombia. These lowland areas in southern Nicaragua and eastern Panama isolate the mountains of Costa Rica and western Panama, creating an island of acceptable habitat for mountainous species. Over time this isolation has allowed many endemic species to evolve.

The Mountains region is generally rural, the steep slopes not lending themselves to human habitation. In traveling through these areas, you will get a peek at the *campesino* lifestyle in Costa Rica. Most people subsist as farmers. Many fruits grow well in the mountainous climate, including *mora*, a relative of our raspberry that is used to make a delicious juice. Ornamental plants are also grown here and exported throughout the world. Other farmers take advantage of the wealth of flowing water to farm rainbow trout, and the region is responsible

Mountains of the Greater Region

for producing most of the country's dairy products, so you will certainly see many cows.

The sites described for this region encompass three of Costa Rica's four mountain ranges. Cordillera de Guanacaste is not covered because high elevations in this range are limited and hard to access, and because the variety of species found there is small when compared to the other mountain ranges farther south.

You will encounter an interesting variety of ecosystems at the sites within the Mountains region. The Monteverde and Santa Elena reserves display quintessential cloud forest, laden with dripping epiphytes. At Km 70 you will find high-elevation oak forest, its ancient trees reaching high into the thin mountain air. Poás Volcano National Park displays an elfin forest with densely packed, short, gnarled trees. Finally, above the tree line atop Cerro de la Muerte, you can see páramo, a wet, stunted ecosystem usually found in the Andes.

Of the 10 sites in the Mountains region, Monteverde Cloud Forest Reserve (4A-1) and Poás Volcano National Park (4B-1) are the best-known tourist destinations. Bosque de Paz Biological Reserve and Lodge (4A-3) and Tapantí National Park (4A-5) are my personal favorites in the Lower Mountains, both offering a large diversity of species and high potential for spotting rarities. Among the Upper Mountain sites, San Gerardo de Dota (4B-4) is the most popular among birders, though I personally prefer the local flavor of Km 70 (4B-3).

Regional Specialties

Highland Tinamou (3) 12:7
*Black Guan (11) 12:5
Buffy-crowned Wood-Partridge (13) 12:16
*Black-breasted Wood-Quail (13) 12:10
Spotted Wood-Quail (13) 12:11
Red-tailed Hawk (47) 17:8
Band-tailed Pigeon (87) 18:1
Ruddy Pigeon (87) 18:5
Maroon-chested Ground-Dove (91) 18:10
*Buff-fronted Quail-Dove (93) 18:19
Barred Parakeet (97) 19:16
Bare-shanked Screech-Owl (107) 20:10
*Costa Rican Pygmy-Owl (109) 20:18
Unspotted Saw-whet Owl (109) 20:13
*Dusky Nightjar (111) 21:19
Violet Sabrewing (123) 23:9
Green Violetear (131) 23:7
*Fiery-throated Hummingbird (123) 24:12

Stripe-tailed Hummingbird (133) 24:17
*Black-bellied Hummingbird (133) 24:21
*Purple-throated Mountain-gem (135) 24:7
*White-throated Mountain-gem (135) 24:6
Magnificent Hummingbird (123) 23:16
*Magenta-throated Woodstar (139) 25:2
*Volcano Hummingbird (139) 25:3
*Scintillant Hummingbird (139) 25:7
Resplendent Quetzal (145) 26:1
*Prong-billed Barbet (153) 28:1
Emerald Toucanet (153) 27:17
Acorn Woodpecker (157) 28:15
Hairy Woodpecker (159) 28:19
*Ruddy Treerunner (163) 29:4
Buffy Tuftedcheek (165) 30:1
Lineated Foliage-gleaner (165) 30:8
*Streak-breasted Treehunter (167) 30:7
Streaked Xenops (169) 29:2
Spot-crowned Woodcreeper (173) 29:9
*Silvery-fronted Tapaculo (187) 32:12
Mountain Elaenia (193) 37:24
*Dark Pewee (203) 36:2
*Ochraceous Pewee (203) 36:3
*Black-capped Flycatcher (207) 36:16
*Golden-bellied Flycatcher (211) 35:9
Barred Becard (217) 33:14
*Yellow-winged Vireo (229) 40:10
Brown-capped Vireo (227) 40:14
Azure-hooded Jay (231) 39:17
*Silvery-throated Jay (231) 39:16
*Ochraceous Wren (243) 38:19
*Timberline Wren (243) 38:16
Gray-breasted Wood-Wren (245) 38:14
*Black-faced Solitaire (249) 39:13
*Black-billed Nightingale-Thrush (247) 38:24
Slaty-backed Nightingale-Thrush (247) 38:23
Ruddy-capped Nightingale-Thrush (247) 38:27
*Sooty Thrush (249) 39:7
Mountain Thrush (251) 39:6
*Black-and-yellow Silky-Flycatcher (253) 39:10
*Long-tailed Silky-Flycatcher (253) 39:15
*Flame-throated Warbler (257) 41:7
Townsend's Warbler (261) 43:8
*Collared Redstart (273) 42:6
*Black-cheeked Warbler (275) 40:17
Three-striped Warbler (275) 40:20
*Wrenthrush (275) 32:13
*Sooty-capped Bush-Tanager (279) 45:14
Flame-colored Tanager (287) 47:9
*Spangle-cheeked Tanager (289) 46:12
Blue Seedeater (297) 49:11
Slaty Finch (299) 49:14
*Peg-billed Finch (299) 49:12
*Slaty Flowerpiercer (299) 49:9
*Yellow-thighed Finch (301) 48:22
*Large-footed Finch (301) 48:20
Chestnut-capped Brush-Finch (301) 48:16
*Volcano Junco (307) 50:11
*Black-thighed Grosbeak (311) 48:7
Elegant Euphonia (325) 45:9
*Golden-browed Chlorophonia (325) 45:10
Yellow-bellied Siskin (325) 50:1

Subregion 4A

The Lower Mountains

The **Prong-billed Barbet** can be found moving through Costa Rican cloud forests in small flocks, searching for berries and fruits. Its song, a series of repeated resonant notes, is sung in unison by the group and is one of the quintessential sounds of the cloud forest. Look for this species at all Lower Mountain sites.

Common Birds to Know

Black Vulture (31) 13:4
Turkey Vulture (31) 13:3
Green Hermit (121) 23:3
*Purple-throated Mountain-gem (135) 24:7
Blue-and-white Swallow (233) 22:20
Gray-breasted Wood-Wren (245) 38:14
Wilson's Warbler (269) 42:4
Slate-throated Redstart (273) 42:7
Common Bush-Tanager (279) 45:13
*Spangle-cheeked Tanager (289) 46:12
Yellow-faced Grassquit (297) 49:6
Rufous-collared Sparrow (307) 50:13

Site 4A-1: Monteverde Cloud Forest Reserve

Birding Time: 5 hours
Elevation: 1,500 meters
Trail Difficulty: 1–2
Reserve Hours: 7 a.m. to 4 p.m.
Entrance Fee: $15 (US 2008)
Bird Guides Available
4×4 Recommended
Trail Map Available on Site

The Monteverde Cloud Forest Reserve is one of Costa Rica's greatest international icons and a world-renowned ecotourism destination. Home to the extravagantly plumaged Resplendent Quetzal, Monteverde is one of Costa Rica's most visited biological reserves, regularly seeing more than 300 visitors a day. The reserve protects about 12,000 acres on the Caribbean and Pacific slopes of the Cordillera de Tilarán and boasts an extensive, well-maintained system of trails leading through pristine forest. The majority of these trails are found right near the Continental Divide. Here you will find Costa Rica's quintessential cloud forest, a type of ecosystem that exists in mountains where clouds consistently pass very low to the ground because of topography and local climate conditions. The constant moisture allows for an incredible diversity of epiphytes, which grow on every exposed surface in the forest. These plants specialize in collecting water out of the moisture-laden air, and they account for the droplets of water that fall constantly from the trees.

The tiny community of Monteverde was originally settled in 1951 by a small group of Quakers who fled the United States in protest of the Korean War. Along with founding the cloud forest reserve, these Quakers soon established a highly successful dairy industry, and today Monteverde cheese and ice cream are some of the best in the country. While the town of Monteverde, true to its Quaker roots, has remained small and quaint, the nearby town of Santa Elena is a tourism boomtown and serves as the commercial center for the area. Many large and elegant hotels have gone up in recent years, along with charming inns, small hostels, and fine restaurants. Canopy tours, skywalks, horseback rides, and a number of other tourist attractions are all available in the area.

Such a popular tourist destination may not seem like a great birding site, but the Monteverde Cloud Forest Reserve has much to recommend it. Its extensive trail system is home to many of Costa Rica's most intriguing birds, including Resplendent Quetzal, Three-wattled Bellbird, and Magenta-throated Woodstar. Furthermore, the reserve offers expert local guides who will know exactly where to find key species. Finally, the cloud forest, itself, is stunningly beautiful and may cause you to completely forget about birds, if only for a minute.

Target Birds

Highland Tinamou (3) 12:7
*Black-breasted Wood-Quail (13) 12:10
*Chiriqui Quail-Dove (93) 18:18
Stripe-tailed Hummingbird (133) 24:17
*Coppery-headed Emerald (133) 24:20
*Magenta-throated Woodstar (139) 25:2
*Orange-bellied Trogon (143) 26:4
Resplendent Quetzal (145) 26:1
Emerald Toucanet (153) 27:17
*Ruddy Treerunner (163) 29:4
Lineated Foliage-gleaner (165) 30:8
*Streak-breasted Treehunter (167) 30:7
Gray-throated Leaftosser (167) 30:10
*Silvery-fronted Tapaculo (187) 32:12
*Three-wattled Bellbird (221) 34:12
Brown-capped Vireo (227) 40:14
Rufous-browed Peppershrike (229) 40:2
Azure-hooded Jay (231) 39:17
*Black-and-yellow Silky-Flycatcher (253) 39:10
White-eared Ground-Sparrow (303) 48:19
*Golden-browed Chlorophonia (325) 45:10

Access

To the town of Santa Elena from Rt. 1: There are three different routes to Santa Elena from Rt. 1. I do not recommend the northern route (not shown on map), which passes through Las Juntas. The road is in poor condition and not well marked.

The middle route is the most straightforward. Turn west off the highway just on the south side of the Río Lagarto (33.0 km north of Rt. 23 and 17.2 km south of Rt. 18) and then proceed straight on this dirt road for 33.2 km, until it ends in the town of Santa Elena. (Driving time from Rt. 1: 85 min)

The southern route leaves the highway 17.4 km north of Rt. 23 on a paved road that heads into the town of Sardinal. After 2.8 km, turn left just after the town soccer field and proceed straight on this road for 16.2 km The road will dead-end at a T-intersection, where you should turn right and continue the final 17.8 km to Santa Elena. (Driving time from Rt. 1: 90 min)

To the Monteverde Cloud Forest Reserve from the town of Santa Elena: Take the road that leaves from the east side of town and proceed straight for 5.2 km (passing the gas station on your left) until the road ends at the Monteverde reserve. (Driving time from Santa Elena: 20 min)

Logistics: The Monteverde/Santa Elena area has a full range of accommodations and restaurants. You should have no problem finding exactly what you want, although it is a good idea to make advanced reservations for the high season.

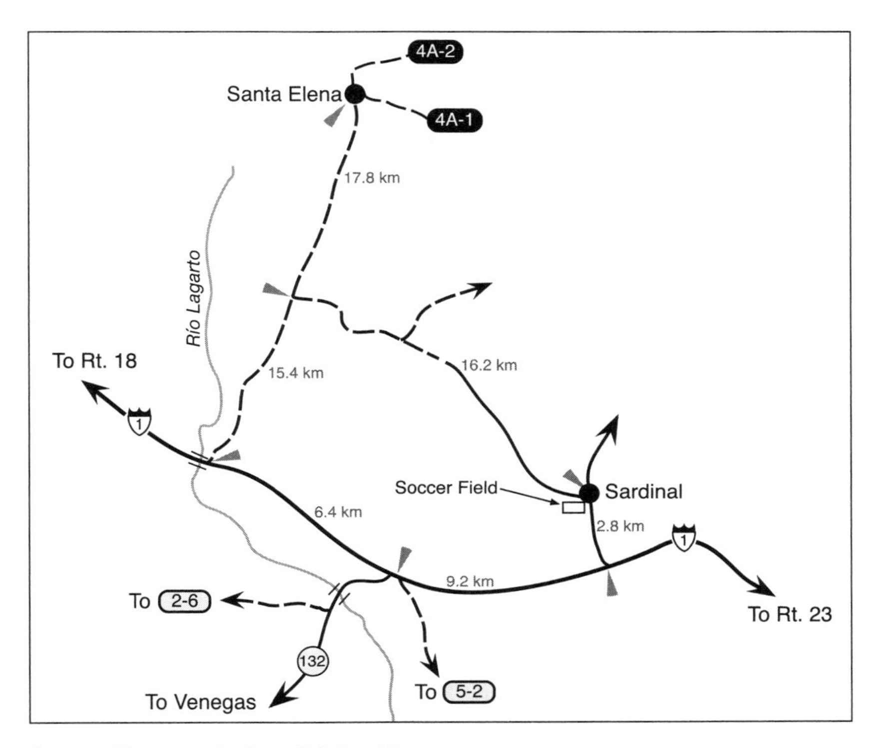

Greater Monteverde Area Driving Map

Birding Sites

I suggest arriving before the park opens and using that time to work the final 0.8 km of road leading to the park, as well as the parking lot area. Highland Tinamou and Prong-billed Barbet can be heard singing, and Resplendent Quetzal, Orange-bellied Trogon, and Azure-hooded Jay are often in the trees above the parking lot before the place gets busy.

As you enter the reserve you will have the opportunity to hire a guide. The guides at Monteverde are very knowledgeable and know exactly where key species like quetzals and bellbirds have been seen recently. Otherwise, be sure to get a map and use it to follow the birding directions below.

Sendero El Camino is the widest and most open trail in the park and has been maintained for vehicles to pass through if needed. Because it is more open than the other trails, seeing birds is easier. You will normally find mixed-species flocks led by Common Bush-Tanagers that might include Spotted Barbtail, Red-faced Spinetail, Collared Redstart, and Spangle-cheeked Tanager. Also, keep an eye out for Orange-billed Nightingale-Thrush and Slaty-backed Nightingale-Thrush foraging on the path.

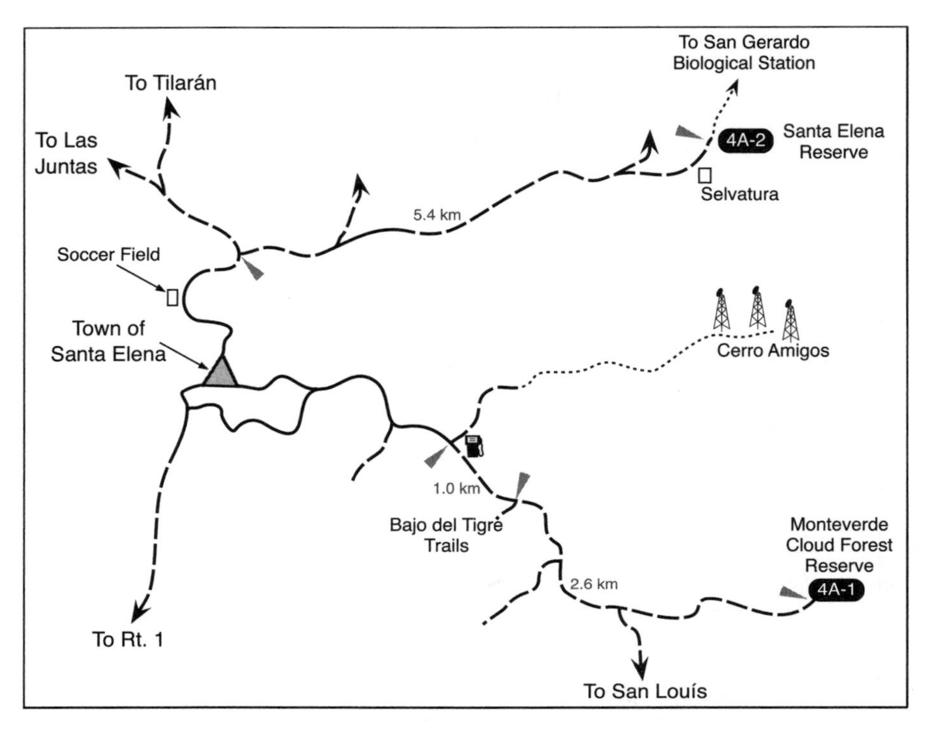

Monteverde Cloud Forest Reserve Area Map. Santa Elena Reserve (4A-2), the Bajo del Tigre Trails, and Cerro Amigos are included.

Sendero La Ventana, which crosses El Camino at its end, leads to a scenic lookout along the Continental Divide where you can see the Pacific Ocean on one side and the thickly forested hills of the Caribbean slope on the other. (Of course, this view depends on the weather being clear, which it rarely is.) Look for Black-and-yellow Silky-Flycatcher, Ruddy Treerunner, and Slaty Flowerpiercer.

Sendero El Río and Sendero Tosi form a small loop around a low wet area, created by Quebrada Quetzal (Quetzal Creek). This area is one of the most productive in the park, especially if it is rainy or windy, because the forest around these trails is well sheltered. Look for Chestnut-capped Brush-Finch and Gray-throated Leaftosser foraging in the undergrowth. Small mixed flocks of Three-striped Warbler, Golden-crowned Warbler, Slaty Antwren, and Plain Antvireo can often be found moving through the understory.

Sendero Bosque Nuboso runs along the southern flank of the Monteverde trail system. Aside from its stunning beauty, the first half of this trail is an especially good place to look for Resplendent Quetzal. The eerie metallic whistles of Black-faced Solitaire, the bubbling, energetic song of Gray-breasted Wood-Wren, and the emphatic call of Silvery-fronted Tapaculo provide the musical background for this lovely walk. Halfway down this trail you can take a shortcut that will connect you with some of the other trails in the park, such as Sendero Wilford Guindon and Sendero El Roble. These are worth exploring if you have the time.

One final "must-see" is the hummingbird garden located just outside of the reserve on the far side of the parking lot. A small gift shop maintains the feeders, and entrance is free. Concrete paths lead up a small embankment to the feeders, where you can find Violet Sabrewing, Purple-throated Mountain-gem, Green-crowned Brilliant, Stripe-tailed Hummingbird, Coppery-headed Emerald, Green Violetear, and Magenta-throated Woodstar. Bananaquit also frequents the feeders.

For the extremely adventurous, the Monteverde reserve maintains two cabins in the Peñas Blancas Valley. The first, Refugio Alemán, is a three-hour hike down the Caribbean Slope. Refugio Eladio, another three hours beyond that, offers even better birding and a fairly extensive trail system. These are not for the faint-of-heart, as you will find yourself completely isolated. The cabins are rustic at best, with mice and bats sharing the space with you. However, the birding is great, and you have the opportunity to find many rarities, including Strong-billed Woodcreeper, Rufous-browed Tyrannulet, Lovely Cotinga, and Yellow-eared Toucanet. Birds for this region are not included in the lists I have provided. For a complete list of the birds in the area, refer to *An Annotated Checklist of the Birds of Monteverde and Peñas Blancas*, compiled by Michael Fogden. If this adventure sounds like your cup of tea, make arrangements with the Monteverde office in advance.

Monteverde Cloud Forest Reserve
Tel: (506) 2645-5122
E-mail: montever@cct.or.cr

Species to Expect (Includes Common Species from the Santa Elena Vicinity)

*Black Guan (11) 12:5
*Black-breasted Wood-Quail (13) 12:10
Black Vulture (31) 13:4
Turkey Vulture (31) 13:3
Broad-winged Hawk (43) 16:13
Barred Forest-Falcon (53) 16:5
Band-tailed Pigeon (87) 18:1
Ruddy Pigeon (87) 18:5
*Chiriqui Quail-Dove (93) 18:18
White-fronted Parrot (99) 19:6
Squirrel Cuckoo (103) 21:7
Groove-billed Ani (103) 21:9
White-collared Swift (117) 22:1
Green Hermit (121) 23:3
Violet Sabrewing (123) 23:9
Green Violetear (131) 23:7
Steely-vented Hummingbird (127) 24:15
Cinnamon Hummingbird (129) 24:11
Stripe-tailed Hummingbird (133) 24:17
***Coppery-headed Emerald (133) 24:20**
***Purple-throated Mountain-gem (135) 24:7**
Green-crowned Brilliant (123) 23:15
*Magenta-throated Woodstar (139) 25:2
Collared Trogon (143) 26:5
*Orange-bellied Trogon (143) 26:4
Resplendent Quetzal (145) 26:1
Blue-crowned Motmot (147) 27:8
***Prong-billed Barbet (153) 28:1**
Emerald Toucanet (153) 27:17
*Hoffmann's Woodpecker (157) 28:16
Smoky-brown Woodpecker (159) 28:13

Golden-olive Woodpecker (159) 28:11
Red-faced Spinetail (163) 30:12
Spotted Barbtail (163) 29:5
***Ruddy Treerunner (163) 29:4**
Lineated Foliage-gleaner (165) 30:8
*Streak-breasted Treehunter (167) 30:7
Gray-throated Leaftosser (167) 30:10
Spotted Woodcreeper (173) 29:20
Plain Antvireo (181) 32:6
Slaty Antwren (183) 32:3
*Silvery-fronted Tapaculo (187) 32:12
Yellow-bellied Elaenia (193) 37:26
Mountain Elaenia (193) 37:24
Olive-striped Flycatcher (195) 36:24
Paltry Tyrannulet (191) 37:10
White-throated Spadebill (189) 37:1
Yellowish Flycatcher (207) 36:19
Bright-rumped Attila (201) 35:6
Dusky-capped Flycatcher (209) 35:21
Great Kiskadee (211) 35:13
Boat-billed Flycatcher (211) 35:12
Social Flycatcher (211) 35:14
*Golden-bellied Flycatcher (211) 35:9
Tropical Kingbird (213) 35:1
*Three-wattled Bellbird (221) 34:12
Yellow-throated Vireo (225) 40:6
Brown-capped Vireo (227) 40:14
Lesser Greenlet (229) 40:7
Rufous-browed Peppershrike (229) 40:2
Brown Jay (231) 39:19
Azure-hooded Jay (231) 39:17
Blue-and-white Swallow (233) 22:20
Rufous-and-white Wren (241) 38:9
Plain Wren (241) 38:17
House Wren (243) 38:18
***Ochraceous Wren (243) 38:19**
Gray-breasted Wood-Wren (245) 38:14
***Black-faced Solitaire (249) 39:13**
Orange-billed Nightingale-Thrush (247) 38:26
Slaty-backed Nightingale-Thrush (247) 38:23
Ruddy-capped Nightingale-Thrush (247) 38:27
Swainson's Thrush (249) 39:2
Mountain Thrush (251) 39:6
Clay-colored Thrush (251) 39:8
*Black-and-yellow Silky-Flycatcher (253) 39:10
Golden-winged Warbler (257) 41:5
Tennessee Warbler (255) 40:22
Black-throated Green Warbler (261) 43:9
Black-and-white Warbler (267) 41:13
Wilson's Warbler (269) 42:4
Slate-throated Redstart (273) 42:7
***Collared Redstart (273) 42:6**
Rufous-capped Warbler (275) 40:16
Three-striped Warbler (275) 40:20
Bananaquit (277) 40:24
Common Bush-Tanager (279) 45:13
Summer Tanager (285) 47:5
Blue-gray Tanager (291) 46:15
***Spangle-cheeked Tanager (289) 46:12**
Yellow-faced Grassquit (297) 49:6
*Yellow-thighed Finch (301) 48:22
White-naped Brush-Finch (301) 48:15
Chestnut-capped Brush-Finch (301) 48:16
White-eared Ground-Sparrow (303) 48:19
Rufous-collared Sparrow (307) 50:13
Great-tailed Grackle (317) 44:16
Baltimore Oriole (321) 44:7
Yellow-throated Euphonia (327) 45:5
*Golden-browed Chlorophonia (325) 45:10

Nearby Birding Opportunities

Bajo del Tigre Trails

The Bajo del Tigre Trails are a wonderful example of how two extremely different habitats can occur at close proximity to one another in Costa Rica. While the trails are just a few kilometers from the Monteverde Cloud Forest Reserve, you won't find cloud forest here. Instead, you will see a mountain dry-forest habitat and corresponding avifauna. Some dry-forest specialties that can be found here include White-fronted Parrot, Plain-capped Starthroat, Canivet's Emerald, Rufous-and-white Wren, and Long-tailed Manakin.

Most of the trails are productive, and the area around the buildings often yields Blue-crowned Motmot and White-eared Ground-Sparrow. When walking the trails, keep an ear open for the loud *BOK!* of the Three-wattled Bellbird, as this is a great place to find this exciting endemic. Other birds to look for include Keel-billed Toucan, Emerald Toucanet, Orange-billed Nightingale-Thrush, Rufous-capped Warbler, Ovenbird, and Red-crowned Ant-Tanager.

Coming from the town of Santa Elena, the Bajo del Tigre Trails are located 1.0 km beyond the gas station and 2.6 km before the Monteverde reserve. You will need to park next to a small guardhouse just off the main road and then walk about 400 meters to get to the ticketing office. (This should be well marked.) The entrance fee is $8 (US 2008) per person, and the park hours are 8 a.m. to 5 p.m. You should plan to spend about two hours birding here.

Cerro Amigos and Other Public Areas

The road that originates next to the gas station can be walked all the way to the top of Cerro Amigos and offers productive birding, but the walk is steep. Expect to find some species more common at higher elevations, such as Slaty Flowerpiercer, Sooty-capped Bush-Tanager, Barred Becard, and Buffy Tuftedcheek.

In the towns of Monteverde and Santa Elena, some noteworthy species can be found along the roadsides and in the gardened grounds of many of the hotels. If your hotel has gardened property, it is probably worth your while to stroll through it. Many of the smaller side roads, which receive less traffic, can be productive as well. Look here for Golden-olive Woodpecker, Emerald Toucanet, Blue-crowned Motmot, Rufous-browed Peppershrike, and Brown-capped Vireo.

Site 4A-2: Santa Elena Reserve

Birding Time: 5 hours
Elevation: 1,650 meters
Trail Difficulty: 1–2
Reserve Hours: 7 a.m. to 4 p.m.
Entrance Fee: $12 (US 2008)
4×4 Recommended
Trail Map Available on Site

Santa Elena Reserve plays second fiddle to the nearby Monteverde reserve, and many visitors are not even aware that it exists. However, this is not a place you should overlook because the birding at Santa Elena is often more productive than that at Monteverde, and a number of interesting species are more easily found here.

The reserve was originally created in 1977, but it was not open to the public until 1992. It protects 765 acres of stunning, bromeliad-laden cloud forest, and on a clear day offers a view of the nearby Arenal Volcano. While the reserve itself is significantly smaller than Monteverde, its trail system is comparable in size and it abuts the immense Children's Eternal Rain Forest, so you can expect a full assortment of flora and fauna. Santa Elena also has a small gift shop and café, which is a good place to have lunch.

Santa Elena's lack of popularity compared with the Monteverde reserve is one of its biggest advantages. During the high season (December through April), when Monteverde has roughly 300 visitors a day, you will greatly appreciate the relative serenity of the Santa Elena Reserve.

Santa Elena is a great location to encounter large mixed-species flocks, which are more easily found here than at the Monteverde Cloud Forest Reserve. It is also home to a slightly different bird population, owing to its higher altitude. Here you will more easily find Wrenthrush, Barred Becard, Ruddy-capped Nightingale-Thrush, Fiery-throated Hummingbird, and Sooty-capped Bush-Tanager.

Target Birds

Highland Tinamou (3) 12:7
*Black-breasted Wood-Quail (13) 12:10
Barred Forest-Falcon (53) 16:5
*Buff-fronted Quail-Dove (93) 18:19
Bare-shanked Screech-Owl (107) 20:10
Stripe-tailed Hummingbird (133) 24:17
*Coppery-headed Emerald (133) 24:20
*Orange-bellied Trogon (143) 26:4
Emerald Toucanet (153) 27:17
*Ruddy Treerunner (163) 29:4
Lineated Foliage-gleaner (165) 30:8
*Streak-breasted Treehunter (167) 30:7
Gray-throated Leaftosser (167) 30:10
Immaculate Antbird (179) 31:10
*Silvery-fronted Tapaculo (187) 32:12
Barred Becard (217) 33:14
*Three-wattled Bellbird (221) 34:12
Azure-hooded Jay (231) 39:17
Ruddy-capped Nightingale-Thrush (247) 38:27
*Wrenthrush (275) 32:13
Sooty-capped Bush-Tanager (279) 45:14
*Golden-browed Chlorophonia (325) 45:10

Access

From the town of Santa Elena: (Refer to site 4A-1 for a map.) Head north out of the town of Santa Elena for 0.4 km and then follow the main paved road as it turns left. The road will pass a soccer field on your left, and in 0.6 km you should turn right, following signs for the Santa Elena Reserve and Selvatura. Follow this road straight (keeping right at any forks) until you come to the Santa Elena Reserve in 5.4 km. (Driving time from the town of Santa Elena: 20 min)

Logistics: The Monteverde/Santa Elena area has a full range of accommodations and restaurants. You should have no problem finding exactly what you want, but it is a good idea to make advanced reservations if you are traveling in the high season.

Birding Sites

As you make your way up to the Santa Elena Reserve, do not hesitate to stop periodically along the roadside if you see or hear birds. You can often find Golden-browed Chlorophonia, Emerald Toucanet, and Ruddy-capped Nightingale-Thrush here. Early in the morning you should listen for Black-breasted Wood-Quail, Highland Tinamou, and Barred Forest-Falcon.

Birding near the parking lot before the park opens might yield Streak-breasted Treehunter or Lineated Foliage-gleaner. There are hummingbird feeders at the park entrance, and while they are not nearly as active as those at the Monteverde reserve, you could find Violet Sabrewing, Green Hermit, Green-crowned Brilliant, and Purple-throated Mountain-gem.

As you enter the reserve and pay your entrance fee, be sure to get a map of the trails, which will help you to understand the birding description that follows. One hike that I find quite productive is a three-hour loop that takes you on the Youth Challenge, Encantado, and Del Bajo trails. To start the loop, take the Youth Challenge Trail, which leads to the right from the main buildings. This area is relatively open, fairly protected from the wind, and tends to attract large mixed flocks, usually led by Common Bush-Tanagers or Three-striped Warblers. In these flocks look for Ruddy Treerunner, Spotted Barbtail, Slaty Antwren, Yellow-thighed Finch, and Spangle-cheeked Tanager.

Turn left onto the Del Bajo Trail, which parallels a small stream. This can be a good place to find Resplendent Quetzal, Tufted Flycatcher, Slaty-backed Nightingale-Thrush, and Chestnut-capped Brush-Finch. Turn right onto Encantado Trail, which often holds displaying Three-wattled Bellbirds during the dry season. The bellbirds prefer to call from the upper reaches of the trees and can be difficult to find, but be persistent, as this is an amazing species to see. Also keep an eye open for White-throated Spadebill, Golden-bellied Flycatcher, and Azure-hooded Jay. The Encantado Trail will eventually loop back to the reserve entrance and parking lot.

Nearby Birding Opportunities

San Gerardo Biological Station

If you are up for a bit of adventure along with excellent birding, the San Gerardo Biological Station is a great place to hike to and stay. The station is located within, and run by, the Children's Eternal Rain Forest Reserve. Encompassing 54,000 acres, this is the largest private reserve in all of Central America, and it dwarfs the famous Monteverde Cloud Forest Reserve by about five to one. (The Bajo del Tigre Trails, described under the "Nearby Birding Opportunities" section of site 4A-1, is also part of this reserve, though the wildlife in that area is totally different.)

To get to San Gerardo Biological Station you must follow the small dirt road that continues past the parking lot of the Santa Elena Reserve. This road is passable only with an all-terrain vehicle, so you will need to backpack in. The road leads down the Caribbean slope, for 2.4 km to the biological station, descending about 400 meters to an altitude of 1,250 meters. This road, which proceeds through beautiful forest, offers excellent birding. Look for Resplendent Quetzal, Black-thighed Grosbeak, Coppery-headed Emerald, Emerald Toucanet, and Immaculate Antbird.

After 2.4 km you have the option of approaching the station either by taking a well-marked trail, which leads off to the left, or by continuing along the road, which loops around the back of the station. At the station you will find a well-maintained building with lodging for 28 people, complete with bedding, bathroom, and a staffed kitchen. If the weather cooperates, a porch affords beautiful views of Arenal Volcano and Lake Arenal. In the afternoons this is a good place from which to scan for raptors, such as Black Hawk-Eagle, Ornate Hawk-Eagle, White Hawk, and Barred Hawk.

An extensive system of trails leads through the forest around San Gerardo Biological Station. A number of difficult-to-find upper-middle-elevation species are regulars along these trails. Look for Scaly-throated Foliage-gleaner, Black-and-white Becard, Gray-throated Leaftosser, Rufous-browed Tyrannulet, Rufous-breasted Antthrush, Rough-legged Tyrannulet, and White-bellied Mountain-gem.

The San Gerardo station is most famous for Bare-necked Umbrellabirds, which form leks in the forest between March and June. The best place to look for this amazing species is along the Tabacón Trail, at the very far end of the loop. Tabacón accesses the lowest-elevation forest in the area, and from here it is possible to find some other lower-elevation species such as Blue-and-gold Tanager, Black-and-yellow Tanager, Tawny-capped Euphonia, and Ocellated Antbird.

You must have reservations to visit the San Gerardo Biological Station, and if you spend the night, three meals will be provided.

Children's Eternal Rain Forest Reserve
Tel: (506) 2645-5200 or 2645-5003
Website: www.acmcr.org
Email: acmcr@acmcr.org

Site 4A-3: Bosque de Paz Biological Reserve and Lodge

Birding Time: 1–2 days
Elevation: 1,500 meters
Trail Difficulty: 1–2
Reserve Hours: N/A
Entrance Fee: Included with overnight
Bird Guides Available
Trail Map Available on Site

Bosque de Paz is a family-run biological reserve that was started in 1989. Its owners, the Gonzales family, opened their doors to small groups of students and researchers in 1994. However, it was not until the 1997 American Birding Association conference, held in Costa Rica, that birders discovered how special this site is. Since then, Bosque de Paz has become one of Costa Rica's most popular destinations among birders.

Bosque de Paz is a shining example of the power of ecotourism, as earnings from the lodge have been used to expand the size of the reserve, which currently stands at 2,470 acres. Bosque de Paz now extends between Poás Volcano National Park and Juan Castro Blanco National Park, creating an important biological corridor between these two large protected areas. The lodge sits along the edge of a rushing mountain stream, tucked between forested hills, and has 10 hotel rooms, allowing for a capacity of about 25 people. The Gonzales have done a wonderful job catering to the needs of birders, who make up about 80% of their clientele. The lodge offers a great trail system, expert guides, delicious meals, and well-furnished, comfortable rooms. What else could a birder ask for?

Most birders would ask for birds, and Bosque de Paz has those in abundance. The reserve is located at 1,500 meters, along the Continental Divide, and is home to many sought-after mountain species such as Resplendent Quetzal and Azure-hooded Jay. The area also seems to attract a number of rarities, and some species, which are extremely difficult to find elsewhere in the country, are legitimate possibilities here. Examples include Great Black-Hawk, Solitary Eagle, Black-banded Woodcreeper, and Blue Seedeater.

Target Birds

*Black Guan (11) 12:5
*Black-breasted Wood-Quail (13) 12:10
Bicolored Hawk (39) 16:8
Great Black-Hawk (45) 13:7
Solitary Eagle (45) 13:8
Maroon-chested Ground-Dove (91) 18:10
*Chiriqui Quail-Dove (93) 18:18
*Buff-fronted Quail-Dove (93) 18:19
Green-fronted Lancebill (135) 23:11
*Black-bellied Hummingbird (133) 24:21
*Magenta-throated Woodstar (139) 25:2
*Scintillant Hummingbird (139) 25:7
*Orange-bellied Trogon (143) 26:4
Resplendent Quetzal (145) 26:1
*Streak-breasted Treehunter (167) 30:7
Strong-billed Woodcreeper (175) 29:18
Black-banded Woodcreeper (169) 29:15
Scaled Antpitta (185) 30:20
Torrent Tyrannulet (201) 36:1
*Dark Pewee (203) 36:2
Barred Becard (217) 33:14
Brown-capped Vireo (227) 40:14
Azure-hooded Jay (231) 39:17
American Dipper (251) 39:11
Ruddy-capped Nightingale-Thrush (247) 38:27
*Long-tailed Silky-Flycatcher (253) 39:15
*Flame-throated Warbler (257) 41:7
Blue Seedeater (297) 49:11
Chestnut-capped Brush-Finch (301) 48:16
*Golden-browed Chlorophonia (325) 45:10

Access

From Rt. 1: Exit off Rt. 1 toward the city of Naranjo. You will enter the downtown area in 2.0 km. Proceed through the town, following signs for Zarcero. At 1.2 km outside Naranjo you will come to a confusing intersection. Fork left, and

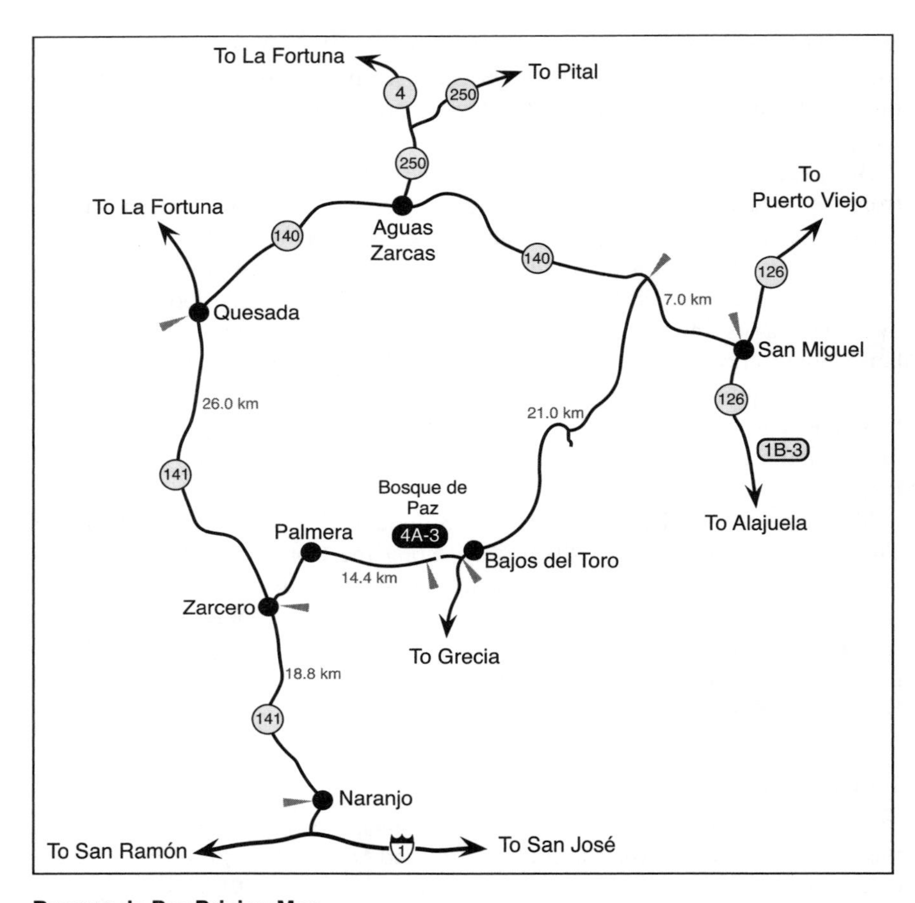

Bosque de Paz Driving Map

then almost immediately fork right. From this point, continue 17.6 km to the Zarcero city park. Turn right at the far corner of the park and follow this road for 14.4 km, passing through the small town of Palmera, to Bosque de Paz. The lodge will be on your left just after the pavement ends. (Driving time from Rt. 1: 60 min)

From Quesada: Take the main road toward Zarcero for 26.0 km and then turn left at the near corner of the Zarcero city park. Follow this road for 14.4 km, passing through the town of Palmera, to Bosque de Paz. The lodge will be on your left just after the pavement ends. (Driving time from Quesada: 60 min)

From the intersection of Rt. 126 and Rt. 140: Take Rt. 140 toward Aguas Zarcas and La Fortuna for 7.0 km. Turn left onto a paved road heading toward Bajos del Toro. In 12.0 km you will come to an intersection, where you should turn right. At 21.0 km from Rt.140, turn right onto a dirt road located just on the far side of the town of Bajos del Toro. Follow this dirt road for 1.6 km to Bosque de Paz, which will be on your right. (Driving time from Rt. 126: 50 min)

Logistics: I highly recommend lodging at Bosque de Paz. The hotel will supply you with three meals a day, but you should bring in anything else you may need.

Bosque de Paz Lodge
Tel: (506) 2234-6676
Website: www.bosquedepaz.com
E-mail: info@bosquedepaz.com

Birding Sites

When you first arrive at Bosque de Paz you will find a wealth of hummingbirds right around the buildings, thanks to feeders and flowering bushes. Look for Violet Sabrewing, Green Violetear, Purple-throated Mountain-gem, Magenta-throated Woodstar, and Magnificent Hummingbird at the feeders. The porterweed bushes in the area should hold Scintillant Hummingbird and Black-bellied Hummingbird. Management also puts out cracked corn near the hummingbird feeders, which attracts Black Guan, Chestnut-capped Brush-Finch, Chiriqui Quail-Dove, and Buff-fronted Quail-Dove. Check the trees around the buildings and parking area for Brown-capped Vireo and Rufous-browed Peppershrike.

Bosque de Paz has a beautiful and extensive system of trails, and describing them all here would be impossible. While all the trails will yield interesting birds, those described below are favorites among birders. Ask at reception for a map of the area.

Sendero El Valle is the most popular birding trail, offering an incredible diversity of birds. The trail loops around regenerating forest in what used to be cattle pasture. Many migrant warblers are attracted to this area, and you might be able to find Golden-winged Warbler, Blackburnian Warbler, Bay-breasted Warbler, and Black-throated Green Warbler during the northern winter. Flame-throated Warbler should also be expected. This is also a good place to look for Dark Pewee, Golden-bellied Flycatcher, and Orange-bellied Trogon. Some rarities that might be found here include Scaled Antpitta, Solitary Eagle, Yellow-bellied Sapsucker, and Blue Seedeater.

Sendero Jaulares, which leads off of Sendero El Valle, passes through similar habitat, and you should expect to find a comparable range of species.

Primary forest is best accessed via Sendero Botánico and Sendero Galería. Along these trails you will have good chances to see Azure-hooded Jay, Slaty-backed Nightingale-Thrush, and Lineated Foliage-gleaner. Black-breasted Wood-Quail and Silvery-fronted Tapaculo are regularly heard but difficult to see.

Some of the best birding can be found during walks along the main road in front of Bosque de Paz. Go right out of the driveway toward Zarcero. Along this first stretch of road, which has old pastures on your left, you may be able to find Long-tailed Silky-Flycatcher, Golden-browed Chlorophonia, and Elegant Euphonia. Keep an eye upward for a soaring Red-tailed Hawk, or possibly a Great Black-Hawk. Less than half a kilometer up the road the pavement begins, and here the birding becomes excellent. There are a variety of forest habitats on either side, and you have good chances of finding Resplendent Quetzal, Red-headed Barbet, White-naped Brush-Finch, Streak-breasted Treehunter, and Emerald Toucanet. Unusual species, such as Peg-billed Finch and Black-banded Woodcreeper, are also possibilities here.

Species to Expect

***Black Guan (11) 12:5**
*Black-breasted Wood-Quail (13) 12:10
Black Vulture (31) 13:4
Turkey Vulture (31) 13:3
Bicolored Hawk (39) 16:8
Great Black-Hawk (45) 13:7
Red-tailed Hawk (47) 17:8
Band-tailed Pigeon (87) 18:1
White-tipped Dove (91) 18:14
*Chiriqui Quail-Dove (93) 18:18
*Buff-fronted Quail-Dove (93) 18:19
White-crowned Parrot (97) 19:7
*Dusky Nightjar (111) 21:19
Black Swift (117) 22:2
White-collared Swift (117) 22:1
Green Hermit (121) 23:3
Green-fronted Lancebill (135) 23:11
Violet Sabrewing (123) 23:9
Green Violetear (131) 23:7
*Black-bellied Hummingbird (133) 24:21
***Purple-throated Mountain-gem (135) 24:7**
Green-crowned Brilliant (123) 23:15
Magnificent Hummingbird (123) 23:16
*Magenta-throated Woodstar (139) 25:2
*Scintillant Hummingbird (139) 25:7
Collared Trogon (143) 26:5
*Orange-bellied Trogon (143) 26:4
Resplendent Quetzal (145) 26:1
Red-headed Barbet (153) 28:2
***Prong-billed Barbet (153) 28:1**
Emerald Toucanet (153) 27:17
Hairy Woodpecker (159) 28:19
Smoky-brown Woodpecker (159) 28:13
Red-faced Spinetail (163) 30:12
Spotted Barbtail (163) 29:5
Lineated Foliage-gleaner (165) 30:8
***Streak-breasted Treehunter (167) 30:7**
Black-banded Woodcreeper (169) 29:15
Spot-crowned Woodcreeper (173) 29:9
Scaled Antpitta (185) 30:20
*Silvery-fronted Tapaculo (187) 32:12
Mountain Elaenia (193) 37:24
Olive-striped Flycatcher (195) 36:24
Slaty-capped Flycatcher (195) 36:26
Paltry Tyrannulet (191) 37:10
Scale-crested Pygmy-Tyrant (197) 37:4
White-throated Spadebill (189) 37:1
Tufted Flycatcher (207) 36:11
***Dark Pewee (203) 36:2**
Yellowish Flycatcher (207) 36:19
Black Phoebe (207) 36:5
Bright-rumped Attila (201) 35:6
Social Flycatcher (211) 35:14
***Golden-bellied Flycatcher (211) 35:9**
Sulphur-bellied Flycatcher (213) 35:10
Tropical Kingbird (213) 35:1
Barred Becard (217) 33:14
Masked Tityra (217) 34:1
*Yellow-winged Vireo (229) 40:10

Brown-capped Vireo (227) 40:14
Rufous-browed Peppershrike (229) 40:2
Brown Jay (231) 39:19
Azure-hooded Jay (231) 39:17
Blue-and-white Swallow (233) 22:20
House Wren (243) 38:18
***Ochraceous Wren (243) 38:19**
Gray-breasted Wood-Wren (245) 38:14
***Black-faced Solitaire (249) 39:13**
Slaty-backed Nightingale-Thrush (247) 38:23
Ruddy-capped Nightingale-Thrush (247) 38:27
Mountain Thrush (251) 39:6
Clay-colored Thrush (251) 39:8
***Long-tailed Silky-Flycatcher (253) 39:15**
***Flame-throated Warbler (257) 41:7**
Tropical Parula (257) 41:3
Black-throated Green Warbler (261) 43:9
Black-and-white Warbler (267) 41:13
Wilson's Warbler (269) 42:4
Slate-throated Redstart (273) 42:7
Golden-crowned Warbler (275) 40:18
*Black-cheeked Warbler (275) 40:17
Three-striped Warbler (275) 40:20
Common Bush-Tanager (279) 45:13
Blue-gray Tanager (291) 46:15
Silver-throated Tanager (291) 46:6
*Spangle-cheeked Tanager (289) 46:12
Scarlet-thighed Dacnis (291) 46:4
Yellow-faced Grassquit (297) 49:6
***Slaty Flowerpiercer (299) 49:9**
***Yellow-thighed Finch (301) 48:22**
White-naped Brush-Finch (301) 48:15
Chestnut-capped Brush-Finch (301) 48:16
Rufous-collared Sparrow (307) 50:13
*Black-thighed Grosbeak (311) 48:7
Bronzed Cowbird (317) 44:15
Elegant Euphonia (325) 45:9
*Golden-browed Chlorophonia (325) 45:10
Yellow-bellied Siskin (325) 50:1

Site 4A-4: La Paz Waterfall Gardens

Birding Time: 3 hours
Elevation: 1,500 meters
Trail Difficulty: 1
Reserve Hours: 8:30 a.m. to 4:00 p.m.
Entrance Fee: $32 (US 2008)
Trail Map Available on Site

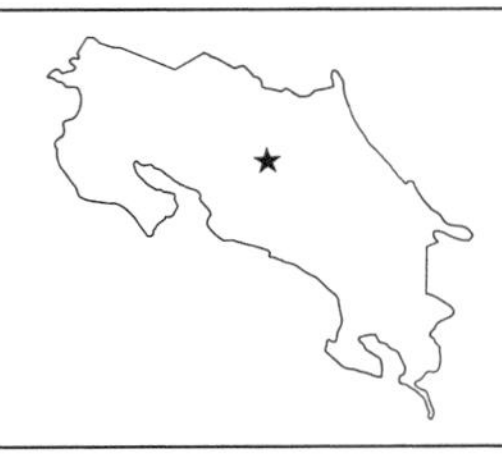

La Paz Waterfall Gardens, which first opened in 2000, sits on 70 acres of forest around the La Paz River Valley. The area offers something for everyone, boasting two restaurants, a butterfly garden, serpentarium, ranarium (frogs), orchid garden, trout pond, aviary, and five magnificent waterfalls. The grounds are kept in immaculate condition, and the army of friendly and helpful employees is always ready to assist in whatever way possible. If you like the feeling of luxury, you will be right at home at La Paz.

Birding this site can be quite productive and relatively easy. The hummingbird garden is always a highlight, with a large variety of easily viewed hummers. Fruit feeders are kept filled and attract many tanagers and sparrows. Access to the forest is provided by three and a half kilometers of paved trails. The whole area can be birded fairly quickly, which means you should plan to take in some of the other attractions offered on site to make the steep entrance fee worthwhile.

Target Birds

Barred Hawk (41) 17:1	*White-bellied Mountain-gem (135) 24:5
Bare-shanked Screech-Owl (107) 20:10	*Scintillant Hummingbird (139) 25:7
Green-fronted Lancebill (135) 23:11	Ochre-breasted Antpitta (185) 30:13
Brown Violetear (131) 23:6	*Silvery-fronted Tapaculo (187) 32:12
Green Thorntail (137) 25:1	Torrent Tyrannulet (201) 36:1
*Black-bellied Hummingbird (133) 24:21	American Dipper (251) 39:11
*Coppery-headed Emerald (133) 24:20	*Sooty-faced Finch (301) 48:21

Access

From Alajuela: (Refer to site 4B-1 for a map.) Start at the Central Park and go four blocks north to a large four-way intersection with a traffic light. Proceed straight across the intersection onto Rt. 130 and follow this road straight for 17.8 km to a T-intersection. Turn right here, and in 2.8 km make another right turn heading toward Vara Blanca. After 5.8 km turn left just before a gas station. La Paz Waterfall Gardens will be on your left in 6.0 km. (Driving time from Alajuela: 50 min)

From the intersection of Rt. 126 and Rt. 140: (Refer to site 4B-1 for a map.) Take Rt. 126 south (uphill) following signs for Heredia. Proceed straight for 19.8 km and La Paz Waterfall Gardens will be on your right. (Driving time from Rt. 140: 30 min)

Logistics: La Paz Waterfall Gardens is easily birded as a stopover between larger sites. However, if you wish to stay in the area, the associated Peace Lodge is one of the nicest hotels in the country, and there are other comfortable hotels in nearby Vara Blanca and Poasito.

Peace Lodge
Tel: (506) 2225-0643
Website: www.waterfallgardens.com/lapaz-peacelodge.html
E-mail: peacelodgereservations@waterfallgardens.com

Birding Sites

Most birders immediately gravitate to the hummingbird garden. Viewing is as easy as it gets, and the variety is exciting. Expect to find Brown Violetear, Green Violetear, Coppery-headed Emerald, Green Thorntail, Black-bellied Hummingbird, Violet Sabrewing, and Purple-throated Mountain-gem.

Three fruit feeding stations have also been set up around the grounds to attract tanagers and other species. One is located immediately behind the reception building, another is behind the hummingbird garden, and the third is near the orchid garden. Birds to look for on the feeders include Silver-throated Tanager, Tawny-capped Euphonia, Yellow-thighed Finch, and Sooty-faced Finch. It is usually worth checking the feeders periodically throughout your time at La Paz, as new groups of birds arrive regularly.

The landscaped grounds around the buildings generally yield a good variety of species. Flycatchers such as Olive-striped Flycatcher and Dark Pewee can be found in the trees. Slaty Flowerpiercer often feeds in the flowering shrubs along with Scintillant Hummingbird.

The trails that lead through the forest will yield many different species. Three-striped Warbler, Gray-breasted Wood-Wren, and Common Bush-Tanager are abundant. You also have a good chance to encounter Spangle-cheeked Tanager, Lineated Foliage-gleaner, Spotted Barbtail, White-throated Spadebill, and Slaty Antwren. Most of the trails run along the river, and while they offer beautiful views, hearing birds over the roar of the water can be quite difficult. The Fern Trail, which leaves from the right edge of the trout pond, is generally quieter and one of the best walks to bird.

Outside the gardens there are a few interesting locations worth checking along Rt. 126. The planted trees immediately across the road from the entrance to La Paz Waterfall Gardens could produce migrant warblers, Yellow-bellied Sapsucker, Rufous-browed Peppershrike, and Ruddy Pigeon.

Following Rt. 126 north toward the Caribbean for 1.4 km will bring you to a large waterfall and a bridge over the La Paz River. There is space to pull off and park on either side of the bridge and you should do so. Scan the rocks in the river for American Dipper and Torrent Tyrannulet.

If you arrive at La Paz before it opens, there is a small road nearby that is worth exploring. From the waterfall gardens you should drive 2.8 km south, toward Vara Blanca, to where a small road forks off to the left. The first half-kilometer of this road contains good habitat that often produces some interesting species. Keep an eye out for Golden-browed Chlorophonia, Brown-capped Vireo, Emerald Toucanet, and Ruddy-capped Nightingale-Thrush.

Species to Expect

*Black Guan (11) 12:5
Black Vulture (31) 13:4
Turkey Vulture (31) 13:3
Ruddy Pigeon (87) 18:5
*Crimson-fronted Parakeet (95) 19:10
White-crowned Parrot (97) 19:7
White-collared Swift (117) 22:1
Green Hermit (121) 23:3
Violet Sabrewing (123) 23:9
Brown Violetear (131) 23:6
Green Violetear (131) 23:7
Green Thorntail (137) 25:1
Rufous-tailed Hummingbird (129) 24:10
***Black-bellied Hummingbird (133) 24:21**
***Coppery-headed Emerald (133) 24:20**
***Purple-throated Mountain-gem (135) 24:7**
Green-crowned Brilliant (123) 23:15
*Scintillant Hummingbird (139) 25:7
Collared Trogon (143) 26:5
*Prong-billed Barbet (153) 28:1
Emerald Toucanet (153) 27:17
Yellow-bellied Sapsucker (161) 28:20
Hairy Woodpecker (159) 28:19
Golden-olive Woodpecker (159) 28:11
Red-faced Spinetail (163) 30:12
Spotted Barbtail (163) 29:5
Lineated Foliage-gleaner (165) 30:8
Tawny-throated Leaftosser (167) 30:11
Wedge-billed Woodcreeper (171) 29:6

Spotted Woodcreeper (173) 29:20
Slaty Antwren (183) 32:3
*Silvery-fronted Tapaculo (187) 32:12
Yellow-bellied Elaenia (193) 37:26
Mountain Elaenia (193) 37:24
Torrent Tyrannulet (201) 36:1
Olive-striped Flycatcher (195) 36:24
Slaty-capped Flycatcher (195) 36:26
Paltry Tyrannulet (191) 37:10
Scale-crested Pygmy-Tyrant (197) 37:4
White-throated Spadebill (189) 37:1
Tufted Flycatcher (207) 36:11
*Dark Pewee (203) 36:2
Tropical Pewee (203) 36:9
Yellowish Flycatcher (207) 36:19
Black Phoebe (207) 36:5
Great Kiskadee (211) 35:13
Social Flycatcher (211) 35:14
*Golden-bellied Flycatcher (211) 35:9
Tropical Kingbird (213) 35:1
Barred Becard (217) 33:14
Rufous-browed Peppershrike (229) 40:2
Brown Jay (231) 39:19
Blue-and-white Swallow (233) 22:20
House Wren (243) 38:18
*Ochraceous Wren (243) 38:19
Gray-breasted Wood-Wren (245) 38:14
American Dipper (251) 39:11
***Black-faced Solitaire (249) 39:13**
Slaty-backed Nightingale-Thrush (247) 38:23
Swainson's Thrush (249) 39:2
Mountain Thrush (251) 39:6
Clay-colored Thrush (251) 39:8
*Long-tailed Silky-Flycatcher (253) 39:15
Golden-winged Warbler (257) 41:5
Tennessee Warbler (255) 40:22
Tropical Parula (257) 41:3
Black-throated Green Warbler (261) 43:9
Black-and-white Warbler (267) 41:13
Louisiana Waterthrush (269) 43:13
Wilson's Warbler (269) 42:4
Slate-throated Redstart (273) 42:7
Three-striped Warbler (275) 40:20
Bananaquit (277) 40:24
Common Bush-Tanager (279) 45:13
Summer Tanager (285) 47:5
Crimson-collared Tanager (287) 47:3
Passerini's Tanager (287) 47:4
Blue-gray Tanager (291) 46:15
Palm Tanager (291) 45:19
Silver-throated Tanager (291) 46:6
Golden-hooded Tanager (289) 46:13
***Spangle-cheeked Tanager (289) 46:12**
Scarlet-thighed Dacnis (291) 46:4
Variable Seedeater (295) 49:3
Yellow-faced Grassquit (297) 49:6
***Slaty Flowerpiercer (299) 49:9**
*Sooty-faced Finch (301) 48:21
***Yellow-thighed Finch (301) 48:22**
Chestnut-capped Brush-Finch (301) 48:16
Rufous-collared Sparrow (307) 50:13
Buff-throated Saltator (309) 48:2
Great-tailed Grackle (317) 44:16
Baltimore Oriole (321) 44:7
***Tawny-capped Euphonia (325) 45:6**
*Golden-browed Chlorophonia (325) 45:10

Site 4A-5: Tapantí National Park

Birding Time: 1 day
Elevation: 1,250 meters
Trail Difficulty: 2
Reserve Hours: 8 a.m. to 4 p.m.
Entrance Fee: $10 (US 2008)
4×4 Recommended
Trail Map Available on Site

Tapantí National Park protects 144,000 acres of forest on the northern edge of the Talamanca Mountain Range. While there are two public entrances to the park, only the principal entrance, located near the towns of Orosí and Puricíl, is described here. This area is a birding gem, providing access to a wide diversity of species in beautiful upper-middle-elevation forest. Steep mountain slopes draped in dense epiphyte-laden trees dominate the landscape, while a rushing mountain river, the Río Grande de Orosí, runs below. A gravel road and a trail system provide access to this beautiful habitat. (The other public entrance to the park, La Esperanza de Tapantí, is located off Rt. 2 and exhibits a much higher-elevation forest. It is described under "Nearby Birding Opportunities" in site 4B-3.)

The location of Tapantí National Park has a lot to recommend it. First, it represents the northern end of contiguous protected land that runs all the way into Panama. This is one of the largest blocks of protected land in Central America, and it provides refuge to a full complement of wildlife. Second, Tapantí is only about an hour's drive from San José, nestled in the picturesque coffee plantations of the Orosí Valley, and makes a wonderful day trip from the Central Valley. On the downside, its proximity to the Central Valley makes it a popular weekend and holiday destination for locals, and it is sometimes difficult to find peace and quiet. The Tapantí region is also one of the rainiest in the country, receiving about seven meters annually, so be sure to come prepared.

Tapantí National Park has the lowest elevation of all the sites in the Mountains region, and you should expect to see a good deal of Caribbean Slope influence on the avifauna. Birds such as White-bellied Mountain-gem, Rufous-breasted Antthrush, and Tawny-capped Euphonia are observed here regularly. While the density of the forest can make this a challenging location for spotting birds, Tapantí always seems to yield rare and exciting species to those who have patience and persistence.

Target Birds

Buffy-crowned Wood-Partridge (13) 12:16
Black Hawk-Eagle (49) 13:9
Barred Parakeet (97) 19:16
Green-fronted Lancebill (135) 23:11
Green Thorntail (137) 25:1
*Black-bellied Hummingbird (133) 24:21
*White-bellied Mountain-gem (135) 24:5
Lineated Foliage-gleaner (165) 30:8
Buff-fronted Foliage-gleaner (165) 30:6
Streaked Xenops (169) 29:2
Strong-billed Woodcreeper (175) 29:18
Black-banded Woodcreeper (169) 29:15
Rufous-rumped Antwren (183) 32:10
Rufous-breasted Antthrush (187) 30:16
Scaled Antpitta (185) 30:20
Ochre-breasted Antpitta (185) 30:13
Torrent Tyrannulet (201) 36:1
*Dark Pewee (203) 36:2
Black-and-white Becard (217) 33:15
Sharpbill (225) 34:7
Azure-hooded Jay (231) 39:17
American Dipper (251) 39:11
White-winged Tanager (285) 47:7
*Sooty-faced Finch (301) 48:21
Elegant Euphonia (325) 45:9
*Golden-browed Chlorophonia (325) 45:10

Access

From Paraíso: (Refer to site 1B-8 for a map.) Head south from the central park on Rt. 224 toward Orosí. Go 10.4 km, passing through the town of Orosí, and then turn right toward the town of Puricíl. Continue straight on this road for 7.4 km, passing through Puricíl and over the Río Grande de Orosí. Here you will come to a fork. Turn right and proceed another 1.6 km to the national park entrance. (Driving time from Paraíso: 30 min)

Logistics: The Kiri Lodge is one of the few lodging options available near the park. Its rooms are small and basic but clean, and the price is reasonable. It has a nice restaurant (with bird feeders) that serves breakfast, lunch, and dinner. Otherwise, there are a few hotels in the town of Orosí, about 20 minutes away.

Kiri Lodge
Tel: (506) 2533-2272
Website: www.kirilodge.net
E-mail: info@kirilodge.net

Birding Sites

Outside the Park

Many interesting birds can be found in the cow pastures and coffee fields just before the national park entrance, and I recommend spending some early-morning time here before the park opens. Start by birding around the bridge that crosses the Río Grande de Orosí. This area often produces Olive-crowned Yellowthroat, Chestnut-headed Oropendola, and Mourning Warbler. Be sure to scan the river for Black Phoebe, Torrent Tyrannulet, and American Dipper.

After the bridge, proceed 0.6 km to a fork in the road. The left fork will take you toward the Kiri Lodge, where you may find Yellow-throated Brush-Finch and Hoffmann's Woodpecker along the roadside. Porterweed bushes around the lodge often attract Green Thorntail and other interesting hummingbirds. At dawn Black-breasted Wood-Quail can be heard singing in the nearby forest.

If you take the right fork, it will lead to the national park entrance in 1.6 km. This stretch of road passes through pastureland and secondary forest, and is often very productive. Taking the time to walk along the road will generally yield species such as Western Wood-Pewee, Dark Pewee, Elegant Euphonia, and Spotted Woodcreeper.

Tapantí National Park

From the park headquarters, where you must pay an entrance fee, a gravel road continues up the valley for 3.8 km. The park trails, which are discussed later, lead off the road at various points. However, the birding found along this road is often the best in the park, and you should spend a large percentage of your time here. Flowering trees will yield many hummingbirds, including White-bellied Mountain-gem, Black-bellied Hummingbird, and Green-fronted Lancebill. Mixed flocks are regularly encountered and contain Lineated Foliage-gleaner, Spangle-cheeked Tanager, and Red-faced Spinetail. Other interesting species to look for from the roadside are White-winged Tanager, Collared Trogon, Red-headed Barbet, Azure-hooded Jay, and Streaked Xenops. Raptors, including Barred Hawk and Black Hawk-Eagle, are regularly seen soaring over the valley. At the very end of the road is a lookout and picnic area. Here you have a good chance of finding Three-striped Warbler, Black-and-yellow Silky-Flycatcher, and Black Guan.

The park also maintains three trails of varying length and difficulty that allow you to go deeper into the forest. The Sendero Oropéndola is a pretty walk that consistently produces interesting birds. It forms a strangely shaped figure eight, the bottom of which runs along the riverbank. Look here for American Dipper and Torrent Tyrannulet. Other birds that can be seen from this trail are Golden-bellied Flycatcher, White-throated Spadebill, Slaty-capped Flycatcher, Scaled Antpitta, and Black-faced Solitaire.

Sendero Árboles Caídos is a strenuous hike with awkward and muddy footing. Nonetheless, it can reward the adventurous birder with sitings of unusual species, including Rufous-breasted Antthrush, Brown-billed Scythebill, and Buff-fronted Foliage-gleaner. Finally, Sendero La Catarata leads to a waterfall, and Sendero La Pava accesses the river. While the scenery on both these trails is nice, the birding is generally not as productive as in the other areas in the park.

Species to Expect

*Black Guan (11) 12:5
*Black-breasted Wood-Quail (13) 12:10
Cattle Egret (23) 5:13
Black Vulture (31) 13:4
Turkey Vulture (31) 13:3
Swallow-tailed Kite (37) 15:2
Broad-winged Hawk (43) 16:13
Ornate Hawk-Eagle (49) 17:11
Spotted Sandpiper (73) 11:8
Band-tailed Pigeon (87) 18:1
Ruddy Pigeon (87) 18:5
***Crimson-fronted Parakeet (95) 19:10**
White-crowned Parrot (97) 19:7
Squirrel Cuckoo (103) 21:7
Bare-shanked Screech-Owl (107) 20:10
Mottled Owl (107) 20:6
Common Pauraque (111) 21:18
White-collared Swift (117) 22:1
Green Hermit (121) 23:3
Green-fronted Lancebill (135) 23:11
Green Thorntail (137) 25:1
Rufous-tailed Hummingbird (129) 24:10
***Black-bellied Hummingbird (133) 24:21**
***White-bellied Mountain-gem (135) 24:5**
***Purple-throated Mountain-gem (135) 24:7**
Green-crowned Brilliant (123) 23:15
Purple-crowned Fairy (125) 23:14
Collared Trogon (143) 26:5
Red-headed Barbet (153) 28:2
***Prong-billed Barbet (153) 28:1**
Smoky-brown Woodpecker (159) 28:13
Golden-olive Woodpecker (159) 28:11
Red-faced Spinetail (163) 30:12
Spotted Barbtail (163) 29:5
*Ruddy Treerunner (163) 29:4

Lineated Foliage-gleaner (165) 30:8
Buff-fronted Foliage-gleaner (165) 30:6
Streaked Xenops (169) 29:2
Olivaceous Woodcreeper (171) 29:7
Wedge-billed Woodcreeper (171) 29:6
Spotted Woodcreeper (173) 29:20
Brown-billed Scythebill (175) 29:14
Slaty Antwren (183) 32:3
Rufous-breasted Antthrush (187) 30:16
Scaled Antpitta (185) 30:20
*Silvery-fronted Tapaculo (187) 32:12
Yellow-bellied Elaenia (193) 37:26
Mountain Elaenia (193) 37:24
Torrent Tyrannulet (201) 36:1
Olive-striped Flycatcher (195) 36:24
Slaty-capped Flycatcher (195) 36:26
Paltry Tyrannulet (191) 37:10
Scale-crested Pygmy-Tyrant (197) 37:4
Common Tody-Flycatcher (197) 37:7
Yellow-olive Flycatcher (195) 37:16
White-throated Spadebill (189) 37:1
Tufted Flycatcher (207) 36:11
***Dark Pewee (203) 36:2**
Western Wood-Pewee (203) 36:7
Eastern Wood-Pewee (203) 36:8
Tropical Pewee (203) 36:9
Yellowish Flycatcher (207) 36:19
Black Phoebe (207) 36:5
Bright-rumped Attila (201) 35:6
Rufous Mourner (201) 34:9
Dusky-capped Flycatcher (209) 35:21
Boat-billed Flycatcher (211) 35:12
Social Flycatcher (211) 35:14
*Golden-bellied Flycatcher (211) 35:9
Tropical Kingbird (213) 35:1
Barred Becard (217) 33:14
Black-and-white Becard (217) 33:15
Yellow-throated Vireo (225) 40:6
Red-eyed Vireo (227) 40:3
Lesser Greenlet (229) 40:7
Brown Jay (231) 39:19
Azure-hooded Jay (231) 39:17
Blue-and-white Swallow (233) 22:20
Plain Wren (241) 38:17
House Wren (243) 38:18
***Ochraceous Wren (243) 38:19**
Gray-breasted Wood-Wren (245) 38:14
American Dipper (251) 39:11
***Black-faced Solitaire (249) 39:13**
Slaty-backed Nightingale-Thrush (247) 38:23
Swainson's Thrush (249) 39:2
Mountain Thrush (251) 39:6
Clay-colored Thrush (251) 39:8
*Black-and-yellow Silky-Flycatcher (253) 39:10
Golden-winged Warbler (257) 41:5
Tennessee Warbler (255) 40:22
Tropical Parula (257) 41:3
Chestnut-sided Warbler (259) 43:4
Black-throated Green Warbler (261) 43:9
Blackburnian Warbler (261) 41:8
Bay-breasted Warbler (263) 41:9
Black-and-white Warbler (267) 41:13
Olive-crowned Yellowthroat (273) 42:11
Wilson's Warbler (269) 42:4
Slate-throated Redstart (273) 42:7
Three-striped Warbler (275) 40:20
Bananaquit (277) 40:24
Common Bush-Tanager (279) 45:13
Summer Tanager (285) 47:5
White-winged Tanager (285) 47:7
Passerini's Tanager (287) 47:4
Blue-gray Tanager (291) 46:15
Palm Tanager (291) 45:19
Silver-throated Tanager (291) 46:6
Speckled Tanager (289) 46:8
Bay-headed Tanager (289) 46:9
Golden-hooded Tanager (289) 46:13
***Spangle-cheeked Tanager (289) 46:12**
Variable Seedeater (295) 49:3
Yellow-faced Grassquit (297) 49:6
*Sooty-faced Finch (301) 48:21
White-naped Brush-Finch (301) 48:15
Rufous-collared Sparrow (307) 50:13
Buff-throated Saltator (309) 48:2
*Black-thighed Grosbeak (311) 48:7
Great-tailed Grackle (317) 44:16
Baltimore Oriole (321) 44:7
Chestnut-headed Oropendola (323) 44:9
Montezuma Oropendola (323) 44:8
Elegant Euphonia (325) 45:9
White-vented Euphonia (327) 45:7
***Tawny-capped Euphonia (325) 45:6**
*Golden-browed Chlorophonia (325) 45:10

Subregion 4B

The Upper Mountains

The **Volcano Junco** is a true high-elevation specialist, consistently found only above the timberline. It usually feeds on the ground, where it is quite accommodating and allows for easy approach. This species is seen most reliably at Irazú Volcano National Park (4B-2), though it is also a regular atop Cerro de la Muerte (4B-5).

Common Birds to Know

Black Vulture (31) 13:4
Turkey Vulture (31) 13:3
Mountain Elaenia (193) 37:24
Blue-and-white Swallow (233) 22:20
Gray-breasted Wood-Wren (245) 38:14
Black-throated Green Warbler (261) 43:9
Wilson's Warbler (269) 42:4
*Sooty-capped Bush-Tanager (279) 45:14
*Slaty Flowerpiercer (299) 49:9
*Yellow-thighed Finch (301) 48:22
Rufous-collared Sparrow (307) 50:13

Site 4B-1: Poás Volcano National Park

Birding Time: 4 hours
Elevation: 2,400 meters
Trail Difficulty: 1–2
Reserve Hours: 8 a.m. to 3:30 p.m.
Entrance Fee: $10 (US 2008)
Trail Map Available on Site

For birders, Poás Volcano National Park offers easy access to high-elevation forest and many of the endemics found there. Although it does not yield as much diversity as some of the Upper Mountain sites farther south, Poás has two significant advantages recommending it. First, it is conveniently situated only 45 minutes from the airport, near other interesting birding destinations, along one of the main routes to the Caribbean coast. Second, it allows you the chance to take in some of Costa Rica's most impressive geology.

The crater of Poás Volcano is one of the world's most accessible. You can stroll up to the edge of the massive 1,300-meter-wide, 300-meter-deep hole and gaze down into an unearthly landscape. Sulfurous gasses vent from the sides of the crater while an acidic lake simmers at the bottom. It is an awesome sight.

If you wish to see the crater, try to arrive as early in the day as possible. Clouds and fog often move in very early, shrouding it from view. On some rainy days the crater is never visible. Good weather has its own price, however, as hundreds of people often come up to visit the park (especially on weekends). But don't let this dissuade you. There are out-of-the-way places to bird that will yield most of the key species and give you some peace and quiet.

Target Birds

Highland Tinamou (3) 12:7
*Buff-fronted Quail-Dove (93) 18:19
*Dusky Nightjar (111) 21:19
*Volcano Hummingbird (139) 25:3
*Sooty Thrush (249) 39:7
*Black-and-yellow Silky-Flycatcher (253) 39:10
*Flame-throated Warbler (257) 41:7
*Wrenthrush (275) 32:13
*Peg-billed Finch (299) 49:12

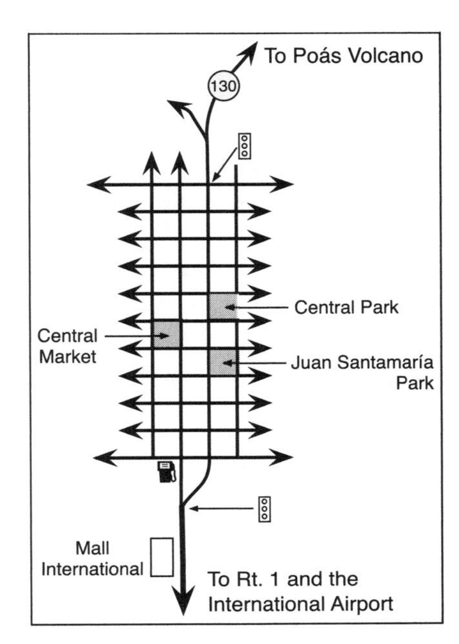

Alajuela City Map

Access

From Alajuela: Start at the Central Park and go four blocks north to a large four-way intersection with a traffic light. Proceed straight across the intersection onto Rt. 130, which you should follow for 17.8 km as it climbs uphill. When the road ends at a T-intersection, turn right, and in 10.2 km you will come to the entrance of the park. (Driving time from Alajuela: 40 min)

From the intersection of Rt. 126 and Rt. 140: Take Rt. 126 heading south into the mountains, following signs for Heredia. After 25.8 km the road will come to a T-intersection in the town of Vara Blanca. Turn right and proceed 5.8 km to another T-intersection. Turn right again, and in 7.4 km you will come to the entrance to the park. (Driving time from Rt. 140: 60 min)

Logistics: There are a number of lodging options available in the nearby towns of Poasito and Vara Blanca. A number of small restaurants are also located here, and there is a snack bar at the park.

Birding Sites

Poás Volcano National Park

When you enter the national park you will pass through a large gate, with rock walls on either side. From here it is 1.4 km to the ticketing station. Feel free to

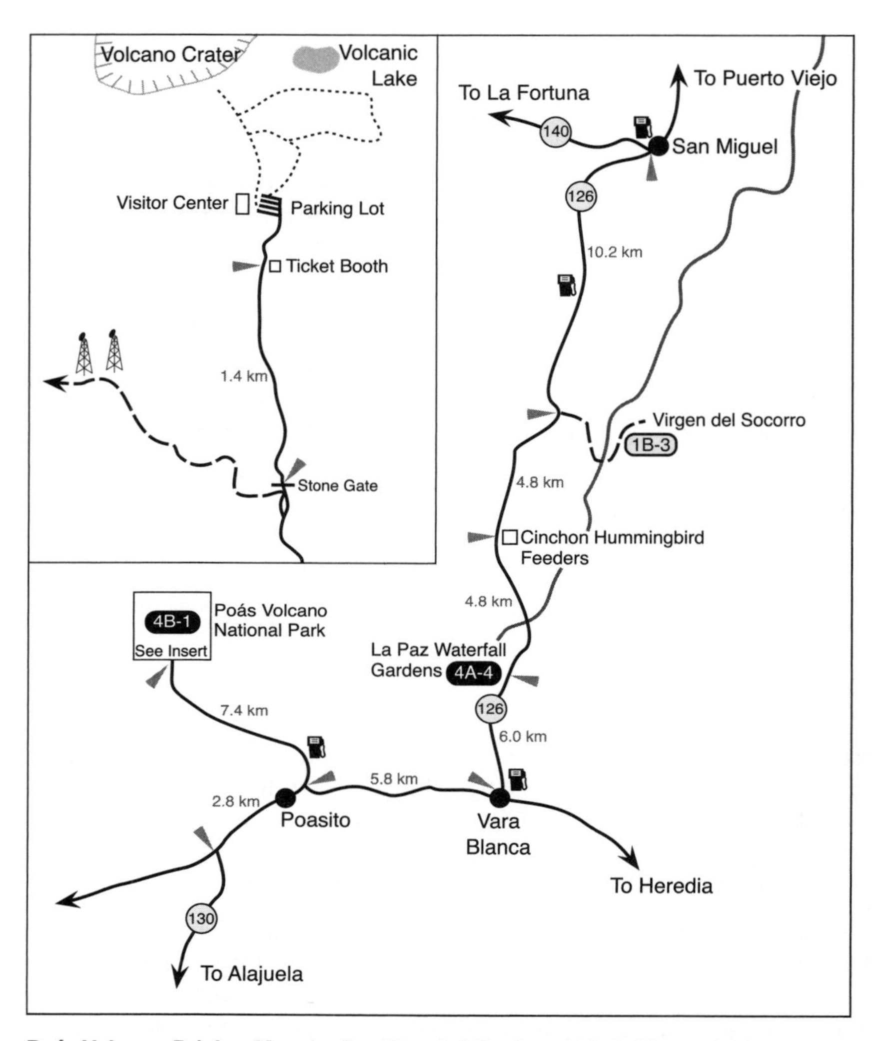

Poás Volcano Driving Map. La Paz Waterfall Gardens (4A-4), Virgen del Socorro (1B-3), and the Cinchona hummingbird feeders are also included. The insert shows roads and trails at Poás Volcano National Park.

bird along the road if traffic isn't too heavy. Just beyond the ticketing station are two small picnic areas, one on each side of the road, which are worth a thorough check. Thrushes, such as Sooty Thrush, Mountain Thrush, and Black-billed Nightingale-Thrush, can be found feeding in the grass. The trees around the clearings often hold Yellow-winged Vireo, Flame-throated Warbler, and Slaty Flowerpiercer. Continue up to the main parking lot, where there is another picnic area that often produces Yellow-thighed Finch and Large-footed Finch.

The park trails are sometimes difficult to bird because of the amount of foot traffic. However, the trail that leads from the top of the parking lot (rather than from the Visitor Center) is usually the quietest and most productive. This is a great place to find Large-footed Finches scratching through the fallen leaves, and possibly a Wrenthrush. You may encounter mixed flocks that could include Ruddy Treerunner, Buffy Tuftedcheek, and Spangle-cheeked Tanager. Black Guan is a regular, and Fiery-throated Hummingbird abounds. After a few hundred meters, you come to a split in the path. Turning right will bring you down a large loop past a volcanic lake, where you will find many of the same species as on the trail you have just been walking. For a shorter walk, continue straight and you will quickly end up on the principal walkway to the crater. This sees the highest amount of foot traffic, and thus birding is difficult. However, you can often find Magnificent Hummingbird and Black-and-yellow Silky-Flycatcher.

Radio Tower Road

A rocky dirt road located just outside the stone entrance gate is a wonderful place to bird, especially if you arrive before the park opens. This can be either walked or driven depending on how comfortable you are leaving your car. The road climbs uphill for about one and a half kilometers to a set of radio towers, and then continues downhill for another few kilometers to some houses. Birding along the roadside can be very productive for many of the high-elevation species. Buff-fronted Quail-Dove, Silvery-fronted Tapaculo, Spot-crowned Woodcreeper, Yellow-thighed Finch, and Wrenthrush can all be found along the road. The section of road around the towers themselves is often an especially good place to find Black-capped Flycatcher, Volcano Hummingbird, Magnificent Hummingbird, and Sooty Thrush.

If you are looking to bird the area a bit more, you might try exploring along the main road on your way back toward Poasito. One productive area is a small dirt road about 4.8 km before the park entrance that leads to Finca Poasito. The lower elevation of this area often produces some different species, such as Ruddy-capped Nightingale-Thrush, Acorn Woodpecker, Flame-colored Tanager, and White-eared Ground-Sparrow.

Species to Expect

*Black Guan (11) 12:5
Black Vulture (31) 13:4
Turkey Vulture (31) 13:3
Red-tailed Hawk (47) 17:8
Band-tailed Pigeon (87) 18:1
Ruddy Pigeon (87) 18:5
*Buff-fronted Quail-Dove (93) 18:19
White-collared Swift (117) 22:1
Green Violetear (131) 23:7
***Fiery-throated Hummingbird (123) 24:12**
Magnificent Hummingbird (123) 23:16
*Volcano Hummingbird (139) 25:3
Emerald Toucanet (153) 27:17
Acorn Woodpecker (157) 28:15
Hairy Woodpecker (159) 28:19
*Ruddy Treerunner (163) 29:4
Buffy Tuftedcheek (165) 30:1

Spot-crowned Woodcreeper (173) 29:9
*Silvery-fronted Tapaculo (187) 32:12
Mountain Elaenia (193) 37:24
Paltry Tyrannulet (191) 37:10
*Black-capped Flycatcher (207) 36:16
Barred Becard (217) 33:14
***Yellow-winged Vireo (229) 40:10**
Blue-and-white Swallow (233) 22:20
Gray-breasted Wood-Wren (245) 38:14
*Black-faced Solitaire (249) 39:13
***Black-billed Nightingale-Thrush (247) 38:24**
***Sooty Thrush (249) 39:7**
Mountain Thrush (251) 39:6
Clay-colored Thrush (251) 39:8
*Black-and-yellow Silky-Flycatcher (253) 39:10
***Flame-throated Warbler (257) 41:7**
Black-throated Green Warbler (261) 43:9
Wilson's Warbler (269) 42:4
*Collared Redstart (273) 42:6
*Black-cheeked Warbler (275) 40:17
*Wrenthrush (275) 32:13
***Sooty-capped Bush-Tanager (279) 45:14**
***Slaty Flowerpiercer (299) 49:9**
***Yellow-thighed Finch (301) 48:22**
***Large-footed Finch (301) 48:20**
Rufous-collared Sparrow (307) 50:13
*Black-thighed Grosbeak (311) 48:7
*Golden-browed Chlorophonia (325) 45:10

Site 4B-2: Irazú Volcano National Park

Birding Time: 5 hours
Elevation: 3,400 meters
Trail Difficulty: 1
Reserve Hours: 8 a.m. to 3:30 p.m.
Entrance Fee: $10 (US 2008)

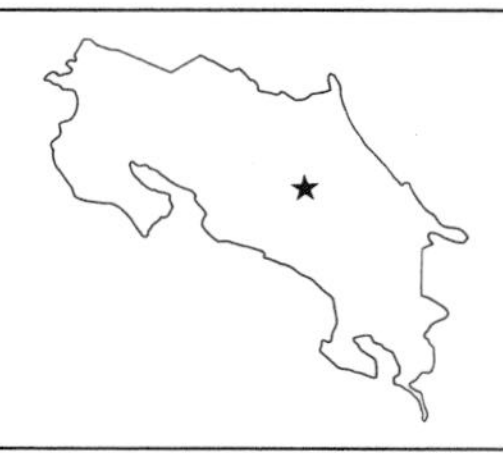

Irazú Volcano is one of Costa Rica's two easily accessible active craters, the other being Poás Volcano. Irazú is located just north of the city of Cartago, but it takes almost a full hour of driving to reach the summit from the city. In 1955 Irazú Volcano National Park was created to protect and preserve the volcanic crater. The park encompasses nearly 5,700 acres of land, although only the summit of the volcano is easily accessible. Irazú is one of Costa Rica's highest peaks, reaching an impressive 3,432 meters (11,260 feet), and you will certainly notice the cool thin air when you reach the top. On a clear day it is even possible to see both the Pacific and the Atlantic Ocean from here! The volcano has five craters, and the main crater is the most impressive, its steep cliffs diving into a green lake at the bottom. The volcano is still quite active, with recent periods of elevated activity during the 1960s and 1990s.

Although the birding within the park is somewhat confined, it does provide the best opportunity in the country for seeing two extreme high-elevation endemics, Timberline Wren and Volcano Junco. There are not many trails in the park, but birding along the roadsides usually yields all the important species, and traffic in the area is minimal. Halfway up the mountain, in an area called the Prusia Sector, are more small roads that are worth birding along. Here you will find a greater variety of birds than at the summit, including some noteworthy species, such as Resplendent Quetzal.

Target Birds

Buffy-crowned Wood-Partridge (13) 12:16
Red-tailed Hawk (47) 17:8
Mourning Dove (89) 18:13
Maroon-chested Ground-Dove (91) 18:10
*Volcano Hummingbird (139) 25:3
Resplendent Quetzal (145) 26:1
Acorn Woodpecker (157) 28:15
*Timberline Wren (243) 38:16
*Sooty Thrush (249) 39:7
*Flame-throated Warbler (257) 41:7
*Volcano Junco (307) 50:11
Yellow-bellied Siskin (325) 50:1

Access

From Cartago: Take Rt. 10 heading east toward Paraíso for 1.2 km. Just beyond a gas station, make a hard left turn. (**From Paraíso:** Take Rt. 10 west toward Cartago for 4.4 km and then fork right just before a gas station.) Follow this road straight through the town of San Rafael de Oreamuno for 1.8 km, at which point the road rapidly bends left about 90 degrees. Immediately after this you will come to a stop sign, where you should turn right. Proceed straight for 0.4 km, and then turn right again. Follow this road straight for 27.2 km and it will take you to the park entrance. (Driving time from Cartago and Paraíso: 50 min)

Logistics: Staying near Irazú is difficult because there is nothing but a few low-end cabins in the vicinity. It is easiest to bird the volcano as a day trip out of San José, or as a travel stop between other sites that offer more comfortable accommodations. The park has a small café, and there are also a number of roadside restaurants before you reach the entrance.

Birding Sites

Irazú Volcano National Park

Purchase your tickets at the entrance, and proceed about one and a half kilometers up the road to a fork and the upper parking lot. Bear right here and follow the road down to the lower parking lot. Check the ground around the small café and gift shop, as this is a reliable place for Volcano Junco.

The concrete path to the crater leads past a large expanse of flat, barren earth. I suggest walking a large loop around this area. The right side will provide views of the crater, while the left side should produce some interesting high-elevation species. The scrubby vegetation here is the easiest place to see Timberline Wren, which should tee up with a little *pishing*. Also, keep an eye out for Flame-throated Warbler, Volcano Hummingbird, Large-footed Finch, Volcano Junco, and Sooty Thrush.

If you wish to spend more time in the park, head back to the upper parking lot. From here you can either walk back toward the ticketing booth or take the left fork and head up to the highest peak. Walking along either road might produce Slaty Flowerpiercer, Black-and-yellow Silky-Flycatcher, Long-tailed Silky-Flycatcher, Black-capped Flycatcher, Fiery-throated Hummingbird, and others.

Prusia Sector Vicinity

A secondary entrance to the park, known as the Prusia Sector, is located 15.2 km before the main park entrance. Historically, this section provided the best birding on Irazú, but it is now unfortunately closed to public use. However, there are places in the surrounding vicinity where the birding is excellent. The Prusia Sector sits at about 2,500 meters, almost 1,000 meters lower than the main park entrance, so a number of different species, including Common Bush-Tanager, Brown-capped Vireo, and Slate-throated Redstart, begin to show up here.

To get to the Prusia Sector area from the main national park entrance, head back down the mountain for 15.2 km and then turn right onto a paved road. About 2.2 km down this road you will enter a small wooded valley that offers good birding for the next 1.2 km. This area often produces Resplendent Quetzal, along with Long-tailed Silky-Flycatcher, Mourning Dove, Acorn Woodpecker, Yellow-bellied Siskin, and Buffy-crowned Wood-Partridge.

A little farther on, the road comes to a fork. The right fork quickly leads to the Prusia Sector Ranger Station, which is now closed to the public. The left fork, however, leads through good birding habitat on its way up to some dairy farms. Birding along this road should yield Rufous-browed Peppershrike and Flame-colored Tanager.

Species to Expect

*Black Guan (11) 12:5
Buffy-crowned Wood-Partridge (13) 12:16
Black Vulture (31) 13:4
Turkey Vulture (31) 13:3
Red-tailed Hawk (47) 17:8
Band-tailed Pigeon (87) 18:1
Ruddy Pigeon (87) 18:5
Mourning Dove (89) 18:13
*Dusky Nightjar (111) 21:19
White-collared Swift (117) 22:1
Vaux's Swift (119) 22:9
Green Violetear (131) 23:7
***Fiery-throated Hummingbird (123) 24:12**
*Purple-throated Mountain-gem (135) 24:7
Magnificent Hummingbird (123) 23:16
***Volcano Hummingbird (139) 25:3**
Collared Trogon (143) 26:5
Resplendent Quetzal (145) 26:1
Emerald Toucanet (153) 27:17
Acorn Woodpecker (157) 28:15
Hairy Woodpecker (159) 28:19
Red-faced Spinetail (163) 30:12
*Ruddy Treerunner (163) 29:4
Spot-crowned Woodcreeper (173) 29:9
*Silvery-fronted Tapaculo (187) 32:12
Mountain Elaenia (193) 37:24
Olive-striped Flycatcher (195) 36:24
Paltry Tyrannulet (191) 37:10
***Black-capped Flycatcher (207) 36:16**
***Yellow-winged Vireo (229) 40:10**
Brown-capped Vireo (227) 40:14

Rufous-browed Peppershrike (229) 40:2
Brown Jay (231) 39:19
Blue-and-white Swallow (233) 22:20
House Wren (243) 38:18
*Ochraceous Wren (243) 38:19
***Timberline Wren (243) 38:16**
Gray-breasted Wood-Wren (245) 38:14
***Black-billed Nightingale-Thrush (247) 38:24**
Ruddy-capped Nightingale-Thrush (247) 38:27
***Sooty Thrush (249) 39:7**
Mountain Thrush (251) 39:6
Clay-colored Thrush (251) 39:8
***Black-and-yellow Silky-Flycatcher (253) 39:10**
***Long-tailed Silky-Flycatcher (253) 39:15**
***Flame-throated Warbler (257) 41:7**
Black-throated Green Warbler (261) 43:9
Black-and-white Warbler (267) 41:13
Wilson's Warbler (269) 42:4
Slate-throated Redstart (273) 42:7
***Collared Redstart (273) 42:6**
***Black-cheeked Warbler (275) 40:17**
Common Bush-Tanager (279) 45:13
***Sooty-capped Bush-Tanager (279) 45:14**
Yellow-faced Grassquit (297) 49:6
***Slaty Flowerpiercer (299) 49:9**
***Yellow-thighed Finch (301) 48:22**
***Large-footed Finch (301) 48:20**
Rufous-collared Sparrow (307) 50:13
***Volcano Junco (307) 50:11**
*Black-thighed Grosbeak (311) 48:7
Yellow-bellied Siskin (325) 50:1
Lesser Goldfinch (325) 50:2

Site 4B-3: Km 70

Birding Time: 1 day
Elevation: 2,750 meters
Trail Difficulty: 2
Reserve Hours: 6:30 a.m. to 8:00 p.m.
Entrance Fee: $7 (US 2008) with guided walk
Bird Guides Available

Seventy kilometers from San José along Rt. 2 you will find two small nature lodges run by the Serrano family. The Serranos first moved here in 1957, living off their land as most people in the area did at that time. They practiced logging, made charcoal, grew mora, and raised dairy cattle. Then, in 1990 Jorge Serrano, the youngest son in the family, began a tourism project, offering home-cooked meals, small cabins, and nature walks. The main attraction was the Resplendent Quetzal, a species easily found in the area. Jorge's project soon gained popularity with birders and nature-lovers alike.

Since 2005 the Serrano family has run two lodges at Km 70. Eddy Serrano now maintains the original Mirador de Quetzales, while Jorge has started the new Paraíso del Quetzal. Both facilities offer small, rustic cabins with hot showers, plenty of blankets, three meals a day, hummingbird feeders, and trails through the surrounding countryside. The birding to be found at each lodge is very similar, and they even share some of the trails. For the sake of brevity I describe only the Paraíso del Quetzal, but Mirador de Quetzales is equally productive.

Many people believe that Km 70, with its wide range of species and habitats, offers the best high-elevation birding in the country. It is the easiest place to find Resplendent Quetzal and regularly produces a number of Costa Rica's more difficult endemic species, including Peg-billed Finch, Silvery-throated Jay, and Ochraceous Pewee.

Target Birds

Spotted Wood-Quail (13) 12:11
Maroon-chested Ground-Dove (91) 18:10
*Buff-fronted Quail-Dove (93) 18:19
*Sulphur-winged Parakeet (95) 19:12
Barred Parakeet (97) 19:16
*Costa Rican Pygmy-Owl (109) 20:18
Unspotted Saw-whet Owl (109) 20:13
*Dusky Nightjar (111) 21:19
*White-throated Mountain-gem (135) 24:6
Resplendent Quetzal (145) 26:1
Buffy Tuftedcheek (165) 30:1
Rough-legged Tyrannulet (191) 37:18
*Ochraceous Pewee (203) 36:3
Barred Becard (217) 33:14
*Silvery-throated Jay (231) 39:16
*Timberline Wren (243) 38:16
*Sooty Thrush (249) 39:7
*Black-and-yellow Silky-Flycatcher (253) 39:10
*Long-tailed Silky-Flycatcher (253) 39:15
*Flame-throated Warbler (257) 41:7
*Wrenthrush (275) 32:13
Flame-colored Tanager (287) 47:9
Slaty Finch (299) 49:14
*Peg-billed Finch (299) 49:12
*Black-thighed Grosbeak (311) 48:7
*Golden-browed Chlorophonia (325) 45:10

Access

From Cartago: (Refer to site 4B-5 for a map.) Take Rt. 2 south toward San Isidro for 48.8 km and then turn right, following signs for both Mirador de Quetzales and Paraíso del Quetzal. (**From San Isidro:** Take Rt. 2 north toward Cartago for 64.0 km and then turn left, following signs for both Mirador de Quetzales and Paraíso del Quetzal.) In 0.8 km you will arrive at the lodges. (Driving time from Cartago: 55 min; from San Isidro: 80 min)

Logistics: Km 70 can be visited as a day trip from San José, or as a travel stop on your way down Rt. 2. It is also a good place to spend the night, and room and board are provided at both lodges. Anything else you may need should be brought in since there are very few stores in the area.

Paraíso del Quetzal
Tel: (506) 2200-0241 or 8390-7894
Website: www.quetzalsparadise.com
E-mail: selvamar@ice.co.cr
(This is a local travel agency)

Mirador de Quetzales
Tel: (506) 2200-5915 or 8381-8456
E-mail: selvamar@ice.co.cr (This is a local travel agency)

Birding Sites

As mentioned above, I only describe the trails of Paraíso del Quetzal. Mirador de Quetzales provides access to some of the same land, and all the same habitat types, so you should be able to find the same species staying there.

Immediately behind the window of the central gathering area and dining room of Paraíso del Quetzal is a hummingbird feeding station that attracts Volcano Hummingbird, Fiery-throated Hummingbird, Magnificent Hummingbird, Green Violetear, and White-throated Mountain-gem. The bushes around the feeders often hold Large-footed Finch, Black-billed Nightingale-Thrush, and Yellow-thighed Finch. Also, keep an eye out for Flame-colored Tanager in the trees above.

Sendero La Zeledonia leads back from the right side of the reception building and loops through a variety of habitats. The first half of the trail leads through old oak forest and crisscrosses along a mountain stream. Here you should look for Buffy Tuftedcheek, Ochraceous Pewee, Barred Becard, Black-cheeked Warbler, and Spangle-cheeked Tanager. Wrenthrush and Silvery-fronted Tapaculo are numerous in the undergrowth, though difficult to see.

The second half of the trail proceeds through old cow pastures and mora fields. These open areas are a great place to find Long-tailed Silky-Flycatcher, Black-and-yellow Silky-Flycatcher, Yellow-bellied Siskin, Sooty Thrush, Acorn Woodpecker, and Flame-throated Warbler. Some more unusual species that can be found in this area are Peg-billed Finch, Slaty Finch, Timberline Wren, and Barred Parakeet. You might also see Red-tailed Hawk or Swallow-tailed Kite circling overhead.

Another area that is great for birding is the avocado grove. This is located behind Mirador de Quetzales, and access is shared by both lodges. Your local guides will take you here to see the Resplendent Quetzal, as many frequent the area, owing to the abundance of fruiting trees. This is the most reliable single location in Costa Rica to find quetzals, and they are present here throughout the year (although scarce in October). Other birds that you might encounter include Black-thighed Grosbeak, Black Guan, Mountain Elaenia, Collared Trogon, and Golden-browed Chlorophonia.

At night it is possible to find Dusky Nightjar near the lodge. If you walk into the forest you may pick up Costa Rican Pygmy-Owl and Bare-shanked Screech-Owl.

Species to Expect

*Black Guan (11) 12:5
Spotted Wood-Quail (13) 12:11
Black Vulture (31) 13:4
Turkey Vulture (31) 13:3
Swallow-tailed Kite (37) 15:2
Band-tailed Pigeon (87) 18:1
Ruddy Pigeon (87) 18:5
Barred Parakeet (97) 19:16
Bare-shanked Screech-Owl (107) 20:10
*Costa Rican Pygmy-Owl (109) 20:18
*Dusky Nightjar (111) 21:19
White-collared Swift (117) 22:1
Green Violetear (131) 23:7

***Fiery-throated Hummingbird (123) 24:12**
*White-throated Mountain-gem (135) 24:6
Magnificent Hummingbird (123) 23:16
***Volcano Hummingbird (139) 25:3**
Collared Trogon (143) 26:5
Resplendent Quetzal (145) 26:1
Emerald Toucanet (153) 27:17
Acorn Woodpecker (157) 28:15
Hairy Woodpecker (159) 28:19
***Ruddy Treerunner (163) 29:4**
Buffy Tuftedcheek (165) 30:1
Spot-crowned Woodcreeper (173) 29:9
*Silvery-fronted Tapaculo (187) 32:12
Mountain Elaenia (193) 37:24
Olive-striped Flycatcher (195) 36:24
Paltry Tyrannulet (191) 37:10
Tufted Flycatcher (207) 36:11
*Ochraceous Pewee (203) 36:3
***Black-capped Flycatcher (207) 36:16**
Barred Becard (217) 33:14
***Yellow-winged Vireo (229) 40:10**
Blue-and-white Swallow (233) 22:20
***Ochraceous Wren (243) 38:19**
*Timberline Wren (243) 38:16
Gray-breasted Wood-Wren (245) 38:14
***Black-billed Nightingale-Thrush (247) 38:24**
***Sooty Thrush (249) 39:7**
Mountain Thrush (251) 39:6
Clay-colored Thrush (251) 39:8
***Black-and-yellow Silky-Flycatcher (253) 39:10**
***Long-tailed Silky-Flycatcher (253) 39:15**
***Flame-throated Warbler (257) 41:7**
Black-throated Green Warbler (261) 43:9
Wilson's Warbler (269) 42:4
***Collared Redstart (273) 42:6**
***Black-cheeked Warbler (275) 40:17**
*Wrenthrush (275) 32:13
***Sooty-capped Bush-Tanager (279) 45:14**
Flame-colored Tanager (287) 47:9
Slaty Finch (299) 49:14
*Peg-billed Finch (299) 49:12
***Slaty Flowerpiercer (299) 49:9**
***Yellow-thighed Finch (301) 48:22**
***Large-footed Finch (301) 48:20**
Rufous-collared Sparrow (307) 50:13
*Black-thighed Grosbeak (311) 48:7
*Golden-browed Chlorophonia (325) 45:10
Yellow-bellied Siskin (325) 50:1

Nearby Birding Opportunities

La Esperanza de Tapantí

La Esperanza de Tapantí is an alternate entrance to Tapantí National Park. The primary entrance to the park, near Orosí, is described in site 4A-5. The Esperanza entrance sits at about 2,600 meters and is located just off Rt. 2, 8.4 km north of the entrance to Km 70. Turn east off the highway and follow a dirt road straight for 2.8 km to the park entrance. (Refer to site 4B-5 for a map.) La Esperanza is a little-known area, and you are unlikely to encounter anyone except the rangers. There is a $10 (US 2008) entrance fee, nonetheless.

Look for Ruddy-capped Nightingale-Thrush and Acorn Woodpecker around the ranger station at the park entrance. Also, listen for the hollow-sounding, rattled, *churit* notes of the Long-tailed Silky-Flycatcher. From the station, begin walking up the road into the forest. The first stretch of road has many of the oldest oak trees and can be an excellent location to find Ochraceous Wren, Buffy Tuftedcheek, and Spot-crowned Woodcreeper. The high, thin, piercing note of the Wrenthrush is commonly heard here, although seeing the secretive little warbler is more difficult. Other birds to look for at La Esperanza include Ochraceous Pewee, Hairy Woodpecker, Yellow-winged Vireo, Yellow-thighed Finch, and Black-thighed Grosbeak.

El Jaular

El Jaular consists of a dirt road that descends through an old-growth oak forest to a small community of houses. The road is gated at the highway, and to get access you must stop at the Abastecedor La Trinidad, pick up the key, and pay a small entrance fee. Spanish is useful here, but the words *El Jaular* and *llave* (key) should be sufficient to get what you want. The *abastecedor* is located on the west side of the highway 7.0 km north of the turnoff for Km 70. Once you get the key, retrace your steps 2.8 km south, and the gated entrance to El Jaular will be on your right. (Refer to site 4B-5 for a map.) Be sure to park inside the gate and out of view from the road, where your car will be safe.

The dirt road starts at about 2,600 meters and then meanders downslope for about 2.2 km through beautiful oak forest. In the undergrowth, look for Yellow-thighed Finch, Large-footed Finch, and Wrenthrush. Flocks of Sooty-capped Bush-Tanagers often include Barred Becard, Spangle-cheeked Tanager, Black-cheeked Warbler, Buffy Tuftedcheek, and Spot-crowned Woodcreeper. Other birds to look for in the forest are Resplendent Quetzal, Golden-browed Chlorophonia, Black Guan, Ochraceous Wren, and Ruddy-capped Nightingale-Thrush.

At the end of the road the forest opens to a few small country homes tucked within cow pastures and trout ponds. This area usually yields a number of different species, including some that are normally associated with lower-mountain and middle-elevation sites. Keep an eye open for Dark Pewee, Black Phoebe, Olive-striped Flycatcher, Yellowish Flycatcher, Tufted Flycatcher, and Scintillant Hummingbird.

Site 4B-4: San Gerardo de Dota

Birding Time: 1–2 days
Elevation: 2,200 meters
Trail Difficulty: 2
Reserve Hours: N/A
Entrance Fee: $5 (US 2008) at Savegre Mountain Lodge for non-guests
Bird Guides Available
4×4 Recommended

San Gerardo de Dota is a small pastoral mountain town situated on the Pacific face of the Talamanca Mountains. The tiny village of about 150 people sits within a breath-taking mountain valley, blanketed with oak forest, along the banks of the rushing Savegre River. Clouds roll in and out, shrouding the scenery in thick fog one moment, then unveiling picturesque vistas the next.

Many of the locals still make their living off the land, farming mora, apples, peaches, and plums as well as rainbow trout. Others have turned to the tourist industry, and a number of hotels and lodges now exist in the area. Most of these cater to birders, as San Gerardo is a favorite stop on the itinerary of many organized birding tours. The management at Savegre Hotel de Montaña estimates that during the high season 80% of their clientele are birders.

The area has gained favor with birders because of the wide range of mountain species that can be found here, the beautiful scenery, and the comfortable accommodations. While San Gerardo is a good place to find some Talamanca Mountain specialties such as White-throated Mountain-gem, Ochraceous Pewee, and Sulphur-winged Parakeet, the real attraction of the area is the Resplendent Quetzal. The quetzal has become a bit of a tourist cliché countrywide, with its image plastered over countless signs. However, when you actually see one, its long tail coverts fluttering in the breeze and its emerald green body glowing with a coppery iridescence, you will understand why the bird enjoys such popularity. San Gerardo de Dota is one of the most reliable places in Costa Rica to find this stunning bird.

Target Birds

Spotted Wood-Quail (13) 12:11
*Sulphur-winged Parakeet (95) 19:12
Barred Parakeet (97) 19:16
Bare-shanked Screech-Owl (107) 20:10
*Costa Rican Pygmy-Owl (109) 20:18
Unspotted Saw-whet Owl (109) 20:13
*Dusky Nightjar (111) 21:19
*White-throated Mountain-gem (135) 24:6
*Volcano Hummingbird (139) 25:3
*Scintillant Hummingbird (139) 25:7
Resplendent Quetzal (145) 26:1
Emerald Toucanet (153) 27:17
Buffy Tuftedcheek (165) 30:1
*Silvery-fronted Tapaculo (187) 32:12
Torrent Tyrannulet (201) 36:1
*Dark Pewee (203) 36:2
*Ochraceous Pewee (203) 36:3
Barred Becard (217) 33:14
*Silvery-throated Jay (231) 39:16
American Dipper (251) 39:11
*Long-tailed Silky-Flycatcher (253) 39:15
*Flame-throated Warbler (257) 41:7
*Wrenthrush (275) 32:13
Flame-colored Tanager (287) 47:9
*Black-thighed Grosbeak (311) 48:7
Elegant Euphonia (325) 45:9
*Golden-browed Chlorophonia (325) 45:10
Yellow-bellied Siskin (325) 50:1

Access

From Cartago: (Refer to site 4B-5 for a map.) Take Rt. 2 south toward San Isidro for 59.0 km and then turn right onto a dirt road. (**From San Isidro:** Take Rt. 2 north toward San José for 54.2 km and then turn left onto a dirt road.) Follow this road downhill for 8 km to the town of San Gerardo de Dota. The road is quite steep, and you should be sure to use as low a gear as possible to save your brakes during the descent. (Driving time from Cartago: 85 min; from San Isidro: 90 min)

Logistics: The area has a number of all-inclusive hotels with varying prices. The upscale Savegre Hotel de Montaña is a popular destination among birders and maintains a very nice trail system. There is hardly a store in the town, however, so be sure to bring in everything you need.

Savegre Hotel de Montaña
Tel: (506) 2740-1028
Website: www.savegre.co.cr
E-mail: savegrehotel@racsa.co.cr

Birding Sites

Savegre Hotel de Montaña

Toward the far end of town, on the left side of the road, you will find Savegre Hotel de Montaña, which offers great opportunities for birders. Just in front of reception is a hummingbird feeding station, which is usually very active and attracts Magnificent Hummingbird, Green Violetear, White-throated Mountain-gem, Scintillant Hummingbird, and Volcano Hummingbird. In the plantings around the buildings you may find Slaty Flowerpiercer and Flame-colored Tanager.

On the hillside behind the hotel are four trails that guests can walk. If you are not a guest, a $5 (US 2008) per person fee will gain you access. These trails, which proceed through primary oak forest, offer some of the most interesting birding in the area. The first trail, Canto de las Aves, runs through new growth and is generally unproductive. I suggest skipping this and focusing your time birding La Quebrada and Los Robles trails. The trailheads for these two trails are about 0.8 km up the hillside, and if you have a 4×4 you can drive directly to them, saving yourself the uphill climb.

La Quebrada trail, about 2.4 km long and relatively flat, is wonderful to bird. You should find mixed flocks of Sooty-capped Bush-Tanager, Black-cheeked Warbler, Ruddy Treerunner, Buffy Tuftedcheek, Spot-crowned Woodcreeper, and Yellow-thighed Finch. Keep your eyes open for other interesting species, including Ochraceous Pewee, Barred Becard, and Streak-breasted Treehunter. Where the trail parallels a small creek, listen for the elusive Wrenthrush and Silvery-fronted Tapaculo. Their songs are always the best clue to their presence.

Los Robles trail, about 6.4 km long, is more of a hike than La Quebrada. It climbs to a higher altitude (2,600 meters), which makes spotting some of the very high-elevation species more likely. Along with the more common species, look for Silvery-throated Jay, Ochraceous Pewee, Highland Tinamou, and Spotted Wood-Quail.

One final trail, called Cerro de la Muerte, connects Savegre with the radio towers described in site 4B-5. This is a serious hike, taking seven to eight hours to complete and covering an extreme altitude gradient. I would not recommend it unless you can arrange to be dropped off at the towers and take the walk back downhill.

San Gerardo

The main road, which runs straight through the community of San Gerardo, is a great area to bird. Along the edges of the dirt road you will find second-growth forest, new growth, and fields with remnant trees. Some of the interesting species that you are likely to find here include Brown-capped Vireo, Flame-colored Tanager, Golden-browed Chlorophonia, Rufous-browed Peppershrike, and Flame-throated Warbler. Fruiting trees often attract Mountain Thrush, Black Guan, and Emerald Toucanet. Keep an eye in the sky, as Long-tailed Silky-Flycatcher, Yellow-bellied Siskin, and Sulphur-winged Parakeet are often seen flying past while Red-tailed Hawk and Swallow-tailed Kite circle overhead.

Just beyond Savegre Hotel de Montaña you will come to a sign advertising Los Ranchos. There is a picnic and camping area in here, and also a short trail that proceeds along a small stream. This is a good place to find species preferring slightly lower elevations. Look for Spangle-cheeked Tanager, Black-faced Solitaire, Ruddy-capped Nightingale-Thrush, Lineated Foliage-gleaner, Yellowish Flycatcher, and Olive-striped Flycatcher.

Another public trail with similar habitat is located 0.4 km beyond Savegre Hotel de Montaña and leads along the Savegre River to a waterfall. The habitat here is open forest, and forest edge, and the scenery is beautiful. You are likely to see many of the species mentioned above.

Though many visitors come to San Gerardo de Dota specifically in search of Resplendent Quetzal, you are certainly not guaranteed to run across one in the normal course of your wandering. This species is quite patchy in its distribution because it is tied to an inconsistent food source, the wild avocado. Avocado trees only occasionally come into fruit, which means that Resplendent Quetzals will congregate around the few trees that are in fruit at any particular time. Luckily, because of the importance of the tourist industry, the locals around San Gerardo are in tune with the movements of their quetzal population. Ask your hotel manager where Resplendent Quetzals are being seen, and they can usually direct you straight to the tree where these amazing birds are feeding.

Species to Expect

***Black Guan (11) 12:5**
Black Vulture (31) 13:4
Turkey Vulture (31) 13:3
Osprey (33) 17:14
Swallow-tailed Kite (37) 15:2
Red-tailed Hawk (47) 17:8
Band-tailed Pigeon (87) 18:1
Ruddy Pigeon (87) 18:5
White-tipped Dove (91) 18:14
*Sulphur-winged Parakeet (95) 19:12
*Costa Rican Pygmy-Owl (109) 20:18
*Dusky Nightjar (111) 21:19
White-collared Swift (117) 22:1
Vaux's Swift (119) 22:9
Green Violetear (131) 23:7
*Fiery-throated Hummingbird (123) 24:12

***White-throated Mountain-gem (135) 24:6**
Magnificent Hummingbird (123) 23:16
***Volcano Hummingbird (139) 25:3**
***Scintillant Hummingbird (139) 25:7**
Collared Trogon (143) 26:5
Resplendent Quetzal (145) 26:1
Emerald Toucanet (153) 27:17
Acorn Woodpecker (157) 28:15
Hairy Woodpecker (159) 28:19
Spotted Barbtail (163) 29:5
***Ruddy Treerunner (163) 29:4**
Buffy Tuftedcheek (165) 30:1
Lineated Foliage-gleaner (165) 30:8
*Streak-breasted Treehunter (167) 30:7
Spot-crowned Woodcreeper (173) 29:9
*Silvery-fronted Tapaculo (187) 32:12
Mountain Elaenia (193) 37:24
Torrent Tyrannulet (201) 36:1
Olive-striped Flycatcher (195) 36:24
Paltry Tyrannulet (191) 37:10
Tufted Flycatcher (207) 36:11
*Dark Pewee (203) 36:2
Yellowish Flycatcher (207) 36:19
*Black-capped Flycatcher (207) 36:16
Black Phoebe (207) 36:5
Boat-billed Flycatcher (211) 35:12
Social Flycatcher (211) 35:14
Tropical Kingbird (213) 35:1
Barred Becard (217) 33:14
***Yellow-winged Vireo (229) 40:10**
Brown-capped Vireo (227) 40:14
Rufous-browed Peppershrike (229) 40:2
Blue-and-white Swallow (233) 22:20
***Ochraceous Wren (243) 38:19**
Gray-breasted Wood-Wren (245) 38:14
***Black-faced Solitaire (249) 39:13**
*Black-billed Nightingale-Thrush (247) 38:24
Ruddy-capped Nightingale-Thrush (247) 38:27
Mountain Thrush (251) 39:6
Clay-colored Thrush (251) 39:8
***Long-tailed Silky-Flycatcher (253) 39:15**
Tennessee Warbler (255) 40:22
***Flame-throated Warbler (257) 41:7**
Black-throated Green Warbler (261) 43:9
Black-and-white Warbler (267) 41:13
Wilson's Warbler (269) 42:4
***Collared Redstart (273) 42:6**
***Black-cheeked Warbler (275) 40:17**
*Wrenthrush (275) 32:13
Common Bush-Tanager (279) 45:13
***Sooty-capped Bush-Tanager (279) 45:14**
Summer Tanager (285) 47:5
Flame-colored Tanager (287) 47:9
Blue-gray Tanager (291) 46:15
Silver-throated Tanager (291) 46:6
***Spangle-cheeked Tanager (289) 46:12**
Yellow-faced Grassquit (297) 49:6
***Slaty Flowerpiercer (299) 49:9**
***Yellow-thighed Finch (301) 48:22**
*Large-footed Finch (301) 48:20
Rufous-collared Sparrow (307) 50:13
*Black-thighed Grosbeak (311) 48:7
Baltimore Oriole (321) 44:7
Elegant Euphonia (325) 45:9
*Golden-browed Chlorophonia (325) 45:10
Yellow-bellied Siskin (325) 50:1

Site 4B-5: Cerro de la Muerte

Birding Time: 1 day
Elevation: 3,000 meters
Trail Difficulty: 1–2
Reserve Hours: N/A
Entrance Fee: N/A

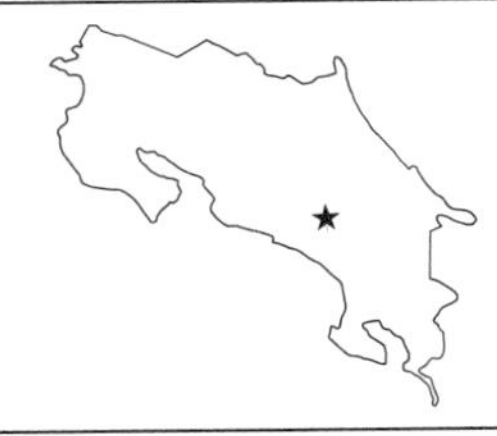

As the Inter-American Highway heads south from the Central Valley toward the South Pacific Slope of Costa Rica, it climbs up and over the Talamanca

Mountain Range. The highway reaches its highest point, about 3,400 meters (11,000 feet), at the top of the mountain known as Cerro de la Muerte. While the mountain's foreboding name (Hill of Death) could easily derive from the danger of the highway that runs over it, the name actually comes from a legend about a group of traveling merchants who died one night from the cold as they were making their way over the mountain toward the markets of San José.

The birding sites described below are located just off the main highway along a 19-kilometer stretch of road at the top of the mountain. Normally, one would not expect to find such productive birding so close to the highway, but because the Cerro is extremely rural, great birds are easily accessible near the main road.

Two factors make Cerro de la Muerte one of the most interesting high-elevation birding areas in the country. First, the mountain offers a wide range of habitats. Oak forest and bamboo thickets are found along the road to Providencia and at La Georgina. New-growth habitat is accessible at Villa Mills, while at the communication towers you experience the low, wet, scrubby habitat known as páramo. Second, the Talamanca Mountains are home to the most diverse mountainous avifauna anywhere in the country, making it possible to see all of Costa Rica's high-elevation specialty species at the Cerro.

Target Birds

Spotted Wood-Quail (13) 12:11
Red-tailed Hawk (47) 17:8
Barred Parakeet (97) 19:16
*Costa Rican Pygmy-Owl (109) 20:18
Unspotted Saw-whet Owl (109) 20:13
*Dusky Nightjar (111) 21:19
*Volcano Hummingbird (139) 25:3
Buffy Tuftedcheek (165) 30:1
*Silvery-fronted Tapaculo (187) 32:12
Barred Becard (217) 33:14
*Silvery-throated Jay (231) 39:16
*Timberline Wren (243) 38:16
*Sooty Thrush (249) 39:7
*Black-and-yellow Silky-Flycatcher (253) 39:10
*Flame-throated Warbler (257) 41:7
*Wrenthrush (275) 32:13
Slaty Finch (299) 49:14
*Peg-billed Finch (299) 49:12
*Volcano Junco (307) 50:11

Access

From Cartago: Follow Rt. 2 south toward San Isidro for 55.0 km, where you will come to the road to Providencia on your right. Refer to the map and "Birding Sites" description for directions to the other locations discussed in this site. (Driving time from Cartago: 60 min)

From San Isidro: Follow Rt. 2 north toward Cartago and San José for 37.8 km to Villa Mills, which will be on your left. Refer to the map and "Birding Sites" description for directions to the other locations discussed in this site. (Driving time from San Isidro: 55 min)

Logistics: Restaurante La Georgina, described in "Birding Sites," offers a few modest cabins in back of the restaurant and is a cheap and convenient lodging option while birding Cerro de la Muerte. Mirador Valle del General, described in "Nearby Birding Opportunities," as well as the nearby sites of San Gerardo de Dota (4B-4) and Km 70 (4B-3), also offer accommodations. There are not, however, any sizable towns or stores in the area, so bring everything you may need with you.

Restaurante La Georgina
Tel: (506) 2770-8043 (Spanish)

Birding Sites

The Road to Providencia (Ojo de Agua)

Just across the highway from Chespiritos Restaurant is a dirt road that leads west to the little town of Providencia. This road can be a very nice area to bird, with good habitat accessible on both sides. Since the road is wide, the viewing is generally good, and easy for a large group. The first two or three kilometers is usually the most productive. Timberline Wren is a possibility, and you should expect to hear Wrenthrush calling from the bamboo thickets. Mixed flocks, usually led by Sooty-capped Bush-Tanagers, are common along the road. Look for Flame-throated Warbler, Yellow-winged Vireo, Ruddy Treerunner, Buffy Tuftedcheek, and Black-cheeked Warbler as well. Of course, there is always the possibility of finding something more exciting, such as a Peg-billed Finch or Silvery-throated Jay.

At the time of this writing, a new park headquarters for the recently created Los Quetzales National Park was under construction at the beginning of the road to Providencia. A trail system through the surrounding forest was in the plans, and I expect this to offer very productive birding. This reserve will be an exciting addition to the many interesting birding areas already accessible along Cerro de la Muerte.

Communication Towers

To get to the communication towers, follow a dirt road that heads west off the highway 12.8 km south of the road to Providencia. The dirt road is not in great condition, and a 4×4 is recommended. About 2 km up the road you will find the towers at the very top of Cerro de la Muerte, 3,450 meters above sea level. This area provides the easiest access to páramo habitat in all of Costa Rica.

The key species here is the Volcano Junco, this area being one of the two easily accessible places in the country where it can be reliably found. I suggest driving or walking the road 2.2 km from the highway to where it forks near some buildings and towers. This is a great place to find the Volcano Junco feeding on the barren earth. If the juncos are not there, follow the left fork around the back of the small hill, to where you might have better luck. Other species to look for

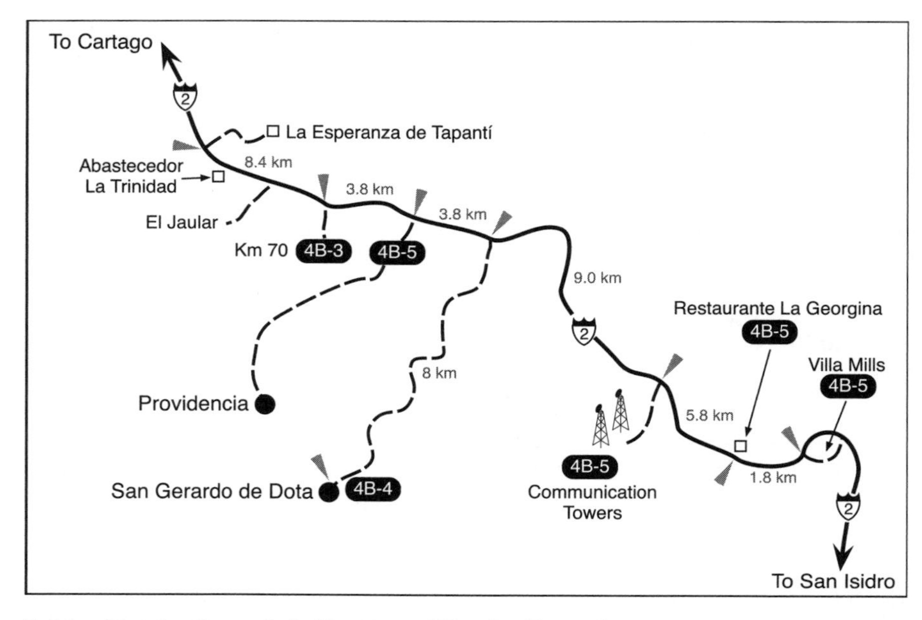

Driving Map for Cerro de la Muerte and Nearby Sites. Cerro de la Muerte (4B-5), San Gerardo de Dota (4B-4), Km 70 (4B-3), El Jaular, and La Esperanza de Tapantí are included.

around the towers are Timberline Wren, Volcano Hummingbird, Black-cheeked Warbler, and Peg-billed Finch.

Restaurante La Georgina

Restaurante La Georgina, located 5.8 km south of the communication towers, offers much more than your average roadside stop. A series of hummingbird feeders set up just behind the window at the back end of the building attract swarms of the energetic little birds, and you can enjoy lunch while watching Fiery-throated Hummingbird, Green Violetear, Magnificent Hummingbird, and Volcano Hummingbird. Walk around the back of the building to where there is some cultivated land beneath a high-tension line. This is a great place to find Large-footed Finch, Sooty Thrush, Black-capped Flycatcher, and Black-billed Nightingale-Thrush.

There are also trails that lead into the forest behind the restaurant. (Keep in mind that the trails are somewhat steep and slick.) The understory of this forest is made up of dense bamboo, and I suggest birding the trail named Sendero Quetzal, as it is the widest and provides the easiest viewing. Wrenthrush and Silvery-fronted Tapaculo are common in this thick and tangled understory. The tall oak trees often hold Hairy Woodpecker, Acorn Woodpecker, and Spot-crowned Woodcreeper. If you get lucky you may spot a Resplendent Quetzal. Mixed flocks moving through the area could yield Barred Becard, Black-cheeked Warbler, Yellow-thighed Finch, and Ruddy Treerunner. Early in the morning you have a good chance of hearing Spotted Wood-Quail, and if you spend the night in the cabins, you may hear Dusky Nightjar and Costa Rican Pygmy-Owl calling outside your window.

Villa Mills

Villa Mills is a small community located just west of Rt. 2, 1.8 km south of Restaurante La Georgina. The area is especially fun to bird because the low, sparse vegetation makes for excellent viewing. Follow a small dirt road for about 300 meters to a widened area on the left, where you can park. (If you come to buildings, you have gone too far.) Lock your car and get out to explore. Heading back toward the highway, you can walk into low open scrub on your left. This is a reliable place to get looks at Timberline Wren, as the little bird will often come into view with a bit of *pishing*. Also, keep an eye open for Magnificent Hummingbird and Green Violetear.

Walking back past your parked car, you will find a trail leading uphill to an open area of new growth, where there is a strange ruined concrete staircase. This is a wonderful place to find Volcano Hummingbird, Slaty Flowerpiercer, Flame-throated Warbler, and Sooty Thrush. You should also encounter Black-capped Flycatcher, Barred Becard, and Yellow-winged Vireo.

Species to Expect

*Black Guan (11) 12:5
Spotted Wood-Quail (13) 12:11
Black Vulture (31) 13:4
Turkey Vulture (31) 13:3
Red-tailed Hawk (47) 17:8
Band-tailed Pigeon (87) 18:1
Ruddy Pigeon (87) 18:5
*Sulphur-winged Parakeet (95) 19:12
Barred Parakeet (97) 19:16
*Costa Rican Pygmy-Owl (109) 20:18
*Dusky Nightjar (111) 21:19
White-collared Swift (117) 22:1
Green Violetear (131) 23:7
***Fiery-throated Hummingbird (123) 24:12**
Magnificent Hummingbird (123) 23:16
***Volcano Hummingbird (139) 25:3**
Resplendent Quetzal (145) 26:1
Acorn Woodpecker (157) 28:15
Hairy Woodpecker (159) 28:19
***Ruddy Treerunner (163) 29:4**
Buffy Tuftedcheek (165) 30:1
Spot-crowned Woodcreeper (173) 29:9
*Silvery-fronted Tapaculo (187) 32:12
Mountain Elaenia (193) 37:24
Paltry Tyrannulet (191) 37:10
***Black-capped Flycatcher (207) 36:16**
Barred Becard (217) 33:14
***Yellow-winged Vireo (229) 40:10**
*Silvery-throated Jay (231) 39:16
Blue-and-white Swallow (233) 22:20
*Timberline Wren (243) 38:16
Gray-breasted Wood-Wren (245) 38:14
***Black-billed Nightingale-Thrush (247) 38:24**
***Sooty Thrush (249) 39:7**
Mountain Thrush (251) 39:6
***Black-and-yellow Silky-Flycatcher (253) 39:10**
*Long-tailed Silky-Flycatcher (253) 39:15

***Flame-throated Warbler (257) 41:7**
Black-throated Green Warbler (261) 43:9
Wilson's Warbler (269) 42:4
***Collared Redstart (273) 42:6**
***Black-cheeked Warbler (275) 40:17**
*Wrenthrush (275) 32:13
***Sooty-capped Bush-Tanager (279) 45:14**
*Peg-billed Finch (299) 49:12
***Slaty Flowerpiercer (299) 49:9**
***Yellow-thighed Finch (301) 48:22**
***Large-footed Finch (301) 48:20**
Rufous-collared Sparrow (307) 50:13
*Volcano Junco (307) 50:11
*Black-thighed Grosbeak (311) 48:7
*Golden-browed Chlorophonia (325) 45:10

Nearby Birding Opportunities

Mirador Valle del General (Vista del Valle)

Following Rt. 2 south from Restaurante La Georgina for 24.0 km (15.4 km north of San Isidro), you will find another small rest stop called Mirador Valle del General. This facility offers a nice, sit-down restaurant, eight mid-range cabins, as well as a short canopy tour. Birders enjoy stopping here to admire the birds coming to the fruit feeders around the restaurant. Expect to see Red-headed Barbet, Silver-throated Tanager, Golden-hooded Tanager, and other colorful species. Swallow-tailed Kites are also easily observed swooping acrobatically over the valley bellow.

The facility also maintains a one-kilometer trail through the surrounding forest that is often productive to bird. At 1,600 meters this is a good place to find some middle-elevation species such as Chestnut-capped Brush-Finch and Golden-crowned Warbler. There is a $3 (US 2008) per person fee to use this trail.

Tel: (506) 2200-5465
Website: www.valledelgeneral.com
E-mail: diego@valledelgeneral.com

Region 5

The Coastline

The **Rufous-necked Wood-Rail** is a shy inhabitant of mangrove swamps around the Gulf of Nicoya. Look for this rare species while on the Mangrove Birding Tour at the Tárcoles River Mouth (5-3).

Regional Map: The Coastline

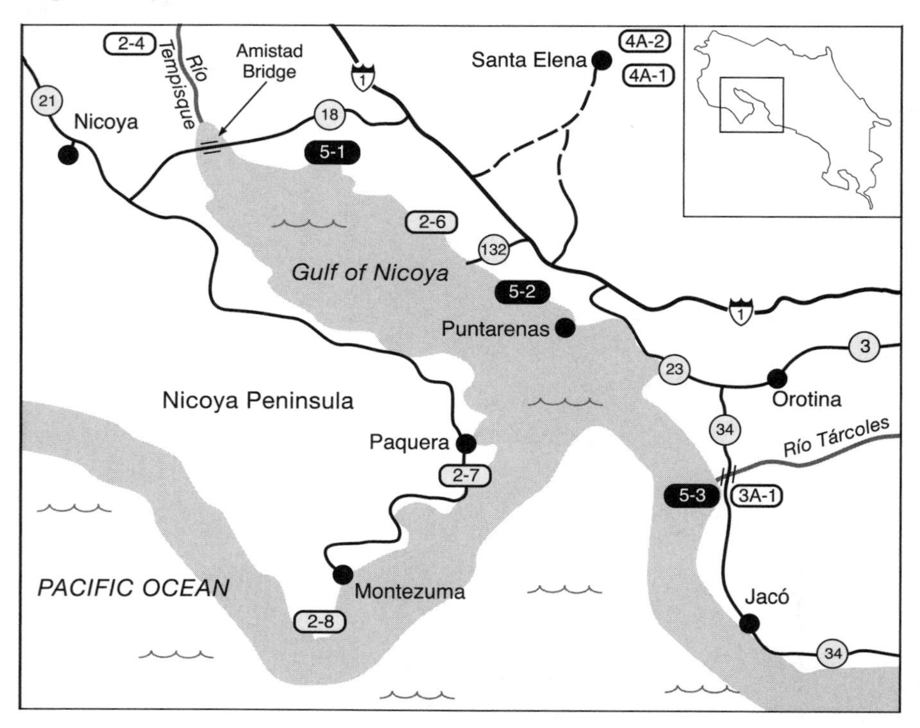

5-1: Colorado
5-2: Chomes
5-3: Tárcoles River Mouth

Introduction

The coastline of Costa Rica is home to many species of birds that do not exist within the interior of the country. Shorebirds, terns, and mangrove-dependent species are found in coastal-specific habitats, such as mangroves, mudflats, and salinas. This book includes three coastline sites, all of which are located along the Pacific Ocean. The Pacific offers better coastal habitats, higher diversity of species, and larger congregations of birds than does the Caribbean, though you will certainly find many of the same species along Caribbean beaches.

A number of distinct habitats are found along the Pacific coast of Costa Rica, several of which warrant some explanation. Salinas are an exciting coastal habitat that may be new to many visitors. These man-made pools are created for the purpose of gathering salt. A system of aqueducts allows seawater into shallow pools at high tide. This water is then allowed to evaporate in the hot tropical sun, leaving its sea salt behind. Most salinas are located in the northern half of the Pacific coast and operate during the dry season, when evaporation is fastest and rain does not slow the process. These pools are a favorite feeding area for many species of shorebirds and herons.

Shrimp pools are another man-made coastal habitat that attracts interesting birds. These pools, of varying depth, harbor shrimp at different stages of life. Shorebirds like to feed where there is exposed mud, while terns, gulls, and herons prefer the deeper pools.

Mangroves, which grow in areas of shallow brackish water physically similar to where salt marshes are found farther north, are the most interesting habitat of the Coastline region. Florally, Costa Rican mangroves are not very diverse environments, with only five species of trees represented: Black Mangrove, White Mangrove, Red Mangrove, Tea Mangrove, and Buttonwood Mangrove. Even though these trees share the common name "mangrove," they are actually unrelated. Instead, they are a wonderful example of convergent evolution, sharing separately evolved adaptive traits that allow them to live in this harsh environment.

One obvious stress of living in brackish, tidal waters is the high salt concentration. Mangrove species deal with high levels of salt in a number of ways: secreting salt through roots or leaves, accumulating excess salt in leaves about to be dropped, actively preventing the uptake of salt, and simply tolerating high levels of salt. These strategies are seen at work to some degree in almost all mangrove species, though each species has its own unique variations.

In addition to providing a crucial environment for aquatic organisms, mangroves are home to a unique avian community. Costa Rica has a handful of birds that are completely dependent on the mangrove habitat, including Mangrove Warbler (a resident race of Yellow Warbler), Mangrove Vireo, Northern Scrub-Flycatcher, Panama Flycatcher, Rufous-necked Wood-Rail, and the endemic Mangrove Hummingbird. I refer to these mangrove-dependent species as "mangrove specialists," and they are certainly one of the most alluring aspects of the Coastline region. The mangroves also provide a home for some unexpected species. Rufous-browed Peppershrike, normally found high in the mountains, and Cinnamon Becard, usually restricted to the Caribbean Slope, are two examples of oddball species that can be found in Pacific mangroves.

Most birders visiting Costa Rica come in search of "tropical birds" such as cotingas, trogons, and antbirds. Coastline sites offer terns, gulls, and shorebirds, most of which are old hat for North American birders. Europeans, however, will find many of these species interesting, and everyone will enjoy the mangrove specialists. Of the three sites presented, the Tárcoles River Mouth (5-3) is exceptional. Aside from great birding, its proximity to Carara National Park (3A-1) makes it a convenient location to incorporate into many itineraries. Chomes (5-2) and Colorado (5-1) are better sites for shorebirds, but access to mangroves is more difficult. It is also worth noting that La Ensenada Wildlife Refuge (2-6) could easily have been included in this region since it offers excellent coastal birding in addition to the opportunities in its dry-forest habitat. Two "Nearby Birding Opportunities," Tivives (described in site 5-3) and the Parrita River mouth (described in site 3A-2), also offer good coastal birding, though they do not warrant a full site description.

Regional Specialties

Blue-footed Booby (15) 1:4
Brown Booby (15) 1:1
Brown Pelican (17) 4:1
Magnificent Frigatebird (19) 1:6
Reddish Egret (25) 5:7
Clapper Rail (59)
Rufous-necked Wood-Rail (55) 6:14
Black-bellied Plover (67) 9:1
Collared Plover (65) 10:3
Snowy Plover (65) 10:4
Wilson's Plover (65) 10:2
Semipalmated Plover (65) 10:5
American Oystercatcher (67) 9:9
Wandering Tattler (73) 10:8
Willet (71) 9:6
Whimbrel (71) 9:14
Long-billed Curlew (71) 9:13
Marbled Godwit (69) 9:8
Ruddy Turnstone (73) 10:6
Surfbird (73) 10:7
Red Knot (77) 11:19
Sanderling (75) 11:11
Dunlin (77) 11:10
Stilt Sandpiper (77) 11:6
Wilson's Phalarope (79) 10:9
Red-necked Phalarope (79) 10:10
Laughing Gull (81) 3:8
Franklin's Gull (81) 3:5
Bonaparte's Gull 3:6
Ring-billed Gull 3:9
Herring Gull 3:10
Brown Noddy (85) 2:8
Bridled Tern (83) 2:2
Least Tern (83) 2:4
Gull-billed Tern (83) 2:1
Caspian Tern (81) 3:1
Black Tern (85) 2:6
Common Tern (83) 2:5
Forster's Tern (82) 2:9
Royal Tern (81) 3:2
Sandwich Tern (83) 3:4
Elegant Tern (81) 3:3
Black Skimmer (85) 3:11
*Mangrove Hummingbird (131) 24:2
Northern Scrub-Flycatcher (193) 37:22
Panama Flycatcher (209) 35:20
Mangrove Vireo (225) 40:11
Yellow (Mangrove) Warbler (259) 42:3

Species to Expect (Applies to all three sites in the region because the variation between sites is minimal. The "Common Birds to Know" are in bold type.)

Black-bellied Whistling-Duck (5) 8:7
Blue-winged Teal (7) 8:5
Brown Pelican (17) 4:1
Neotropic Cormorant (17) 4:4
Magnificent Frigatebird (19) 1:6
Bare-throated Tiger-Heron (21) 5:16
Great Blue Heron (23) 5:6
Great Egret (23) 5:14
Snowy Egret (23) 5:10
Little Blue Heron (23) 5:9
Tricolored Heron (25) 5:8
Cattle Egret (23) 5:13
Green Heron (25) 6:2
Yellow-crowned Night-Heron (27) 5:3
White Ibis (29) 4:8
Roseate Spoonbill (27) 4:7
Wood Stork (19) 4:6
Black Vulture (31) 13:4
Turkey Vulture (31) 13:3
Osprey (33) 17:14
Common Black-Hawk (45) 13:6
Yellow-headed Caracara (53) 15:9

Black-bellied Plover (67) 9:1
Collared Plover (65) 10:3
Wilson's Plover (65) 10:2
Semipalmated Plover (65) 10:5
Black-necked Stilt (69) 9:12
Spotted Sandpiper (73) 11:8
Greater Yellowlegs (71) 11:2
Willet (71) 9:6
Lesser Yellowlegs (71) 11:3
Upland Sandpiper (73) 11:17
Whimbrel (71) 9:14
Marbled Godwit (69) 9:8
Ruddy Turnstone (73) 10:6
Red Knot (77) 11:19
Sanderling (75) 11:11
Semipalmated Sandpiper (75) 11:13
Western Sandpiper (75) 11:12
Least Sandpiper (75) 11:14
Stilt Sandpiper (77) 11:6
Short-billed Dowitcher (79) 9:3
Laughing Gull (81) 3:8
Franklin's Gull (81) 3:5
Least Tern (83) 2:4
Gull-billed Tern (83) 2:1
Caspian Tern (81) 3:1
Royal Tern (81) 3:2
Sandwich Tern (83) 3:4
Elegant Tern (81) 3:3
Black Skimmer (85) 3:11
Lesser Nighthawk (113) 21:12
*Mangrove Hummingbird (131) 24:2
Ringed Kingfisher (149) 27:1
Belted Kingfisher (149) 27:2
Amazon Kingfisher (149) 27:4
Green Kingfisher (149) 27:5
American Pygmy Kingfisher (149) 27:6
Streak-headed Woodcreeper (173) 29:8
Northern Scrub-Flycatcher (193) 37:22
Panama Flycatcher (209) 35:20
Brown-crested Flycatcher (209) 35:18
Cinnamon Becard (215) 33:11
Mangrove Vireo (225) 40:11
Rufous-browed Peppershrike (229) 40:2
Mangrove Swallow (233) 22:23
Yellow Warbler (259) 42:2
Yellow (Mangrove) Warbler (259) 42:3
Prothonotary Warbler (267) 42:1
Northern Waterthrush (269) 43:14
Great-tailed Grackle (317) 44:16

Site 5-1: Colorado

Birding Time: 3 hours
Elevation: Sea level
Trail Difficulty: 2
Reserve Hours: N/A
Entrance Fee: N/A

Colorado is a coastal town along the shores of the Gulf of Nicoya. Because the coastline here is mostly mudflats, the area has been passed over by the developers who have ravaged much of the Pacific Coast beaches. The area is relatively poor, and many of Colorado's residents make their living as fishermen, while others harvest sea salt from the salinas.

Colorado is one of the best sites in Costa Rica for shorebirds. Especially at high tide, when the mudflats in the Gulf of Nicoya are covered with water, large numbers of shorebirds, terns, and gulls come to rest and feed in the salinas of Colorado. The viewing here is easy; you can walk between the various pools and get quite close to many of the birds. Keep in mind that the salinas are private property, and while the owners are friendly toward birders, be considerate and be sure to ask permission of someone in the area before entering.

Target Birds

Least Grebe (9) 7:4
Hook-billed Kite (35) 16:9
Common Black-Hawk (45) 13:6
Zone-tailed Hawk (45) 14:2
Peregrine Falcon (51) 15:5
American Golden-Plover (67) 9:2
Wilson's Plover (65) 10:2
American Oystercatcher (67) 9:9
Upland Sandpiper (73) 11:17
Marbled Godwit (69) 9:8
Surfbird (73) 10:7
Red Knot (77) 11:19
Baird's Sandpiper (77) 11:15
Pectoral Sandpiper (75) 11:7
Stilt Sandpiper (77) 11:6
Long-billed Dowitcher (79) 9:4
Wilson's Phalarope (79) 10:9
Franklin's Gull (81) 3:5
Least Tern (83) 2:4
Gull-billed Tern (83) 2:1
Caspian Tern (81) 3:1
Black Tern (85) 2:6
Common Tern (83) 2:5
Elegant Tern (81) 3:3
Black Skimmer (85) 3:11
Yellow (Mangrove) Warbler (259) 42:3
Spot-breasted Oriole (319) 44:1

Access

From the intersection of Rt. 1 and Rt. 18: Head west on Rt. 18 for 10.2 km and then turn left. (**From the Amistad Bridge along Rt. 18:** Head east on Rt. 18 for 17.0 km and then turn right.) Follow this road for 4.8 km to the town of Colorado. (Driving time from Rt. 1: 15 min; from the Amistad Bridge: 20 min)

Logistics: Colorado is best birded as a quick stop while traveling between more extensive sites. There are no accommodations in the area, although there are a few local restaurants.

Birding Sites

There are two large sets of salinas near Colorado, one on each side of town. Both can be very productive. To reach the east salinas, take your first left upon entering Colorado and go 0.8 km. At this point another dirt road will merge in from the right. Immediately after this, turn right down a driveway and proceed 0.4 km to a small farmhouse, where you can ask permission to bird the area. You should see hundreds of shorebirds feeding in the pools. Western Sandpiper, Short-billed Dowitcher, Whimbrel, and Willet are among the more common species, although you may encounter Marbled Godwit, Stilt Sandpiper, and Red Knot. Also possible are White Ibis, Wood Stork, Little Blue Heron, Black-bellied Whistling-Duck, and Blue-winged Teal. Some scrubby mangroves surround the salinas, and a bit of *pishing* should bring out Yellow (Mangrove) Warbler and maybe a Northern Scrub-Flycatcher or Panama Flycatcher.

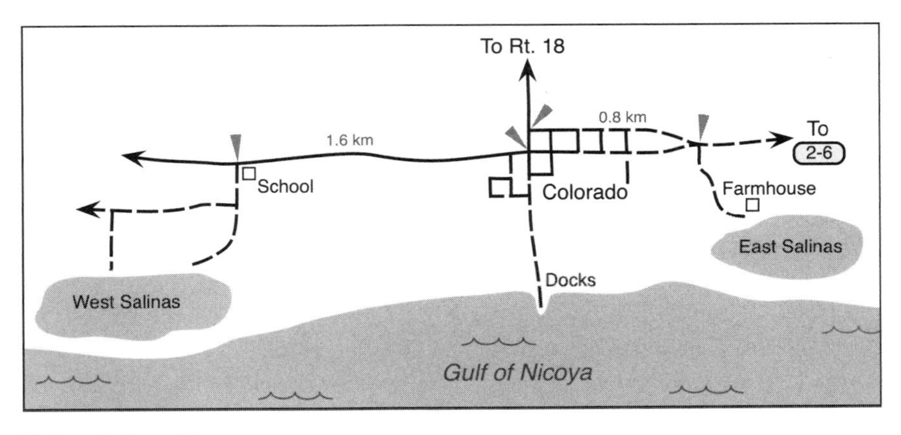

Colorado Area Map

To reach the west salinas, take your first right onto a paved road as you enter Colorado. Follow this road for 1.6 km to a school on your left. Turn left onto the dirt road immediately after the school, and follow this straight for 0.8 km to the salinas. There are usually workers in the area, and you should be sure to ask them for permission before you enter. This is another great place to find many of the afore-mentioned shorebirds, and is especially good for gulls and terns. Keep an eye open for Laughing Gull, Royal Tern, Elegant Tern, Gull-billed Tern, Sandwich Tern, and Black Skimmer.

The fishing docks, straight through town, can be a nice place to stop when the tide is low. Many shorebirds and herons, such as Tricolored Heron, Little Blue Heron, and Great Egret, can be seen from here.

One final place to check is a small seasonal pond along the north side of Rt. 18, 0.8 km east of the turn for Colorado. When filled with water, the pond is home to breeding Least Grebes and affords good views.

Site 5-2: Chomes

Birding Time: 3 hours
Elevation: Sea level
Trail Difficulty: 1
Reserve Hours: N/A
Entrance Fee: N/A
4×4 Recommended

Conveniently located near the Inter-American Highway (Rt. 1) on the eastern side of the Gulf of Nicoya, the town of Chomes offers some of the best coastal birding in the country. The attraction is an extensive series of shrimp ponds on the south side of town, as well as excellent mangroves and mudflats nearby. The area is productive all day long, although the concentrations of birds vary. At high tide the birds are pushed up into the man-made pools, while at low tide they move out into the mudflats. Chomes makes a great travel stop on your way along the Inter-American Highway, but to bird the area thoroughly you will certainly want to allow a few hours.

Target Birds

Reddish Egret (25) 5:7
Hook-billed Kite (35) 16:9
Common Black-Hawk (45) 13:6
Zone-tailed Hawk (45) 14:2
Peregrine Falcon (51) 15:5
Double-striped Thick-knee (63) 9:10
American Golden-Plover (67) 9:2
Wilson's Plover (65) 10:2
American Oystercatcher (67) 9:9
Upland Sandpiper (73) 11:17
Long-billed Curlew (71) 9:13
Marbled Godwit (69) 9:8
Surfbird (73) 10:7
Red Knot (77) 11:19
Baird's Sandpiper (77) 11:15
Pectoral Sandpiper (75) 11:7
Stilt Sandpiper (77) 11:6
Long-billed Dowitcher (79) 9:4
Wilson's Phalarope (79) 10:9
Franklin's Gull (81) 3:5
Least Tern (83) 2:4
Gull-billed Tern (83) 2:1
Caspian Tern (81) 3:1
Black Tern (85) 2:6
Common Tern (83) 2:5
Elegant Tern (81) 3:3
Black Skimmer (85) 3:11
*Mangrove Hummingbird (131) 24:2
Northern Scrub-Flycatcher (193) 37:22
Panama Flycatcher (209) 35:20
Mangrove Vireo (225) 40:11
Yellow (Mangrove) Warbler (259) 42:3

Access

From the intersection of Rt. 1 and Rt. 23: Head north on Rt. 1 for 26.4 km and take a left onto a dirt road just before Rt. 132. (**From the intersection of Rt. 1 and Rt. 18:** Head south on Rt. 1 for 23.8 km and then take a right onto a dirt road just beyond Rt. 132.) Follow this road straight for 9.8 km, when it enters the outskirts of Chomes. Continue straight for a few blocks until you come to a T-intersection. Here turn left, go a few hundred meters to another T-intersection and then turn right. This road may look extremely small and unused, but it leads almost immediately to the shrimp pools. (Driving time from Rt. 1: 20 min)

Logistics: Chomes is a small town without much to offer beyond a restaurant or two. Plan to stay elsewhere.

Birding Sites

On your way into and out of the Chomes area, look for Double-striped Thick-knee standing in the open fields. When you approach the pools, keep an eye out for any workers. The ponds are private land, and it is polite to ask permission to bird the area. This is never a problem, as the people are always quite friendly.

The best time to bird the area is at high tide, when the birds are pushed out of the Gulf of Nicoya and into the pools. If the tides cooperate, mornings or afternoons are best because the midday heat haze can make viewing difficult. The pools are filled with varying amounts of water, which has a large impact on the bird species found in and around them. Many pools will be mostly muddy, and these can be great places to find shorebirds. Look for Semipalmated Sandpiper, Least Sandpiper, Western Sandpiper, and Semipalmated Plover. Pools with a shallow layer of water could yield Black-necked Stilt and Short-billed Dowitcher, Stilt Sandpiper, and Red Knot. Large waders such as Great Blue Heron, White Ibis, Wood Stork, and Roseate Spoonbill are also found around the pools, and the deepest ones attract terns. Look for Royal Tern, Elegant Tern, Sandwich Tern, and Gull-billed Tern. Laughing Gull and Magnificent Frigatebird often cruise overhead, while Mangrove Swallows swoop low over the water.

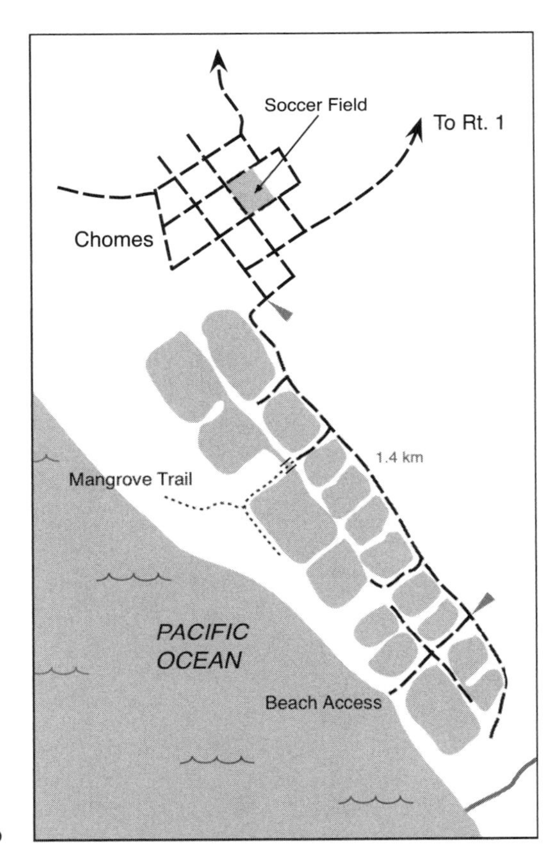

Chomes Area Map

Sections of mangrove line the edges of many of the pools, and you should check the larger patches for specialists of this habitat. The best access point is a trail that runs for a few hundred meters into the mangrove (refer to map). Prothonotary Warbler and Yellow (Mangrove) Warbler should be easy to find, and you may be able to spot Northern Scrub-Flycatcher, Mangrove Hummingbird, or Panama Flycatcher. Be careful with the identification of the latter because Brown-crested Flycatcher is also common in the area.

The beach is the final place to check, and it is often swarming with birds. Low tide exposes vast mudflats that provide habitat for a range of species. Look for the afore-mentioned shorebirds, although you might be able to find Marbled Godwit and Collared Plover as well. Many species of heron feed along the tidal waters, and Brown Pelicans roost on the mud. There are often large flocks of terns and gulls, which could include Black Skimmer, Franklin's Gull, Caspian Tern, and Least Tern. Use caution here, however, as American Crocodiles sometimes swim out of the nearby river and bask on the mud.

Site 5-3: Tárcoles River Mouth

Birding Time: 2–4 hours
Elevation: Sea level
Trail Difficulty: 2
Reserve Hours: N/A
Entrance Fee: N/A

The Tárcoles River Mouth offers excellent coastal birding. You can find a variety of shorebirds, terns, and gulls, as well as all of the mangrove specialties. It is an especially appealing location because of its proximity to Carara National Park (3A-1), and birding the two areas in conjunction will yield an amazing variety of species. The town of Tárcoles is certainly not one of Costa Rica's most pristine, as garbage removal in the area seems to be sporadic at best. This problem is compounded by the fact that the Tárcoles River drains much of the Central Valley, resulting in trash scattered along its banks. However, as anyone who has birded a sewage treatment plant or landfill can attest, birds do not really care about aesthetics. The area consistently yields interesting species.

Target Birds

Reddish Egret (25) 5:7
Boat-billed Heron (27) 5:2
Plumbeous Kite (37) 14:8
Crane Hawk (41) 14:4
Common Black-Hawk (45) 13:6
Zone-tailed Hawk (45) 14:2
Rufous-necked Wood-Rail (55) 6:14
Double-striped Thick-knee (63) 9:10
Southern Lapwing (67)
Collared Plover (65) 10:3
Wilson's Plover (65) 10:2
Red Knot (77) 11:19
Sanderling (75) 11:11
White-rumped Sandpiper (77) 11:16

Least Tern (83) 2:4
Caspian Tern (81) 3:1
Elegant Tern (81) 3:3
Black Skimmer (85) 3:11
Scarlet Macaw (99) 19:1
Lesser Nighthawk (113) 21:12
*Mangrove Hummingbird (131) 24:2
American Pygmy Kingfisher (149) 27:6
Northern Scrub-Flycatcher (193) 37:22
Panama Flycatcher (209) 35:20
Mangrove Vireo (225) 40:11
Rufous-browed Peppershrike (229) 40:2
Yellow (Mangrove) Warbler (259) 42:3
Prothonotary Warbler (267) 42:1

Access

From the headquarters of Carara National Park: (Refer to site 3A-1 for a map.) Head south on Rt. 34 for 2.8 km and then turn right onto a dirt road that heads into the town of Tárcoles. After 1.0 km the road comes to a T-intersection and you should take a right. Follow this straight for 2.2 km until you come to the Tárcoles River and a small boat landing used by Crocodile Tours. (Driving time from Carara National Park: 10 min)

Logistics: Refer to site 3A-1.

Birding Sites

The Tárcoles River Mouth

The Crocodile Tours boat landing is a good place to start your birding. Scan the banks of the river for herons, including Little Blue Heron, Yellow-crowned Night-Heron, and Great Blue Heron. Mangrove Swallows can almost always be seen swooping low over the river.

As you pull out of the boat landing parking lot, take a right. You will soon pass a field on your right that is often flooded. This is a good place to find White Ibis, Tricolored Heron, Ringed Kingfisher, and others. Continue down the road and then take your first right. Proceed until this road terminates at the Tárcoles River and the Tarcol Lodge, which sits along the banks. If the river level is low, you may find Semipalmated Plover, Least Sandpiper, Spotted Sandpiper, Willet, and Whimbrel feeding in the exposed mud. Here you will find another boat landing, the launch site for the Mangrove Birding Tour described below. There is a small patch of mangroves on the far side of the Tarcol Lodge that could yield Prothonotary Warbler, Panama Flycatcher, and Streak-headed Woodcreeper, although other mangroves in the area are generally much more productive.

From the Tarcol Lodge retrace your steps about 100 meters. Take your first right and you will arrive at an overgrown soccer field. Park your car in an out-of-the-way place, do not leave any valuables inside, and be sure to lock up. Walk across the soccer field to the beach. Yellow-headed Caracara and Common Black-Hawk often patrol the area around the soccer field. Walk up the beach toward the river mouth, and you will soon come to a patch of mangroves. This area can be extremely productive and you have a good chance of finding Panama

Flycatcher, Northern Scrub-Flycatcher, Yellow (Mangrove) Warbler, Prothonotary Warbler, and Mangrove Vireo.

If you continue walking the beach, you will eventually come to the river mouth, which is the best place for shorebirds, terns, and gulls. Be aware that this is a healthy hike (about one and a half kilometers in sand) but worth the effort. There is usually a large flock of terns roosting by the river mouth. The majority of these birds are Royal Terns and Elegant Terns, but you may also be able to find Sandwich Tern, Least Tern, Caspian Tern, and Black Skimmer. Shorebirds also congregate here, and you should see Whimbrel, Least Sandpiper, Black-bellied Plover, Wilson's Plover, Collared Plover, and Short-billed Dowitcher among others. Lesser Nighthawk likes to roost in the beach grasses.

Mangrove Birding Tour

The Mangrove Birding Tour is a two-and-a-half-hour boat trip along the Tárcoles River. The boat holds up to 30 people and is manned by knowledgeable guides. If you want to experience the beauty of the unique mangrove habitat, and see many of the specialty species found there, the Mangrove Birding Tour is the best way to do it. The boat will take you through otherwise inaccessible mangrove canals that are home to many species of birds. Boat-billed Herons are seen roosting in the vegetation near the canal. Rufous-browed Peppershrike sings constantly, and you should see Mangrove Vireo, Panama Flycatcher, Northern Scrub-Flycatcher, Mangrove Hummingbird, and Cinnamon Becard. You will usually see Common Black-Hawk circling overhead, and occasionally a Plumbeous Kite. If you are really lucky, you may happen across a Rufous-necked Wood-Rail winding its way among the tangled mangrove roots.

The boat also takes you up the principal river channel, which passes by fields and farmland. This is a great area to see Roseate Spoonbill, Wood Stork, Double-striped Thick-knee, Southern Lapwing, Scarlet Macaw, and Zone-tailed Hawk.

Tours are run daily at 6 a.m., 9 a.m., 12 p.m., and 3 p.m. The cost is $25 per person (US 2008) for groups smaller than 10 people and $15 per person for large groups. Be sure to make a reservation in advance.

Mangrove Birding Tour
Tel: (English) (506) 2643-1983, (506) 8889-8815, (506) 2433-8278
(Spanish) (506) 8888-2004, (506) 2637-0472

Nearby Birding Opportunities

Tivives

Tivives is another area that provides access to mangroves as well as beach and a river mouth. While Tivives does produce some interesting species, it is generally not as productive as the three primary coastal sites presented in this guide.

To get to Tivives from the intersection of Rt. 34 and Rt. 23, take Rt. 23 toward Puntarenas for 12.4 km. After this, turn left onto a dirt road and go straight for 6.0 km to a small gatehouse at the entrance to Tivives. Politely ask the gatekeeper if you can look for birds around the mangroves and river, and he should let you pass without hassle. Just after the gatehouse you will come to a T-intersection, where you should turn left. This road parallels the coast for about 1.5 km and you should park at the end, when you come to a river mouth.

Walking out to the beach will give you a good view of the river mouth, which often has a number of birds foraging in and around the waters. Snowy Egret and Roseate Spoonbill often wade in the surf, Neotropic Cormorant and Brown Pelican hunt the converging waters, and many sandpipers can be found along the shoreline. Look for Least Sandpiper, Collared Plover, Wilson's Plover, Willet, Whimbrel, Ruddy Turnstone, and others.

Walking up along the edge of the river will give you access to some mangroves, though you should use caution because there are crocodiles that live in the river. The mangroves are home to Prothonotary Warbler, Rufous-browed Peppershrike, Mangrove Hummingbird, Panama Flycatcher, and others.

Region 6
The Central Valley

While found throughout the country, the **White-tailed Kite** is frequently seen in elegant flight over farm fields and pastures in the Central Valley. Look for this species at the University of Costa Rica (6-2) and Lankester Gardens (6-3).

Regional Map: The Central Valley

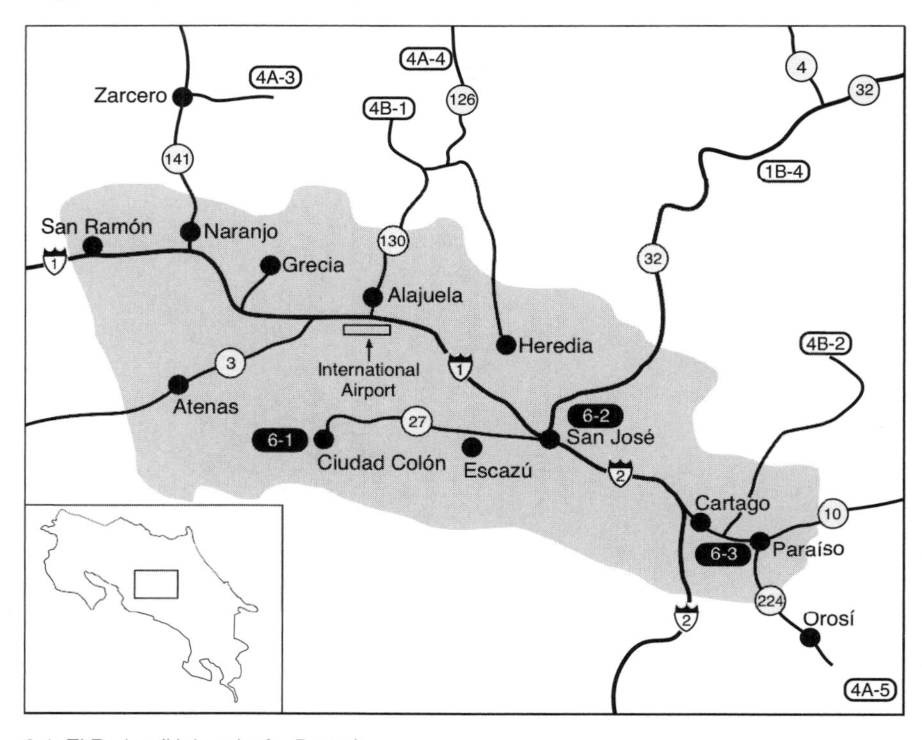

6-1: El Rodeo (University for Peace)
6-2: The University of Costa Rica
6-3: Lankester Gardens

Introduction

The Central Valley sits in the center of Costa Rica, between the Cordillera Central to the north and the Cordillera de Talamanca to the south. The area is home to about half the population of Costa Rica. Eight of the country's 10 largest cities are located in the Central Valley, and San José, the country's capital, has over 328,000 residents.

One factor that makes the Central Valley such a popular place to live is its ideal climate. Most of the valley lies between 800 and 1,300 meters, which helps to alleviate the overbearing heat of the tropical lowlands. The temperatures in the valley reach a high of around 29°C (85°F) on most days and are comfortably cool at night. Very few buildings use air-conditioning because the natural temperatures are so comfortable. Instead, much of the architecture is able to utilize open-air designs. The area is also relatively flat (by Costa Rican standards), making it conducive to large-scale human settlement.

The Central Valley is also the center for one of Costa Rica's largest industries, coffee. The hills around the Central Valley have ideal coffee-growing conditions and produce some of the best in the world. In the past, coffee and bananas were the two principal players in Costa Rica's economy, making this area especially important. While both are still significant today, other industries are developing quickly as Costa Rica's economy diversifies. The past few decades have seen a wave of foreign investment in Costa Rica, as large corporations like Intel, Proctor & Gamble, and Hewlett Packard have set up major operations within the Central Valley.

Because the Central Valley spans both slopes, it is not an ecologically cohesive region. The Continental Divide runs roughly between the cities of San José and Cartago, and there is quite a difference in habitat and birdlife between the west side of the valley and the east. To make matters more confusing, the Central Valley is a low pass between slopes, and a number of species cross over to the opposite slope from where they are usually found.

While the Central Valley does not provide the most exciting birding, many visitors may find themselves staying in the area for various reasons. Each of the three sites described can be thoroughly birded in a few hours and make very nice morning trips from wherever you may be staying in the Central Valley. If you are looking to find the few Central Valley specialty species, Lankester Gardens (6-3) and the University of Costa Rica (6-2) are your best options. However, I find El Rodeo (6-1) to be a more pleasant overall birding experience.

Other sites that make good day trips from the Central Valley include Braulio Carrillo National Park (1B-4), Carara National Park (3A-1), La Paz Waterfall Gardens (4A-4), Tapantí National Park (4A-5), Poás Volcano National Park (4B-1), Irazú Volcano National Park (4B-2), Km 70 (4B-3), and Tárcoles River Mouth (5-3).

Regional Specialties

White-throated Flycatcher (205) 36:13
Sedge Wren (243) 38:1
Prevost's Ground-Sparrow (303) 48:18
White-eared Ground-Sparrow (303) 48:19

Common Birds to Know

Black Vulture (31) 13:4
Turkey Vulture (31) 13:3
Red-billed Pigeon (87) 18:2
White-tipped Dove (91) 18:14
*Crimson-fronted Parakeet (95) 19:10
Rufous-tailed Hummingbird (129) 24:10
*Hoffmann's Woodpecker (157) 28:16
Great Kiskadee (211) 35:13
Boat-billed Flycatcher (211) 35:12
Social Flycatcher (211) 35:14
Tropical Kingbird (213) 35:1
Blue-and-white Swallow (233) 22:20
Plain Wren (241) 38:17
Clay-colored Thrush (251) 39:8
Blue-gray Tanager (291) 46:15
Rufous-collared Sparrow (307) 50:13
Great-tailed Grackle (317) 44:16
Baltimore Oriole (321) 44:7

Site 6-1: El Rodeo (University for Peace)

Birding Time: 4 hours
Elevation: 850 meters
Trail Difficulty: 1
Reserve Hours: 8 a.m. to 4 p.m.
Entrance Fee: $1 (US 2008)

El Rodeo is a small protection zone near the town of Ciudad Colón, and is my favorite birding site in the Central Valley. Although it is only 30 minutes from San José, you feel quite removed from the hubbub of the big city. One of the best birding areas at El Rodeo is the University for Peace, a small school run under the auspices of the United Nations that seeks to educate students from around the globe in peace, tolerance, and acceptance. The university maintains a small forest preserve on campus and a system of trails lead through it. These trails are open to the public for a nominal entrance fee, and the habitat, which is primarily secondary forest, is some of the best in the area. The countryside around the University for Peace is blanketed with rows of coffee bushes, and the trees and shrubs within and around the small fields attract a variety of birds.

The habitat found at El Rodeo is reminiscent of North Pacific dry forest, and many of the usual suspects can be found here. Long-tailed Manakin, Canivet's Emerald, Plain-capped Starthroat, and many other North Pacific specialties can be seen, as well as the more ubiquitous disturbed-habitat species.

Target Birds

Gray-headed Chachalaca (11) 12:1
Striped Cuckoo (101) 21:5
Lesser Ground-Cuckoo (103) 21:11
Tropical Screech-Owl (107) 20:15
Ferruginous Pygmy-Owl (109) 20:17
Canivet's Emerald (127) 24:13
Blue-throated Goldentail (129) 24:9
Cinnamon Hummingbird (129) 24:11
Plain-capped Starthroat (135) 23:19
Ruby-throated Hummingbird (131) 25:9
Blue-crowned Motmot (147) 27:8
*Fiery-billed Aracari (155) 27:16
Keel-billed Toucan (155) 27:18
Barred Antshrike (175) 31:3
Northern Beardless-Tyrannulet (191) 37:19
White-winged Becard (217) 33:13
Rose-throated Becard (217) 33:12
Long-tailed Manakin (223) 33:4
Rufous-breasted Wren (239) 38:8
Red-crowned Ant-Tanager (277) 47:10
Olive Sparrow (303) 50:15
Blue Grosbeak (313) 48:11

Access

From Sabana Park (San José): Take Rt. 27, which starts on the southeast corner of the park, west toward the towns of Escazú and Ciudad Colón. Follow the highway straight for 18.6 km until you enter the town of Ciudad Colón. Here the

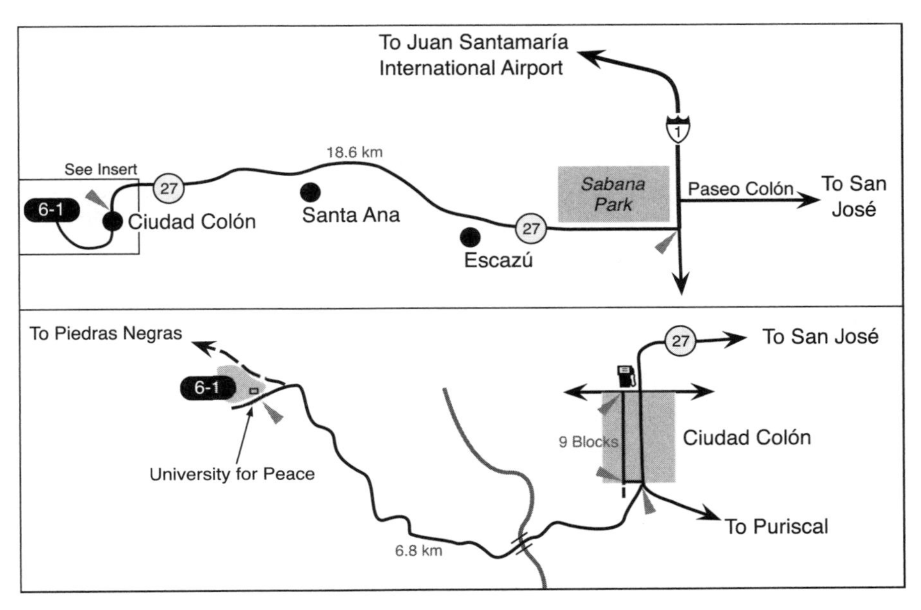

El Rodeo Driving Map. The bottom half of this map shows detail around Ciudad Colón and the University for Peace.

road turns into a one-way street going the wrong way, and you should turn right at a gas station. Go one block, turn left, and proceed straight for nine blocks to the end of town. Turn left again, go one block, and take a right, following signs for the University for Peace. Stay on the paved road and you will arrive at the University in 6.8 km. (Driving time from Sabana Park: 25 min)

Logistics: El Rodeo is best visited as a morning excursion from San José. If you want to stay closer, there are some nice hotels in the nearby towns of Santa Ana and Escazú.

Birding Sites

The road leading from Ciudad Colón to the University for Peace offers some excellent birding, although the traffic can be slightly frustrating. One of the most productive sections to bird is the first kilometer after crossing a small river about 1.0 km from Ciudad Colón. The road winds up a hillside covered with new-growth vegetation, and you can often find many interesting birds. Look for Barred Antshrike, Gray-headed Chachalaca, Canivet's Emerald, Violaceous Trogon, and Northern Beardless-Tyrannulet. The area can also be good for migrants and winter residents, including Baltimore Oriole, Yellow-throated Vireo, Black-and-white Warbler, and Mourning Warbler.

Another interesting area is 3.6 km from Ciudad Colón where there are some overgrown fields. This habitat often yields Painted Bunting, Indigo Bunting, Gray-crowned Yellowthroat, and Ruby-throated Hummingbird.

You should stop and bird anywhere along the road that catches your fancy. Brown Jay, Montezuma Oropendola, Blue-crowned Motmot, and Crimson-fronted Parakeet are all common along the roadside, and Short-tailed Hawk is often seen soaring overhead.

Just before arriving at the University for Peace, you will come to a dirt road that forks off to the right and heads toward the community of Piedras Negras. This is also a nice road to bird, especially since the traffic is minimal. A little ways down you will find a hedge of bushes that attract many species of hummingbirds when flowering. Blue-throated Goldentail, Cinnamon Hummingbird, and Steely-vented Hummingbird can be found here. As you continue down the road, also look for Rufous-breasted Wren, Stripe-headed Sparrow, Olive Sparrow, and Keel-billed Toucan.

To get to the university trails, continue along the paved road 0.4 km past the university's main entrance. You will come to a small guardhouse on your right with some ponds behind it. Park your car here and pay the entrance fee. Since neither the ponds nor the surrounding manicured grounds tend to be very productive, your time is best spent walking the trails that run through the forest in back. Keep an eye open for Red-crowned Ant-Tanager, Blue-crowned Motmot, Long-tailed Manakin, Rose-throated Becard, Masked Tityra, and Yellow-throated Euphonia. At dawn and dusk you will probably hear Ferruginous Pygmy-Owl calling, and at night Mottled Owl and Tropical Screech-Owl are possibilities.

Species to Expect

Little Tinamou (3) 12:17
Black-bellied Whistling-Duck (5) 8:7
Gray-headed Chachalaca (11) 12:1
Black Vulture (31) 13:4
Turkey Vulture (31) 13:3
White-tailed Kite (37) 15:1
Broad-winged Hawk (43) 16:13
Short-tailed Hawk (43) 16:11
Red-billed Pigeon (87) 18:2
White-winged Dove (89) 18:12
Inca Dove (89) 18:11
White-tipped Dove (91) 18:14
***Crimson-fronted Parakeet (95) 19:10**
Orange-chinned Parakeet (97) 19:14
Squirrel Cuckoo (103) 21:7
Black-billed Cuckoo (101) 21:1
Striped Cuckoo (101) 21:5
Lesser Ground-Cuckoo (103) 21:11
Groove-billed Ani (103) 21:9
Tropical Screech-Owl (107) 20:15
Ferruginous Pygmy-Owl (109) 20:17
White-collared Swift (117) 22:1
Vaux's Swift (119) 22:9
Canivet's Emerald (127) 24:13
Blue-throated Goldentail (129) 24:9
Steely-vented Hummingbird (127) 24:15
Rufous-tailed Hummingbird (129) 24:10
Cinnamon Hummingbird (129) 24:11
Plain-capped Starthroat (135) 23:19
Ruby-throated Hummingbird (131) 25:9
Violaceous Trogon (141) 26:8
Blue-crowned Motmot (147) 27:8
Green Kingfisher (149) 27:5
Keel-billed Toucan (155) 27:18
***Hoffmann's Woodpecker (157) 28:16**
Lineated Woodpecker (161) 27:13
Pale-billed Woodpecker (161) 27:14
Northern Barred-Woodcreeper (169) 29:19

Streak-headed Woodcreeper (173) 29:8
Barred Antshrike (175) 31:3
Northern Beardless-Tyrannulet (191) 37:19
Yellow-bellied Elaenia (193) 37:26
Ochre-bellied Flycatcher (199) 36:25
Paltry Tyrannulet (191) 37:10
Common Tody-Flycatcher (197) 37:7
Yellow-olive Flycatcher (195) 37:16
Brown-crested Flycatcher (209) 35:18
Great Kiskadee (211) 35:13
Boat-billed Flycatcher (211) 35:12
Social Flycatcher (211) 35:14
Sulphur-bellied Flycatcher (213) 35:10
Piratic Flycatcher (193) 35:8
Tropical Kingbird (213) 35:1
White-winged Becard (217) 33:13
Rose-throated Becard (217) 33:12
Masked Tityra (217) 34:1
Long-tailed Manakin (223) 33:4
Yellow-throated Vireo (225) 40:6
Red-eyed Vireo (227) 40:3
Yellow-green Vireo (227) 40:5
Lesser Greenlet (229) 40:7
Brown Jay (231) 39:19
Blue-and-white Swallow (233) 22:20
Rufous-naped Wren (239) 38:2
Rufous-breasted Wren (239) 38:8
Plain Wren (241) 38:17
House Wren (243) 38:18
Swainson's Thrush (249) 39:2
Clay-colored Thrush (251) 39:8
Tennessee Warbler (255) 40:22
Yellow Warbler (259) 42:2
Chestnut-sided Warbler (259) 43:4
Black-and-white Warbler (267) 41:13
Ovenbird (267) 43:12
Mourning Warbler (271) 42:14
Gray-crowned Yellowthroat (273) 42:13
Wilson's Warbler (269) 42:4
Canada Warbler (269) 42:8
Rufous-capped Warbler (275) 40:16
Red-crowned Ant-Tanager (277) 47:10
Summer Tanager (285) 47:5
Scarlet Tanager (287) 47:8
Blue-gray Tanager (291) 46:15
Red-legged Honeycreeper (293) 46:2
Blue-black Grassquit (297) 49:7
Variable Seedeater (295) 49:3
Yellow-faced Grassquit (297) 49:6
Olive Sparrow (303) 50:15
Stripe-headed Sparrow (303) 50:7
Grayish Saltator (309) 48:3
Buff-throated Saltator (309) 48:2
Blue-black Grosbeak (311) 48:10
Indigo Bunting (313) 48:12
Painted Bunting (313) 48:13
Baltimore Oriole (321) 44:7
Montezuma Oropendola (323) 44:8
Yellow-throated Euphonia (327) 45:5

Site 6-2: The University of Costa Rica

Birding Time: 3 hours
Elevation: 1,200 meters
Trail Difficulty: 1
Reserve hours: N/A
Entrance Fee: N/A

It seems that within every city, birders have found a location where a little bit of nature still clings amid the sea of concrete and the turmoil of city life. In San José this area is the University of Costa Rica (UCR). If you are staying in the city, this is the perfect place to spend a few morning hours searching out birds. While the area does not offer extensive forests, or many rarities, it can produce

30 or more species in a morning of birding. It is also the most reliable site described in this guide to find Prevost's Ground-Sparrow, which is restricted to the Central Valley.

The University of Costa Rica is the country's largest school, enrolling about 39,000 students and offering 45 fields of study. It is also the most prestigious school in the country and regarded as one of the best universities in Latin America. The campus, located on the east side of San José in an area called San Pedro, is easily accessible by taxi from anywhere in the city. The birding sites below describe the main campus itself, as well as the Instalaciones Deportivas (sports facilities) and a secondary campus known as the Ciudad de la Investigación.

Because of its urban location, UCR is one site where you should take extra precautions. While there are many police and campus security officers who patrol the area, you should still be alert to possible dangers. In particular, I do not suggest arriving to bird before dawn.

Target Birds

White-tailed Kite (37) 15:1
Gray-necked Wood-Rail (55) 6:13
*Crimson-fronted Parakeet (95) 19:10
Tropical Screech-Owl (107) 20:15
Ferruginous Pygmy-Owl (109) 20:17
Blue-crowned Motmot (147) 27:8
Golden-olive Woodpecker (159) 28:11
Masked Tityra (217) 34:1
Rufous-browed Peppershrike (229) 40:2
White-naped Brush-Finch (301) 48:15
Prevost's Ground-Sparrow (303) 48:18
White-eared Ground-Sparrow (303) 48:19
Montezuma Oropendola (323) 44:8

Access

From Mall San Pedro (San José): Take Rt. 39 north toward Guadalupe for 0.4 km and then turn right into the UCR campus. Drive past the guard station, where you will receive a ticket (which you must return when exiting), and take your first left. Follow this road straight for 0.8 km to a parking lot on your right, in front of the Escuela de Biología and Escuela de Música. The birding directions start from this point, and I suggest leaving your car here and walking to the different areas.

Logistics: The university is located on the east side of San José, so you should not encounter any unusual logistical concerns here.

Birding Sites

UCR Campus

Start at the parking lot in front of the Escuela de Biología and Escuela de Música. Here you can often find Tropical Screech-Owls roosting. Their favorite perch is in the far right tree on the first median strip in front of the music building.

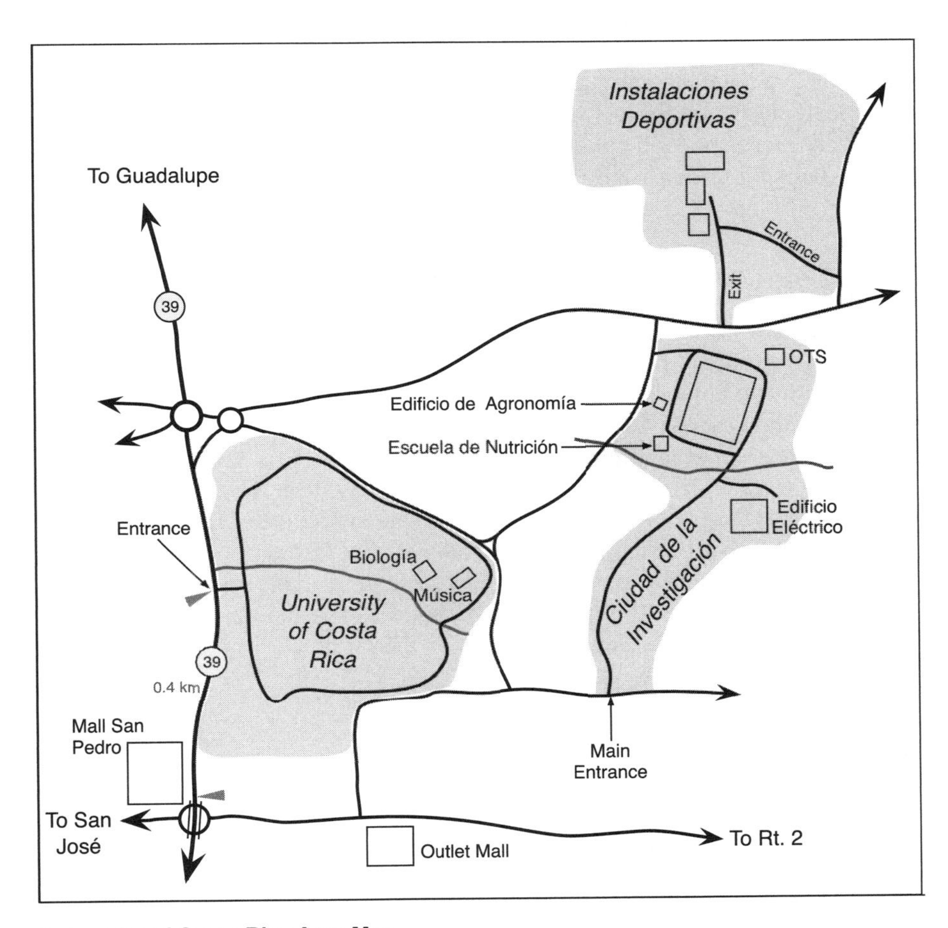

University of Costa Rica Area Map

From here proceed around the right side of the biology building to where you will find a small biological reserve in back. The reserve is fenced and locked, and the best way to bird the area is to walk the perimeter, as well as along the stream that crosses through campus. Look for Masked Tityra, White-eared Ground-Sparrow, Ferruginous Pygmy-Owl, and Piratic Flycatcher. The stream might yield a Gray-necked Wood-Rail. If you get really lucky, you might see Bat Falcon perching atop the campus buildings early in the morning.

Ciudad de la Investigación

The Ciudad de la Investigación is easiest to enter on foot, near the Edificio de Agronomía. You can drive a car in the main entrance, though you will need to tell a guard where you are going, and I find it easier just to walk from the UCR campus.

Start birding this area behind the Escuela de Nutrición, where there is some new-growth habitat along a stream. Follow the stream toward the large blue and white Edificio Eléctrico. Just north of this building there is similar habitat, and both these areas are good places to find Red-billed Pigeon, White-winged Dove, Brown Jay, and Grayish Saltator. This area is also the best place to look for Prevost's Ground-Sparrow and White-tailed Kite.

Instalaciones Deportivas

The Instalaciones Deportivas (sports facilities) are the most productive area in UCR, and you should plan on spending at least an hour and a half exploring it. On foot you can enter the Instalaciones Deportivas through either the entrance or the exit, though with a car you must use the entrance. It is best to walk along the perimeter of the facilities, where there is scrubby new growth as well as some older trees. Early in the morning it is common to hear the repetitive tooting of Ferruginous Pygmy-Owl. You may also be able to find Blue-crowned Motmot, White-naped Brush-Finch, Melodious Blackbird, and Golden-olive Woodpecker. Recently, Montezuma Oropendolas have made a home in the area, building their many hanging nests in a large tree near the exit. You may also find Bronzed Cowbird and Eastern Meadowlark feeding in the grass.

Species to Expect

Black Vulture (31) 13:4
Turkey Vulture (31) 13:3
White-tailed Kite (37) 15:1
Gray-necked Wood-Rail (55) 6:13
Rock Pigeon (85)
Red-billed Pigeon (87) 18:2
White-winged Dove (89) 18:12
Inca Dove (89) 18:11
White-tipped Dove (91) 18:14
***Crimson-fronted Parakeet (95) 19:10**
Orange-chinned Parakeet (97) 19:14
White-crowned Parrot (97) 19:7
Squirrel Cuckoo (103) 21:7
Groove-billed Ani (103) 21:9
Tropical Screech-Owl (107) 20:15
Ferruginous Pygmy-Owl (109) 20:17
Vaux's Swift (119) 22:9
Green-breasted Mango (129) 23:13
Steely-vented Hummingbird (127) 24:15
Rufous-tailed Hummingbird (129) 24:10
Blue-crowned Motmot (147) 27:8
***Hoffmann's Woodpecker (157) 28:16**
Great Kiskadee (211) 35:13
Boat-billed Flycatcher (211) 35:12
Social Flycatcher (211) 35:14
Sulphur-bellied Flycatcher (213) 35:10
Tropical Kingbird (213) 35:1
Masked Tityra (217) 34:1
Brown Jay (231) 39:19
Blue-and-white Swallow (233) 22:20
Plain Wren (241) 38:17
House Wren (243) 38:18
Clay-colored Thrush (251) 39:8
Tennessee Warbler (255) 40:22
Yellow Warbler (259) 42:2
Chestnut-sided Warbler (259) 43:4
Rufous-capped Warbler (275) 40:16
Blue-gray Tanager (291) 46:15
Palm Tanager (291) 45:19
White-naped Brush-Finch (301) 48:15
Prevost's Ground-Sparrow (303) 48:18
White-eared Ground-Sparrow (303) 48:19

Rufous-collared Sparrow (307) 50:13
Grayish Saltator (309) 48:3
Eastern Meadowlark (315) 50:16
Melodious Blackbird (315) 52:6
Great-tailed Grackle (317) 44:16
Bronzed Cowbird (317) 44:15
Baltimore Oriole (321) 44:7
Montezuma Oropendola (323) 44:8

Site 6-3: Lankester Gardens

Birding Time: 2.5 hours
Elevation: 1,300 meters
Trail Difficulty: 1
Reserve Hours: 8:30 a.m. to 4:30 p.m.
Entrance Fee: $5 (US 2008)
Trail Map Available on Site

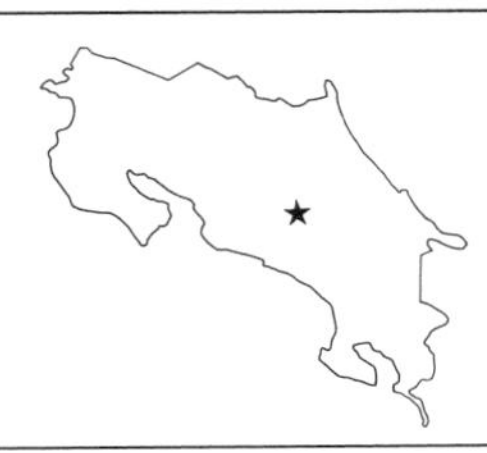

Lankester Gardens is a 27-acre botanical reserve located just east of Cartago. In 1973 it was established as an institution devoted to research, conservation, and education in tropical botany. Today, Lankester Gardens is operated by the University of Costa Rica, with continuing dedication to science and conservation. Orchids hold special interest among the many botanical collections here. The garden has over 1,000 species of orchids in cultivation, and these are the focal point of the facility's research.

Lankester Gardens is a small oasis set among fields, cattle pastures, and residential neighborhoods outside the city of Cartago. Large walls block out most of this surrounding world, creating a serene garden/forest environment. You will be able to happily spend a few hours strolling the trails in search of birds, as well as appreciating the lush tropical botany. There is interesting birding to be found in the fields just outside the garden walls as well. Here you can see two specialty birds for the Cartago area—Sedge Wren and White-throated Flycatcher. Also, a small pond behind the gardens can yield interesting waterfowl.

Target Birds

American Wigeon (7) 8:6
Northern Shoveler (7) 8:2
Northern Pintail (7) 8:1
Least Grebe (9) 7:4
White-tailed Kite (37) 15:1
Sora (59) 6:12
Common Moorhen (61) 7:2
American Coot (61) 7:1
Wilson's Snipe (79) 9:5
Blue-crowned Motmot (147) 27:8
Golden-olive Woodpecker (159) 28:11
Scaled Antpitta (185) 30:20
Lesser Elaenia (193) 37:25
White-throated Flycatcher (205) 36:13
Sedge Wren (243) 38:1
Orange-billed Nightingale-Thrush (247) 38:26
Tropical Mockingbird (253)
White-naped Brush-Finch (301) 48:15
White-eared Ground-Sparrow (303) 48:19
Lesser Goldfinch (325) 50:2

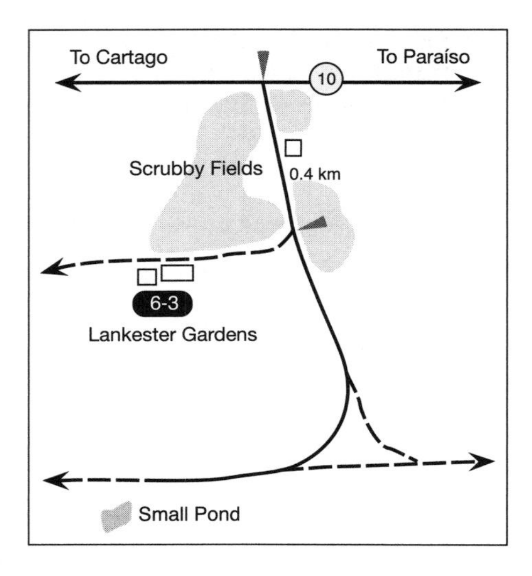

Lankester Gardens Area Map

Access

From Cartago: Head east on Rt. 10 toward Paraíso for 3.8 km and then turn right onto a paved road. (**From Paraíso:** Head west on Rt. 10 toward Cartago for 1.8 km and then turn left onto a paved road.) After only 0.4 km, fork right onto a dirt road, and the gardens will be on your left almost immediately. (Driving time from Cartago: 10 min; from Paraíso: 5 min)

Logistics: Because Lankester Gardens only takes a few hours to bird thoroughly, it is best visited as a morning excursion from San José. Alternatively you can bird the area while traveling between more extensive sites.

Birding Sites

As you enter Lankester Gardens, be sure to pay your entrance fee and pick up a map. The manicured area around the orchid greenhouse, just in front of the entrance and small gift shop, is one of the easiest places to spot birds. Golden-olive Woodpecker, Streak-headed Woodcreeper, Social Flycatcher, and Grayish Saltator are among the birds frequenting the area. As you walk the trails into the forest, keep an eye open for White-eared Ground-Sparrow and Orange-billed Nightingale-Thrush feeding among the fallen leaves. Rufous-capped Warbler and White-breasted Wood-Wren are also likely. Near the heliconia garden, keep an eye out for hummingbirds, which feed in the beautiful bracts. Violet Sabrewing and Steely-vented Hummingbird are possibilities too.

Birding outside the walls of the garden can be interesting as well. The road that leads between Lankester Gardens and Rt. 10 proceeds through wet scrubby fields around some farm buildings. Walking along the roadside should produce

Eastern Meadowlark, Yellow-faced Grassquit, White-collared Seedeater, Plain Wren, and Gray-crowned Yellowthroat as well as two specialty birds for the area—Sedge Wren and White-throated Flycatcher.

If you continue straight on the paved road for 1.4 km, you will come to a small pond, which is visible off to the left. Blue-and-white Swallows and Gray-breasted Martins often swoop over the surface, while White-throated Flycatchers can be seen in the trees around the water. The pond itself often yields interesting waterfowl such as Black-bellied Whistling-Duck, Blue-winged Teal, Northern Shoveler, American Coot, and Common Moorhen.

Species to Expect

Black-bellied Whistling-Duck (5) 8:7
Blue-winged Teal (7) 8:5
Least Grebe (9) 7:4
Snowy Egret (23) 5:10
Green Heron (25) 6:2
Black Vulture (31) 13:4
Turkey Vulture (31) 13:3
Broad-winged Hawk (43) 16:13
Common Moorhen (61) 7:2
American Coot (61) 7:1
Red-billed Pigeon (87) 18:2
Inca Dove (89) 18:11
Ruddy Ground-Dove (89) 18:7
White-tipped Dove (91) 18:14
*Crimson-fronted Parakeet (95) 19:10
Groove-billed Ani (103) 21:9
Vaux's Swift (119) 22:9
Violet Sabrewing (123) 23:9
Steely-vented Hummingbird (127) 24:15
Rufous-tailed Hummingbird (129) 24:10
Blue-crowned Motmot (147) 27:8
***Hoffmann's Woodpecker (157) 28:16**
Golden-olive Woodpecker (159) 28:11
Streak-headed Woodcreeper (173) 29:8
Yellow-bellied Elaenia (193) 37:26
Paltry Tyrannulet (191) 37:10
Common Tody-Flycatcher (197) 37:7
Olive-sided Flycatcher (203) 36:4
White-throated Flycatcher (205) 36:13
Great Kiskadee (211) 35:13
Boat-billed Flycatcher (211) 35:12
Social Flycatcher (211) 35:14
Tropical Kingbird (213) 35:1
Gray-breasted Martin (235) 22:15
Blue-and-white Swallow (233) 22:20
Southern Rough-winged Swallow (235) 22:19
Barn Swallow (237) 22:12
Plain Wren (241) 38:17
House Wren (243) 38:18
Sedge Wren (243) 38:1
White-breasted Wood-Wren (245) 38:15
Orange-billed Nightingale-Thrush (247) 38:26
Swainson's Thrush (249) 39:2
Clay-colored Thrush (251) 39:8
Tennessee Warbler (255) 40:22
Chestnut-sided Warbler (259) 43:4
Black-and-white Warbler (267) 41:13
Gray-crowned Yellowthroat (273) 42:13
Rufous-capped Warbler (275) 40:16
Summer Tanager (285) 47:5
Blue-gray Tanager (291) 46:15
Palm Tanager (291) 45:19
Blue-black Grassquit (297) 49:7
White-collared Seedeater (295) 49:2
Yellow-faced Grassquit (297) 49:6
White-naped Brush-Finch (301) 48:15
White-eared Ground-Sparrow (303) 48:19
Rufous-collared Sparrow (307) 50:13
Grayish Saltator (309) 48:3
Buff-throated Saltator (309) 48:2
Black-headed Saltator (309) 48:1
Eastern Meadowlark (315) 50:16
Great-tailed Grackle (317) 44:16
Bronzed Cowbird (317) 44:15
Baltimore Oriole (321) 44:7
Montezuma Oropendola (323) 44:8
Yellow-throated Euphonia (327) 45:5
House Sparrow (307) 50:17

Appendix

The **Pied Puffbird** is an uncommon inhabitant of the Caribbean Lowlands. Selva Bananito Lodge (1A-5) is a good place to look for this species, as it is often seen perched high atop a dead snag or sallying for dragonflies, beetles, and other insects.

Outdated English Names 1989–2007	Current English Names Used in This Guide 2008
Spot-bellied Bobwhite (lumped with)	Crested Bobwhite
Rufous-fronted Wood-Quail	Black-eared Wood-Quail
Olivaceous Cormorant	Neotropic Cormorant
Green-backed Heron	Green Heron
Chestnut-bellied Heron	Agami Heron
American Swallow-tailed Kite	Swallow-tailed Kite
Black-shouldered Kite	White-tailed Kite
Black-chested Hawk	Barred Hawk
Mangrove Black-Hawk (lumped with)	Common Black-Hawk
Bay-winged Hawk	Harris's Hawk
Common Gallinule	Common Moorhen
Lesser Golden Plover	American Golden Plover
Common Snipe	Wilson's Snipe
Gray-fronted Dove	Gray-headed Dove
Common Barn Owl	Barn Owl
Andean Pygmy-Owl	Costa Rican Pygmy-Owl
Least Pygmy-Owl	Central American Pygmy-Owl
Common Potoo (split into)	Common Potoo (in the wet forest) and Northern Potoo (in the dry forest)
Band-rumped Swift	Costa Rican Swift
Long-tailed Hermit	Long-billed Hermit
Little Hermit	Stripe-throated Hermit
Fork-tailed Emerald (split into)	Canivet's Emerald (in the North Pacific) and Garden Emerald (in the South Pacific)
Crowned Woodnymph	Violet-crowned Woodnymph
Beryl-crowned Hummingbird	Charming Hummingbird
Red-footed Plumeleteer	Bronze-tailed Plumeleteer
Gray-tailed Mountain-Gem	White-throated Mountain-gem
Striped Foliage-gleaner	Striped Woodhaunter
Spectacled Foliage-gleaner	Scaly-throated Foliage-gleaner
Barred Woodcreeper	Northern Barred-Woodcreeper
Buff-throated Woodcreeper	Cocoa Woodcreeper
Slaty Antshrike	Western Slaty-Antshrike
Spectacled Antpitta	Streak-chested Antpitta
Fulvous-breasted Antpitta	Thicket Antpitta
Zeledon's Tyrannulet	Rough-legged Tyrannulet
Mistletoe Tyrannulet	Paltry Tyrannulet
Scrub Flycatcher	Northern Scrub-Flycatcher
Thrush-like Manakin	Thrush-like Schiffornis
Gray-headed Manakin	Gray-headed Piprites
Whistling Wren	Scaly-breasted Wren
Sooty Robin	Sooty Thrush

Outdated English Names 1989–2007	Current English Names Used in This Guide 2008
Mountain Robin	Mountain Thrush
Pale-vented Robin	Pale-vented Thrush
Clay-colored Robin	Clay-colored Thrush
White-throated Robin	White-throated Thrush
Mangrove Warbler (lumped with)	Yellow Warbler
Zeledonia	Wrenthrush
Olive Tanager	Carmiol's Tanager
Scarlet-rumped Tanager (split into)	Passerini's Tanager (in the Caribbean) and Cherrie's Tanager (in the Pacific)
Pink-billed Seed-Finch	Nicaraguan Seed-Finch
Yellow-throated Brush-Finch	White-naped Brush-Finch
Black-headed Brush-Finch	Stripe-headed Brush-Finch
Northern Oriole	Baltimore Oriole
Blue-hooded Euphonia	Elegant Euphonia

Recent Invasive Species

The following are new residents that have established themselves in Costa Rica since the printing of Stiles and Skutch's guide (1989).

Clapper Rail: This species has been found inhabiting mangrove swamps around the Gulf of Nicoya. It has been most consistently found along the banks of the Tempisque River near the Amistad Bridge.

Southern Lapwing: Now present throughout the Caribbean and Pacific lowlands, this invasive species from South America is likely to become more numerous in years to come.

Brown-throated Parakeet: This invader from Panama into the South Pacific Lowlands is likely to continue its spread northward.

Veraguan Mango: Because this species looks almost identical to the Green-breasted Mango, its status in Costa Rica remains uncertain. There have been several reports of a mango species in the South Pacific Slope of Costa Rica, which could be either Veraguan Mango spreading north from Panama or Green-breasted Mango spreading south.

Mouse-colored Tyrannulet: This rare and recently discovered species occurs in the Coto Brus Valley of the South Pacific Slope.

Rusty-margined Flycatcher: This species is currently moving into the South Pacific corner of Costa Rica from Panama.

Tropical Mockingbird: Breeding pairs have been observed at a number of locations throughout the country, a clear indication that this bird is now establishing itself in Costa Rica. It is expected to become more widespread in the future.

Melodious Blackbird: Described in the "Hypothetical and Recent Additions" section of Stiles and Skutch's guide, this species has since become common around human settlement throughout the country.

Shiny Cowbird: This is another southern invasive from Panama, and it was recently recorded on the Caribbean coast near Puerto Viejo de Talamanca.

Crested Oropendola: This Panamanian invasive is now found in the very southern portion of the Pacific Slope. It is expected to continue expanding northward.

Tricolored Munia: A single population of munias was discovered near Filadelfia, at the northern part of the Nicoya Peninsula. Since this is originally an Asian species, the population was most likely established from escaped birds.

Where to Find the Endemics and Other Sought-After Species

Sites mentioned in the list below are not necessarily the only locations where a species can be found. Also, many of the following species are quite rare, even at the optimal locations. Asterisks indicate endemic species.

Highland Tinamou: Your best chances of finding this species are at 4A-1 and 4A-2.

Slaty-breasted Tinamou: This tinamou is restricted to the northern half of the Caribbean Slope. It is most reliably found at 1A-4.

Masked Duck: This rare duck could show up in wetlands throughout the country, but the most reliable place to find it is the San Joaquín Marsh (described in 3B-3). Your best chances are during the dry season, December through April.

Plain Chachalaca: This species is most readily found at 2-5, although it also occurs at 2-7 and 2-8.

***Black Guan:** Relatively common throughout the Mountains region, this bird is most easily observed at 4A-3, where large groups are attracted to a feeding station.

Great Curassow: Because of overhunting, this large bird is regularly found only near expansive forest, such as that found at 1A-4, 1B-2, 2-1, 2-3, and 3A-6.

Buffy-crowned Wood-Partridge: This species is confined to a section of mountains in the middle of the country. It is best found at 4B-2 and 4A-5.

Marbled Wood-Quail: This species is found throughout the South Pacific region, and your best chances of finding it are at 3A-5 and 3B-3.

***Black-eared Wood-Quail:** This rare endemic of the Caribbean Slope can occasionally be found at 1A-5, 1B-4, and 1B-5.

***Black-breasted Wood-Quail:** While relatively abundant within its elevational range, this species is best found at 1B-8, 4A-1, 4A-2, and 4A-3.

Spotted Wood-Quail: Confined to high elevations, this species can be found at 4B-3, 4B-4, and 4B-5.

Tawny-faced Quail: This very rare species has been reported from 1A-2.

Pinnated Bittern: This rare bittern is most frequently reported from El Tigre Fields (described in 1A-4), although it could show up at wetland sites such as 1A-1 and 2-3.

Least Bittern: This species is best looked for at 2-3.

Rufescent Tiger-Heron: This difficult-to-find species is confined to the Caribbean Lowlands and is best looked for at 1A-3.

Fasciated Tiger-Heron: Preferring fast-moving, rocky rivers, this species can be found at 1B-2 and 3A-5 among other sites.

Agami Heron: Sites 1A-2 and 1A-3 are the best places to look for this striking heron, although it is also occasionally found at 1A-4.

Boat-billed Heron: This heron can be easily seen roosting during the day at 1A-1, 2-8, 3A-1, and 3A-5.

Green Ibis: This species is a regular at 1A-1 and 1A-3.

Jabiru: This massive bird is best found at 2-3, 2-4, and 1A-1, especially during the dry season, December through April.

Lesser Yellow-headed Vulture: This vulture can reliably be found only at 1A-1.

King Vulture: This impressive bird can show up in lowlands across the country, although it is more common near expansive forests. It is most abundant on the Osa Peninsula, as at site 3A-5.

Gray-headed Kite: This species is uncommon at many locations, though it is a regular at 1A-4.

Hook-billed Kite: This species is often found near bodies of water, as at sites 1A-3, 2-3, and 2-6.

Pearl Kite: This invasive can be found at 3B-1 and at the Coastal Plains (described in 3A-5).

Snail Kite: This raptor is found most reliably at 2-3, but can also be seen at 2-4 and 1A-1.

Plumbeous Kite: Look for this bird at 1A-1 and 5-3, and migrating over 1A-7 in August.

Black-collared Hawk: This hawk is most reliably found at 1A-1.

Tiny Hawk: This raptor could occur at most sites in the Caribbean and South Pacific regions, though it is not regular at any single location.

Bicolored Hawk: This rare bird is most frequently found at middle elevations. Sites 1B-7, 3B-4, 3B-6, and 4A-3 are good places to look.

Crane Hawk: This species is most likely found at 1A-3, 2-3, and 3A-1.

Semiplumbeous Hawk: This species is confined to the Caribbean Lowlands. The best place to check is 1A-4.

Great Black-Hawk: This hawk is best looked for at 4A-3, as well as at 1B-5, 1B-8, and 2-3.

Solitary Eagle: Look for this rare bird at 4A-3. Sites 3B-4 and 3B-6 are also good possibilities.

Crested Eagle: This very rare raptor is confined to expansive forests in the Caribbean and South Pacific slopes. It could occur at 1A-3, 1B-5, and 3A-6.

Harpy Eagle: This extremely rare species in Costa Rica is best looked for at 3A-6.

Black-and-white Hawk-Eagle: Look for this very rare raptor at 1B-1, 1B-4, 3A-6, and 3B-4.

Black Hawk-Eagle: This species is most likely to be found at 1A-7, 3A-3, 3A-7, and 4A-5, but it could occur at many sites in the Caribbean and South Pacific slopes.

Ornate Hawk-Eagle: This large raptor can be found near expansive wet forest in the Caribbean and South Pacific slopes, such as at sites 1B-1, 1B-4, 1B-8, 3B-4, and 3B-6.

Slaty-backed Forest-Falcon: Confined to the Caribbean Lowlands, this falcon is most reliably found at 1A-7.

Red-throated Caracara: This caracara has become extremely rare in recent years. It is possible to spot at 3A-6 and 1B-5.

Rufous-necked Wood-Rail: Confined to mangroves, this rare bird is best looked for at 5-3.

Uniform Crake: Site 3A-5 is perhaps the most reliable location for viewing this bird. This crake can also be found at 1A-4, 1A-7, and 3A-7. You are most likely to find it during the dry season, December through April.

Sungrebe: Site 1A-1 is the best place to find this unique species, although 1A-2, 1A-3, and 1A-4 are also good places to look.

Sunbittern: Since this bird prefers rocky rivers, 1B-7 is a good place to find it, although it occurs elsewhere.

Limpkin: This bird is locally common at 1A-1 and 2-3.

Double-striped Thick-knee: This strange-looking bird is reliably found at 2-3 and 5-3.

Southern Lapwing: Sites 5-3 and 3B-2 and the San Isidro Sewage Ponds (described in 3B-1) are good places to look for this invasive species.

Wattled Jacana: This Panamanian invasive sometimes shows up at 3A-7 and the San Joaquín Marsh (described in 3B-3).

Maroon-chested Ground-Dove: This extremely rare species could show up at 4A-3, 4B-2, and 4B-3.

Gray-headed Dove: This species is most likely found at 1A-1 and 2-5.

***Chiriqui Quail-Dove:** This endemic can be found at 4A-1, 4A-3, and many other lower-mountain and middle-elevation sites.

Purplish-backed Quail-Dove: Sites 1B-4, 1B-5, and 1B-7 are good places to find this species.

***Buff-fronted Quail-Dove:** This dove is most easily observed at 4A-3.

Violaceous Quail-Dove: This is probably the rarest quail-dove in Costa Rica. Look for this bird at 1B-2 and 2-2.

***Sulphur-winged Parakeet:** Sites 3B-4, 4B-3, and 4B-4 are especially good locations to find this endemic.

***Crimson-fronted Parakeet:** This parakeet is common throughout the Caribbean and South Pacific slopes and is abundant in the Central Valley, as at site 6-2.

Great Green Macaw: This bird can be encountered at 1A-2, 1A-3, and 1A-4.

Scarlet Macaw: This macaw is most abundant on the Osa Peninsula, as at sites 3A-4, 3A-5, and 3A-6. Isolated populations also occur at 1A-2, 3A-1, and 2-3.

Barred Parakeet: This tiny parakeet is most likely found at 4B-3, 4B-4, and 4B-5.

***Red-fronted Parrotlet:** This rare endemic could show up at many Caribbean Middle-Elevation sites, though Silent Mountain (described in 1B-7) is the most reliable location.

Yellow-naped Parrot: This parrot is reliably found at 2-1 and 2-3.

Mangrove Cuckoo: Look for this cuckoo at 1A-1 and 2-3. It is most often encountered during the dry season, December through April.

Pheasant Cuckoo: This very rare species is regularly reported from 3B-4.

Lesser Ground Cuckoo: Sites 2-2 and 2-5 are good places to see this skulking bird.

Rufous-vented Ground-Cuckoo: This very rare species is occasionally found at the Arial Tram (described in site 1B-4). There are also reports of this bird at 1B-1 and 2-2.

Tropical Screech-Owl: Look for this species at 3B-1, 3B-3, 6-1, and 6-2.

***Costa Rican Pygmy-Owl:** This high-elevation endemic is reliably found at 4B-3, 4B-4, and 4B-5.

Central American Pygmy-Owl: Look for this tiny owl at 1A-2, 1A-5, and 1A-7.

Striped Owl: This species can be found at 1A-1, 1A-7, 3A-1, and 3A-2.

Unspotted Saw-whet Owl: This extremely rare owl could be encountered at 4B-3 and 4B-5.

Short-tailed Nighthawk: This nighthawk's distinctive silhouette is often seen flying over 1A-4 at dusk and dawn.

***Dusky Nightjar:** This high-elevation endemic is found regularly at all Upper Mountain sites.

Great Potoo: Look for this species at 1A-1, 1A-4, and 1A-7.

Common Potoo: This Potoo is most likely encountered in the South Pacific region, as at sites 3A-2, 3B-1, and 3B-3.

Northern Potoo: Look for this bird at 2-5, 2-6, and 2-7.

***Costa Rican Swift:** This species is commonly encountered throughout the South Pacific Slope.

White-tipped Sicklebill: This striking hummingbird is most reliably found at 3A-5.

Green-fronted Lancebill: This bird is found at many Lower Mountain and Caribbean Middle-Elevation sites, although 4A-5 is the most reliable.

Black-crested Coquette: This impressive little hummingbird is most easily found at 1B-2.

***White-crested Coquette:** This endemic could show up at most South Pacific sites, although it is most consistent at 3A-3 and 3A-5. December and January are good months to look for this species.

Green Thorntail: The Cinchona hummingbird feeders (described in 1B-3) are a great place to find this bird. It can also be found at 1B-7, 1B-8, 4A-4, and 4A-5.

***Garden Emerald:** This bird can be looked for at 3B-1, 3B-3, and 3B-4.

***Fiery-throated Hummingbird:** This endemic is common at 4B-1, 4B-2, 4B-3, and 4B-5.

***Charming Hummingbird:** This charming bird is found at all South Pacific Lowland sites, although 3A-5 is the most reliable.

***Mangrove Hummingbird:** Confined to mangroves, this hummingbird is best found at Rincón (described in 3A-5), 2-6, and 5-3.

***Snowy-bellied Hummingbird:** This bird is a regular at all sites in the South Pacific Middle-Elevations.

Stripe-tailed Hummingbird: This hummingbird is easily found at 4A-1.

***Black-bellied Hummingbird:** This endemic is abundant at 4A-4, but 4A-3 and 4A-5 are also good places to look.

***White-tailed Emerald:** Look for this endemic at 3B-3, 3B-4, 3B-5, and 3B-6.

***Coppery-headed Emerald:** This hummingbird is easily found at 4A-1 and 4A-4, as well as the Cinchona hummingbird feeders (described in 1B-3).

***Snowcap:** This striking bird is most easily observed at 1B-7 and 1B-8.

***White-bellied Mountain-gem:** Look for this species at the Cinchona hummingbird feeders (described in 1B-3) as well as at 4A-5.

***Purple-throated Mountain-gem:** A common species at all Lower Mountain sites.

***White-throated Mountain-gem:** This species is most easily seen at 4B-3 and 4B-4.

***Magenta-throated Woodstar:** Look for this impressive endemic at 4A-1 and 4A-3.

***Volcano Hummingbird:** This hummingbird is found regularly at all Upper Mountain sites.

***Scintillant Hummingbird:** This tiny bird can be seen at 4A-3 and 4B-4.

***Baird's Trogon:** This South Pacific endemic is regularly found at 3A-1, 3A-3, 3A-4, 3A-5, 3A-6, and 3A-7.

***Orange-bellied Trogon:** This species is most easily found at 4A-1 and 4A-3.

***Lattice-tailed Trogon:** Sites 1B-4 and 1B-5 are good locations to search for this bird.

Resplendent Quetzal: Site 4B-3 is the most reliable location in the country to find this sought-after species, although it can be difficult to find in October. Another very reliable site is 4B-4, and during the breeding season, March to June, 4A-1 and 4A-2 are also excellent locations.

Tody Motmot: This tiny motmot is best found at 1B-1, although it also occurs at 2-2.

Keel-billed Motmot: This difficult species is occasionally seen at 1B-2 as well as at the nearby Coter Lake Eco-Lodge.

Green-and-rufous Kingfisher: Site 1A-3 is the only location in this guide where this species is regularly found.

Pied Puffbird: Site 1A-5 is the best place to look for this species, although it can also be found at 1A-2, 1A-4, and 1A-7.

Lanceolated Monklet: This rare species is usually found near streams at middle-elevations of the Caribbean Slope. It has been reported from 1B-3, 1B-4, 1B-7, and 1B-8.

White-fronted Nunbird: Sites 1A-3 and 1B-1 are the best areas to search for this bird.

Great Jacamar: This very rare species is occasionally reported from 1A-5.

***Prong-billed Barbet:** This bird is regular at all Lower Mountain sites.

Emerald Toucanet: This species occurs throughout the Mountains region, although 4A-1, 4A-2, and 4B-4 are perhaps the most reliable locations to find it.

***Fiery-billed Aracari:** This endemic toucan can be found at most sites in the South Pacific Slope, but it is especially common at 3B-3.

***Yellow-eared Toucanet:** This beautiful toucan can be found at 1B-1, 1B-4, 1B-5, 1B-8, and 2-2.

Olivaceous Piculet: This species is uncommon at most sites in the South Pacific Slope. However, 3A-3 and 3B-1 are two of the most likely locations.

***Golden-naped Woodpecker:** Check for this uncommon bird at all South Pacific Lowland sites.

***Hoffmann's Woodpecker:** This woodpecker is abundant in the North Pacific and Central Valley regions.

Red-rumped Woodpecker: This woodpecker is most easily seen at 3A-5, but it could also occur at 3A-6 and 3A-7.

***Rufous-winged Woodpecker:** This bird occurs throughout the Caribbean and South Pacific slopes, although it is most likely found at 1A-4, 1B-4, 1B-7, and 3A-7.

Chestnut-colored Woodpecker: The best place to look for this striking woodpecker is at 1A-4.

***Ruddy Treerunner:** This endemic is easily found at all Upper Mountain sites as well as at 4A-1 and 4A-2.

Striped Woodhaunter: Site 1B-4 is a reliable location to find this species.

Scaly-throated Foliage-gleaner: This rare ovenbird could be encountered at 1B-3, 3B-4, and 3B-6.

Buff-fronted Foliage-gleaner: Sites 1B-8, 3B-4, and 4A-5 are the best locations to search for this rare bird.

Ruddy Foliage-gleaner: Confined to a tiny portion of the country, this species is most easily found at 3B-3.

***Streak-breasted Treehunter:** This endemic is best found at 4A-2 and 4A-3.

Streaked Xenops: This bird is perhaps best looked for at 4A-5.

Gray-throated Leaftosser: This species is most easily found at 4A-1 and 4A-2.

Scaly-throated Leaftosser: Rare throughout most of its range, this bird is reliably found at 3A-5.

Long-tailed Woodcreeper: This woodcreeper is found at 1A-2, 3A-1, 3A-5, and 3A-7.

Strong-billed Woodcreeper: This rare bird is occasionally reported from site 4A-3 and Silent Mountain (described in 1B-7). Refugio Eladios (described in 4A-1) is another good place to search.

Black-banded Woodcreeper: This rare woodcreeper can be found at 4A-3 and should also be looked for at 1B-8 and 4A-5.

Fasciated Antshrike: This bird should be looked for at 1A-4 and 1A-5.

Great Antshrike: This antshrike is most easily found at 1A-4 and 3A-5.

***Black-hooded Antshrike:** This endemic is abundant in the South Pacific Lowlands.

***Streak-crowned Antvireo:** Look for this bird at 1A-2, 1B-1, 1B-2, and 1B-4.

Spot-crowned Antvireo: This species has an extremely limited range in Costa Rica, and 1A-7 is the only location described in this guide where you have a chance of finding it.

Rufous-rumped Antwren: This is a very rare species in Costa Rica that has been reported from 1B-3 and 4A-5. However, it is consistently found at Silent Mountain (described in 1B-7).

Bare-crowned Antbird: This bird is most easily found at 1B-2, but it also occurs at 1A-1 and 1A-4.

Dull-mantled Antbird: This species of the Caribbean Middle-Elevations is most easily observed at 1B-4, 1B-7, and 1B-8.

Ocellated Antbird: Normally accompanying army ants, this species is best found at 1A-4.

Black-headed Antthrush: This skulking bird is most easily found at 1B-8.

Rufous-breasted Antthrush: San Gerardo Biological Station (described in 4A-2) is the easiest place to find this antthrush. Also look for it at 4A-5.

***Black-crowned Antpitta:** This rare endemic is best searched for at 1B-4 and 1B-5.

Scaled Antpitta: Look for this species at 1B-8, 4A-3, and 4A-5.

Streak-chested Antpitta: This bird is most likely encountered at 3A-1, 3A-4, and 3A-6.

***Thicket Antpitta:** While this antpitta can be found at many Caribbean Slope sites, it is most abundant at 1B-2.

Ochre-breasted Antpitta: This rare species of middle elevations should be searched for at 1B-8, 3B-4, 3B-6, 4A-4, and 4A-5.

***Silvery-fronted Tapaculo:** This often heard, though seldom seen, endemic is common at all sites in the Mountains region, being especially abundant at 4A-1 and 4A-2.

Yellow-crowned Tyrannulet: This uncommon flycatcher occurs at many South Pacific Slope sites, though the most reliable are 3A-7, 3B-1, and 3B-3.

Greenish Elaenia: Look for this bird at 3A-1 and 3B-3.

Lesser Elaenia: This species can be found at 3A-7, 3B-1, and 6-3.

Torrent Tyrannulet: This cute flycatcher is almost always found near rushing rivers. Look for it at 1B-3, 3B-4, 4A-3, 4A-4, 4A-5, and 4B-4.

Sepia-capped Flycatcher: This rare flycatcher can be found at 3B-5.

Rufous-browed Tyrannulet: This species is most likely found at Silent Mountain (described in 1B-7), San Gerardo Biological Station (described in 4A-2), and 1B-8.

Rough-legged Tyrannulet: This rare flycatcher could turn up at a variety of locations. Among the more reliable spots are Silent Mountain (described in site 1B-7) and at 3B-3 and 4B-3.

Black-capped Pygmy-Tyrant: This flycatcher can be found at 1A-2 and 1A-4.

Black-headed Tody-Flycatcher: This bird is most easily found at 1A-7, 1B-3, and 1B-7.

Golden-crowned Spadebill: Look for this flycatcher at 3A-4, 3A-5, and 3A-6.

Royal Flycatcher: This dramatic species can sometimes be found at 1A-2 and 3A-1.

Black-tailed Flycatcher: This species, which is extremely similar to the Sulphur-rumped Flycatcher, can be found at 3A-5, 3A-6, and 3A-7.

***Tawny-chested Flycatcher:** This flycatcher is a reliable find at 1B-7, although it also occurs at 1A-4 and 1B-8.

***Dark Pewee:** This endemic flycatcher can be easily found at 4A-3, 4A-5, and 4B-4.

***Ochraceous Pewee:** This species is most reliably found at 4B-3, but it also occurs at 4B-4.

White-throated Flycatcher: Look for this bird at 6-3.

***Black-capped Flycatcher:** This flycatcher is easily found at all Upper Mountain sites.

Nutting's Flycatcher: The best chance you will have finding this bird is at 2-5.

Rusty-margined Flycatcher: This Panamanian invasive is currently best found at 3A-7.

***Golden-bellied Flycatcher:** This flycatcher can be seen at all Lower Mountain sites.

Fork-tailed Flycatcher: Look for this distinctive species at 3A-7 and 3B-1.

Thrush-like Schiffornis: A regular at 1B-8, this bird can also be found at 1B-1, 3A-1, and 3A-6 among other sites.

***Gray-headed Piprites:** This rare endemic should be looked for at 1A-2, 1B-4, and 1B-5.

Speckled Mourner: Sites 1A-4 and 1A-2 offer your best chances of finding this rare species.

Black-and-white Becard: This rare becard could be found at 1B-3, 1B-8, and 4A-5.

Lovely Cotinga: Site 1B-8 is a good place to search for this hard-to-find species. Especially during the month of March, Silent Mountain (described in 1B-7) is another excellent place to look. Sites 1B-1 and 1B-2 also hold potential.

***Turquoise Cotinga:** Sites 3A-5, 3B-1, and 3B-4 are especially good places to look for this colorful endemic.

***Yellow-billed C nga:** Between December and April, Rincón (described in 3A-5) is a reliable location to find this rare endemic.

***Snowy Cotinga:** Site 1A-5 is a very good spot to find this cotinga, although 1A-2, 1A-4, and 1A-7 are also good possibilities.

***Bare-necked Umbrellabird:** This strange bird is most easily found at 1A-4 and at the San Gerardo Biological Station (described in 4A-2), where males display on leks from March to June.

***Three-wattled Bellbird:** While it could show up at many sites, this striking bird is most easily found at 2-6 in January and at 4A-1, 4A-2, 3B-4, and 3B-6 during the breeding season (March to June).

***Orange-collared Manakin:** This manakin is easily found at 3A-1, 3A-3, 3A-5, and 3B-1. The males display on leks most actively from February through June.

Lance-tailed Manakin: With a limited range in Costa Rica, this manakin can be seen only at 3B-5 and 3B-6. The males display on leks most actively from July through September.

White-crowned Manakin: This bird can be found on leks at 1B-5, 1B-7, and 1B-8.

Sharpbill: Silent Mountain (described in 1B-7) is, perhaps, the most reliable location for finding this rare species, although it also occurs at 1B-1, 1B-4, 1B-8, and 4A-5.

***Yellow-winged Vireo:** This vireo is common at all Upper Mountain sites.

Scrub Greenlet: This bird is reliably found at 3A-5 and 3A-7.

Black-chested Jay: Site 1A-7 had always been the only reliable location to find this species. Recently it was reported from 3B-5 and could expand its range into the South Pacific region in the future.

***Silvery-throated Jay:** Look for this endemic jay at 4B-3, 4B-4, and 4B-5.

Rock Wren: This species can be looked for at 2-2.

***Black-throated Wren:** This wren can be found at many Caribbean Slope sites, but it is most abundant at 1A-4, 1B-2, and 1B-7.

***Black-bellied Wren:** Look for this endemic wren at 3A-3, 3A-5, and 3A-7.

***Riverside Wren:** This endemic is easily found throughout most of the South Pacific Slope.

***Stripe-breasted Wren:** This bird is quite common on the Caribbean Slope.

***Ochraceous Wren:** Look for this endemic wren at all Lower Mountain sites as well as at 4B-3 and 4B-4.

Sedge Wren: This Central Valley specialty can be found only at 6-3.

***Timberline Wren:** Most reliably found at 4B-2, this wren is also a regular at 4B-3 and 4B-5.

Song Wren: This odd-looking wren is best found at 1B-1, 1B-2, and 1B-8.

American Dipper: This bird, which inhabits rushing mountain rivers, is often found at 1B-3, 4A-3, 4A-4, 4A-5, and 4B-4.

***Black-faced Solitaire:** This species is common at all Lower Mountain sites as well as at 4B-4.

***Black-billed Nightingale-Thrush:** This thrush is abundant at 4B-1, 4B-2, 4B-3, and 4B-5.

***Sooty Thrush:** This endemic is easily found at 4B-2, 4B-3, and 4B-5.

Tropical Mockingbird: Look for this invasive at sites 1B-2 and 6-3.

***Black-and-yellow Silky-Flycatcher:** This species is locally common at 4B-1, 4B-2, 4B-3, and 4B-5.

***Long-tailed Silky-Flycatcher:** This bird is found most regularly at 4B-3 and 4B-4.

***Flame-throated Warbler:** This pretty endemic warbler can be found at all Upper Mountain sites as well as at 4A-3.

Masked Yellowthroat: This species has a limited range in Costa Rica, but it can be found at 3B-3, 3B-4, and 3B-6.

***Collared Redstart:** This common endemic is most abundant at 4A-1, 4A-2, and all Upper Mountain sites.

***Black-cheeked Warbler:** This warbler is found at all Upper Mountain sites.

***Wrenthrush:** This skulking warbler is easily heard at 4B-3 and 4B-5.

***Sooty-capped Bush-Tanager:** This species is abundant at all Upper Mountain sites.

Ashy-throated Bush-Tanager: This uncommon tanager can be found in the middle-elevations of the Caribbean Slope at 1B-4, 1B-5, 1B-7, and 1B-8.

***Black-and-yellow Tanager:** This bird can most likely be found at 1B-3, 1B-4, 1B-5, and 1B-8.

Rosy Thrush-Tanager: This tanager can be found at 3B-2.

***White-throated Shrike-Tanager:** Look for this tanager at 1B-4, 3A-3, 3A-4, 3A-6, and 3A-7.

***Sulphur-rumped Tanager:** This rare endemic is possible to find at 1A-7.

***Black-cheeked Ant-Tanager:** This endemic can be found only at sites on or near the Osa Peninsula. Of these, 3A-5 is the most reliable, where the tanager comes to a feeding station.

***Cherrie's Tanager:** This common species occurs throughout the South Pacific Slope.

***Blue-and-gold Tanager:** This uncommon bird is most easily found at 1B-5, but it should also be looked for at 1B-3, 1B-4, and 1B-8.

***Plain-colored Tanager:** This bird is best found at 1A-5, although it is also a likely occurrence at 1A-4 and 1A-7.

Rufous-winged Tanager: The most likely area for this uncommon species seems to be Coter Lake Eco-Lodge (described in 1B-2), but it could occur at 1B-1, 1B-2, and 1A-7.

***Spangle-cheeked Tanager:** This tanager is regularly encountered at all Lower Mountain sites.

Yellow-bellied Seedeater: Look for this nomadic bird at 3A-7 and at the Coastal Plains (described in 3A-5).

Ruddy-breasted Seedeater: This species is often found at 3A-7, 1A-1, and the Coastal Plains (described in 3A-5).

***Nicaraguan Seed-Finch:** This bird is often found in El Tigre Fields (described in 1A-4) and at sites 1A-1 and 1A-2.

Blue Seedeater: This rare species could show up at 3B-4, 3B-6, and 4A-3.

Slaty Finch: This rare species has occurred at 4B-3 and 4B-5.

***Peg-billed Finch:** This uncommon endemic is most likely seen at 4B-5, although 4B-3 and 4B-4 also hold potential.

***Slaty Flowerpiercer:** This bird is common at all Upper Mountain sites as well as 4A-3 and 4A-4.

Wedge-tailed Grass-Finch: This bird can be looked for at 3B-2.

***Sooty-faced Finch:** This species is most easily found at 1B-3 and 4A-4.

***Yellow-thighed Finch:** This finch is common throughout the Mountains region, with the exception of 4A-5.

***Large-footed Finch:** This species is easily encountered at all Upper Mountain sites.

Stripe-headed Brush-Finch: This uncommon species occurs throughout the South Pacific Middle-Elevations.

Prevost's Ground-Sparrow: Look for this Central Valley specialty at 6-2.

White-eared Ground-Sparrow: This sparrow is most easily found at 6-2 and 6-3, but it also occurs at 4A-1.

***Volcano Junco:** This high-elevation specialist is a reliable find at 4B-2, although 4B-5 is another good place to look.

***Black-thighed Grosbeak:** This bird is found throughout the Mountains region and is especially likely at 4B-4.

Red-breasted Blackbird: Look for this blackbird at 1A-5, 3A-7, the Coastal Plains (described in 3A-5), and El Tigre Fields (described in 1A-4).

***Nicaraguan Grackle:** Look for this endemic grackle at 1A-1.

Yellow-tailed Oriole: This oriole is found most regularly at 1A-4.

Spot-breasted Oriole: Site 2-6 is the best place to look.

Crested Oropendola: Look for this Panamanian invasive at 3B-3.

***Yellow-crowned Euphonia:** This species is common in the Caribbean and South Pacific lowlands.

***Spot-crowned Euphonia:** Found throughout the South Pacific Slope, this endemic is especially abundant at 3A-5 and 3A-7.

***Tawny-capped Euphonia:** This euphonia is common at most middle-elevation sites in the Caribbean Slope.

***Golden-browed Chlorophonia:** This beautiful little bird is found throughout the Mountains region.

Sites
1A-4 La Selva Biological Station
1B-7 Rancho Naturalista
2-3 Palo Verde National Park
3A-5 Bosque del Río Tigre
3B-3 Las Cruces Biological Station
4A-3 Bosque de Paz
4B-4 San Gerardo de Dota
5-2 Chomes

Regions
Caribbean Lowlands
Caribbean Middle-Elevations
North Pacific Slope
South Pacific Lowlands
South Pacific Middle-Elevations
Lower Mountains
Upper Mountains
Coastline

Abundance Codes
C— Common: Generally recorded at least once every 2–3 days
U— Uncommon: Generally recorded at least once every 2–3 weeks
R— Rare: Generally recorded at least once a year

Note: Seasonal residents have been rated according to their abundance while present.

	Tinamous—TINAMIDAE	La Selva	Rancho Naturalista	Palo Verde	Bosque DRT	Las Cruces	Bosque DP	San Gerardo	Chomes
	Highland Tinamou *Nothocercus bonapartei*						U	R	
	Great Tinamou *Tinamus major*	C	U		R	C	R		
	Little Tinamou *Crypturellus soui*	C	U		C	C	R		
	Thicket Tinamou *Crypturellus cinnamomeus*			C					C
	Slaty-breasted Tinamou *Crypturellus boucardi*	U							
	Ducks—ANATIDAE								
	Black-bellied Whistling-Duck *Dendrocygna autumnalis*	R		C					C
	Fulvous Whistling-Duck *Dendrocygna bicolor*			U					
	Muscovy Duck *Cairina moschata*	R		U					R
	American Wigeon *Anas americana*			R					
	Blue-winged Teal *Anas discors*	R		C					C
	Cinnamon Teal *Anas cyanoptera*			R					

Species	La Selva	Rancho Naturalista	Palo Verde	Bosque DRT	Las Cruces	Bosque DP	San Gerardo	Chomes
Northern Shoveler *Anas clypeata*			R					R
Northern Pintail *Anas acuta*			R					
Green-winged Teal *Anas crecca*								
Ring-necked Duck *Aythya collaris*			R					
Lesser Scaup *Aythya affinis*			R					
Masked Duck *Nomonyx dominicus*			R					
Guans, Chachalacas, and Curassow—CRACIDAE								
Plain Chachalaca *Ortalis vetula*			R					
Gray-headed Chachalaca *Ortalis cinereiceps*	U	C		C	U	R		
Crested Guan *Penelope purpurascens*	C	R	R	R	C	R		
Black Guan *Chamaepetes unicolor*						C	C	
Great Curassow *Crax rubra*	R		U	R	R	R		
Quails—ODONTOPHORIDAE								
Buffy-crowned Wood-Partridge *Dendrortyx leucophrys*						R	R	
Crested Bobwhite *Colinus cristatus*			U					C
Marbled Wood-Quail *Odontophorus gujanensis*				C	C			
Black-eared Wood-Quail *Odontophorus melanotis*								
Black-breasted Wood-Quail *Odontophorus leucolaemus*						C		
Spotted Wood-Quail *Odontophorus guttatus*					R		U	
Tawny-faced Quail *Rhynchortyx cinctus*								
Grebes—PODICIPEDIDAE								
Least Grebe *Tachybaptus dominicus*			U	C				
Pied-billed Grebe *Podilymbus podiceps*								
Shearwaters—PROCELLARIIDAE								
Pink-footed Shearwater *Puffinus creatopus*								

	Species	La Selva	Rancho Naturalista	Palo Verde	Bosque DRT	Las Cruces	Bosque DP	San Gerardo	Chomes
	Wedge-tailed Shearwater *Puffinus pacificus*								
	Audubon's Shearwater *Puffinus lherminieri*								
	Storm-Petrels—HYDROBATIDAE								
	Wilson's Storm-Petrel *Oceanites oceanicus*								
	Leach's Storm-Petrel *Oceanodroma leucorhoa*								
	Band-rumped Storm-Petrel *Oceanodroma castro*								
	Wedge-rumped Storm-Petrel *Oceanodroma tethys*								
	Black Storm-Petrel *Oceanodroma melania*								
	Markham's Storm-Petrel *Oceanodroma markhami*								
	Least Storm-Petrel *Oceanodroma microsoma*								
	Tropicbird—PHAETHONTIDAE								
	Red-billed Tropicbird *Phaethon aethereus*								
	Boobies—SULIDAE								
	Blue-footed Booby *Sula nebouxii*								
	Brown Booby *Sula leucogaster*								
	Pelican—PELECANIDAE								
	Brown Pelican *Pelecanus occidentalis*			R					C
	Cormorant—PHALACROCORACIDAE								
	Neotropic Cormorant *Phalacrocorax brasilianus*	C		C	C				C
	Anhinga—ANHINGIDAE								
	Anhinga *Anhinga anhinga*	C		C	U				U
	Frigatebird—FREGATIDAE								
	Magnificent Frigatebird *Fregata magnificens*			R	U				C
	Herons, Egrets, and Bitterns—ARDEIDAE								
	Pinnated Bittern *Botaurus pinnatus*			R					
	Least Bittern *Ixobrychus exilis*			R					
	Rufescent Tiger-Heron *Tigrisoma lineatum*								

	Species	La Selva	Rancho Naturalista	Palo Verde	Bosque DRT	Las Cruces	Bosque DP	San Gerardo	Chomes
☐	Fasciated Tiger-Heron *Tigrisoma fasciatum*	R	R		U				
☐	Bare-throated Tiger-Heron *Tigrisoma mexicanum*	R		C	R				U
☐	Great Blue Heron *Ardea herodias*	U		C	R		R		C
☐	Great Egret *Ardea alba*	R	R	C	U				C
☐	Snowy Egret *Egretta thula*	R	U	C	C				C
☐	Little Blue Heron *Egretta caerulea*	U	R	C	C				C
☐	Tricolored Heron *Egretta tricolor*	R		C	R				C
☐	Reddish Egret *Egretta rufescens*								R
☐	Cattle Egret *Bubulcus ibis*	C	C	C	C	U	C		C
☐	Green Heron *Butorides virescens*	U		C	C				C
☐	Agami Heron *Agamia agami*	R							
☐	Black-crowned Night-Heron *Nycticorax nycticorax*	R		C					U
☐	Yellow-crowned Night-Heron *Nyctanassa violacea*	R		C	U				C
☐	Boat-billed Heron *Cochlearius cochlearius*	R		C	C	C			
	Ibis and Spoonbill—THRESKIORNITHIDAE								
☐	White Ibis *Eudocimus albus*			C	C				C
☐	Glossy Ibis *Plegadis falcinellus*			U					
☐	Green Ibis *Mesembrinibis cayennensis*	U							
☐	Roseate Spoonbill *Platalea ajaja*	R		C	R				U
	Storks—CICONIIDAE								
☐	Jabiru *Jabiru mycteria*			U					R
☐	Wood Stork *Mycteria americana*	R		C	R				C
	New World Vultures—CATHARTIDAE								
☐	Black Vulture *Coragyps atratus*	C	C	C	C	C	C	C	C
☐	Turkey Vulture *Cathartes aura*	C	C	C	C	C	C	C	C

	Species	La Selva	Rancho Naturalista	Palo Verde	Bosque DRT	Las Cruces	Bosque DP	San Gerardo	Chomes
	Lesser Yellow-headed Vulture *Cathartes burrovianus*			R					
	King Vulture *Sarcoramphus papa*	U	R	R	C	U	R		R

Hawks, Eagles, and Kites—ACCIPITRIDAE

	Species	La Selva	Rancho Naturalista	Palo Verde	Bosque DRT	Las Cruces	Bosque DP	San Gerardo	Chomes
	Osprey *Pandion haliaetus*	U	R	C	R	R		C	U
	Gray-headed Kite *Leptodon cayanensis*	U	U	U	U	R			
	Hook-billed Kite *Chondrohierax uncinatus*	U	R	U	R				U
	Swallow-tailed Kite *Elanoides forficatus*	U	C		C	U	R	C	
	Pearl Kite *Gampsonyx swainsonii*				R				
	White-tailed Kite *Elanus leucurus*	R		R		U			
	Snail Kite *Rostrhamus sociabilis*			C					
	Double-toothed Kite *Harpagus bidentatus*	C	R	U	C	U			
	Mississippi Kite *Ictinia mississippiensis*	R							
	Plumbeous Kite *Ictinia plumbea*	R							U
	Black-collared Hawk *Busarellus nigricollis*								
	Northern Harrier *Circus cyaneus*			R					R
	Tiny Hawk *Accipiter superciliosus*	R	R		R				
	Sharp-shinned Hawk *Accipiter striatus*		R			R	R	R	
	Cooper's Hawk *Accipiter cooperii*		R			R	R	R	
	Bicolored Hawk *Accipiter bicolor*	R	R		R	R	R		
	Crane Hawk *Geranospiza caerulescens*	R		R	R				R
	Barred Hawk *Leucopternis princeps*	R	U		R	R	U		
	Semiplumbeous Hawk *Leucopternis semiplumbeus*	U							
	White Hawk *Leucopternis albicollis*	R	R		C	R			

		La Selva	Rancho Naturalista	Palo Verde	Bosque DRT	Las Cruces	Bosque DP	San Gerardo	Chomes
☐	Common Black-Hawk *Buteogallus anthracinus*	R		U	C				C
☐	Great Black-Hawk *Buteogallus urubitinga*	R		R	R	R	U		
☐	Harris's Hawk *Harpyhaliaetus solitarius*			U					U
☐	Solitary Eagle *Harpyhaliaetus solitarius*						U		
☐	Roadside Hawk *Buteo magnirostris*	R	U	C	C	C			C
☐	Broad-winged Hawk *Buteo platypterus*	C	C	U	C	U	C	U	U
☐	Gray Hawk *Buteo nitidus*	R		U	R				C
☐	Short-tailed Hawk *Buteo brachyurus*	R	C	U	C	C			R
☐	Swainson's Hawk *Buteo swainsoni*	U	R	R		R		R	R
☐	White-tailed Hawk *Buteo albicaudatus*			R					R
☐	Zone-tailed Hawk *Buteo albonotatus*	R	R	U	U	U			U
☐	Red-tailed Hawk *Buteo jamaicensis*		R			R	U	C	
☐	Crested Eagle *Morphnus guianensis*								
☐	Harpy Eagle *Harpia harpyja*								
☐	Black-and-white Hawk-Eagle *Spizaetus melanoleucus*	R							
☐	Black Hawk-Eagle *Spizaetus tyrannus*	R	R		U	U			
☐	Ornate Hawk-Eagle *Spizaetus ornatus*	U					U	R	
	Falcons and Caracaras—FALCONIDAE								
☐	Barred Forest-Falcon *Micrastur ruficollis*	R	U		U	C	R	R	
☐	Slaty-backed Forest-Falcon *Micrastur mirandollei*	R							
☐	Collared Forest-Falcon *Micrastur semitorquatus*	U		U	U	U	U		
☐	Red-throated Caracara *Ibycter americanus*								
☐	Crested Caracara *Caracara cheriway*			C	R				C
☐	Yellow-headed Caracara *Milvago chimachima*			R	C	C			R

		La Selva	Rancho Naturalista	Palo Verde	Bosque DRT	Las Cruces	Bosque DP	San Gerardo	Chomes
	Laughing Falcon *Herpetotheres cachinnans*	U	U	C	C	C	R		C
	American Kestrel Falco sparverius			U		R			U
	Merlin *Falco columbarius*			R					
	Bat Falcon *Falco rufigularis*	U			C	C	R	R	
	Peregrine Falcon *Falco peregrinus*	R		U					U
	Rails, Gallinule, and Coot—RALLIDAE								
	Ocellated Crake *Micropygia schomburgkii*								
	White-throated Crake *Laterallus albigularis*	C	C	U	C	U			
	Gray-breasted Crake *Laterallus exilis*	R							
	Clapper Rail *Rallus longirostris*								
	Rufous-necked Wood-Rail *Aramides axillaris*								R
	Gray-necked Wood-Rail *Aramides cajanea*	U		R	C	C			
	Uniform Crake *Amaurolimnas concolor*	R			U				
	Sora *Porzana carolina*			R					
	Yellow-breasted Crake *Porzana flaviventer*			R					
	Paint-billed Crake *Neocrex erythrops*								
	Spotted Rail *Pardirallus maculatus*			R					
	Purple Gallinule *Porphyrio martinica*	R		C	C				U
	Common Moorhen *Gallinula chloropus*			C					
	American Coot *Fulica americana*			C					U
	Sungrebe—HELIORNITHIDAE								
	Sungrebe *Heliornis fulica*	R							
	Sunbittern—EURYPYGIDAE								
	Sunbittern *Eurypyga helias*	U	U			U	R		

Species	La Selva	Rancho Naturalista	Palo Verde	Bosque DRT	Las Cruces	Bosque DP	San Gerardo	Chomes
Limpkin—ARAMIDAE								
Limpkin *Aramus guarauna*			C					R
Thick-knee—BURHINIDAE								
Double-striped Thick-knee *Burhinus bistriatus*			C					U
Plovers and Lapwing—CHARADRIIDAE								
Southern Lapwing *Vanellus chilensis*								
Black-bellied Plover *Pluvialis squatarola*			U					C
American Golden-Plover *Pluvialis dominica*								R
Collared Plover *Charadrius collaris*								C
Snowy Plover *Charadrius alexandrinus*								R
Wilson's Plover *Charadrius wilsonia*			R					C
Semipalmated Plover *Charadrius semipalmatus*								C
Killdeer *Charadrius vociferus*			R					
Oystercatcher—HAEMATOPODIDAE								
American Oystercatcher *Haematopus palliatus*								U
Stilt—RECURVIROSTRIDAE								
Black-necked Stilt *Himantopus mexicanus*			C					C
Jacanas—JACANIDAE								
Northern Jacana *Jacana spinosa*	U		C	C				C
Wattled Jacana *Jacana jacana*								
Sandpipers and Allies—SCOLOPACIDAE								
Spotted Sandpiper *Actitis macularius*	C	C	C	C				C
Solitary Sandpiper *Tringa solitaria*	R		R	R				U
Wandering Tattler *Tringa incana*								U
Greater Yellowlegs *Tringa melanoleuca*			U	C				C
Willet *Tringa semipalmata*			U					C
Lesser Yellowlegs *Tringa flavipes*			U					C

		La Selva	Rancho Naturalista	Palo Verde	Bosque DRT	Las Cruces	Bosque DP	San Gerardo	Chomes
	Upland Sandpiper *Bartramia longicauda*								R
	Whimbrel *Numenius phaeopus*			U					C
	Long-billed Curlew *Numenius americanus*								
	Marbled Godwit *Limosa fedoa*								C
	Ruddy Turnstone *Arenaria interpres*								C
	Surfbird *Aphriza virgata*								R
	Red Knot *Calidris canutus*								R
	Sanderling *Calidris alba*								U
	Semipalmated Sandpiper *Calidris pusilla*			U					C
	Western Sandpiper *Calidris mauri*			U					C
	Least Sandpiper *Calidris minutilla*			U					C
	White-rumped Sandpiper *Calidris fuscicollis*								R
	Baird's Sandpiper *Calidris bairdii*								R
	Pectoral Sandpiper *Calidris melanotos*			R					U
	Dunlin *Calidris alpina*								
	Stilt Sandpiper *Calidris himantopus*			R					C
	Buff-breasted Sandpiper *Tryngites subruficollis*								R
	Short-billed Dowitcher *Limnodromus griseus*			U					C
	Long-billed Dowitcher *Limnodromus scolopaceus*			R					R
	Wilson's Snipe *Gallinago delicata*			R					
	Wilson's Phalarope *Phalaropus tricolor*								U
	Red-necked Phalarope *Phalaropus lobatus*								R
	Red Phalarope *Phalaropus fulicarius*								

	Gulls, Terns, and Skimmer—LARIDAE	La Selva	Rancho Naturalista	Palo Verde	Bosque DRT	Las Cruces	Bosque DP	San Gerardo	Chomes
	Sabine's Gull *Xema sabini*								
	Bonaparte's Gull *Larus philadelphia*								
	Laughing Gull *Larus atricilla*			R					C
	Franklin's Gull *Larus pipixcan*			R					U
	Ring-billed Gull *Larus delawarensis*								
	Herring Gull *Larus argentatus*								
	Brown Noddy *Anous stolidus*								
	Bridled Tern *Onychoprion anaethetus*								
	Least Tern *Sternula antillarum*								U
	Gull-billed Tern *Gelochelidon nilotica*			R					C
	Caspian Tern *Hydroprogne caspia*								U
	Black Tern *Chlidonias niger*								R
	Common Tern *Sterna hirundo*								U
	Forster's Tern *Sterna forsteri*								R
	Royal Tern *Thalasseus maxima*			R					C
	Sandwich Tern *Thalasseus sandvicensis*			R					C
	Elegant Tern *Thalasseus elegans*			R					C
	Black Skimmer *Rynchops niger*								U

Jaegers—STERCORARIIDAE

	Species	La Selva	Rancho Naturalista	Palo Verde	Bosque DRT	Las Cruces	Bosque DP	San Gerardo	Chomes
	Pomarine Jaeger *Stercorarius pomarinus*								
	Parasitic Jaeger *Stercorarius parasiticus*								

Pigeons and Doves—COLUMBIDAE

	Species	La Selva	Rancho Naturalista	Palo Verde	Bosque DRT	Las Cruces	Bosque DP	San Gerardo	Chomes
	Rock Pigeon *Columba livia*		R						
	Pale-vented Pigeon *Patagioenas cayennensis*	U			C				

		La Selva	Rancho Naturalista	Palo Verde	Bosque DRT	Las Cruces	Bosque DP	San Gerardo	Chomes
	Scaled Pigeon *Patagioenas speciosa*	R				U			
	Red-billed Pigeon *Patagioenas flavirostris*	U	C	C			U		C
	Band-tailed Pigeon *Patagioenas fasciata*		R			R	C	C	
	Ruddy Pigeon *Patagioenas subvinacea*		U			U	U	C	
	Short-billed Pigeon *Patagioenas nigrirostris*	C	C		C	C	U		
	White-winged Dove *Zenaida asiatica*			C					C
	Mourning Dove *Zenaida macroura*			R					R
	Inca Dove *Columbina inca*			C					C
	Common Ground-Dove *Columbina passerina*			C					C
	Plain-breasted Ground-Dove *Columbina minuta*			U					R
	Ruddy Ground-Dove *Columbina talpacoti*	C	U	U	C	C			C
	Blue Ground-Dove *Claravis pretiosa*	U		R	C	U			R
	Maroon-chested Ground-Dove *Claravis mondetoura*								
	White-tipped Dove *Leptotila verreauxi*	U	C	C	C	C	C	R	C
	Gray-headed Dove *Leptotila plumbeiceps*								
	Gray-chested Dove *Leptotila cassini*	C	C		C	C			
	Olive-backed Quail-Dove *Geotrygon veraguensis*	U							
	Chiriqui Quail-Dove *Geotrygon chiriquensis*					R	U	U	
	Purplish-backed Quail-Dove *Geotrygon lawrencii*		U						
	Buff-fronted Quail-Dove *Geotrygon costaricensis*						C	R	
	Violaceous Quail-Dove *Geotrygon violacea*								
	Ruddy Quail-Dove *Geotrygon montana*		U		U	R			
	Parrots, Macaws, and Allies—PSITTACIDAE								
	Sulphur-winged Parakeet *Pyrrhura hoffmanni*		U					C	

		La Selva	Rancho Naturalista	Palo Verde	Bosque DRT	Las Cruces	Bosque DP	San Gerardo	Chomes
	Crimson-fronted Parakeet *Aratinga finschi*	U	C		C	C	U		
	Olive-throated Parakeet *Aratinga nana*	C							
	Orange-fronted Parakeet *Aratinga canicularis*			C					C
	Brown-throated Parakeet *Aratinga pertinax*								
	Great Green Macaw *Ara ambiguus*	U							
	Scarlet Macaw *Ara macao*			C	C				
	Barred Parakeet *Bolborhynchus lineola*					R	U	U	
	Orange-chinned Parakeet *Brotogeris jugularis*	C		C	C	C			C
	Red-fronted Parrotlet *Touit costaricensis*		R						
	Brown-hooded Parrot *Pyrilia haematotis*	C	C		C	U	R		
	Blue-headed Parrot *Pionus menstruus*					C			
	White-crowned Parrot *Pionus senilis*	C	C		C	C	C	U	
	White-fronted Parrot *Amazona albifrons*			C					C
	Red-lored Parrot *Amazona autumnalis*	C			C	R			
	Mealy Parrot *Amazona farinosa*	C			C				
	Yellow-naped Parrot *Amazona auropalliata*			C					U

Cuckoos—CUCULIDAE

		La Selva	Rancho Naturalista	Palo Verde	Bosque DRT	Las Cruces	Bosque DP	San Gerardo	Chomes
	Squirrel Cuckoo *Piaya cayana*	C	C	U	C	C	U		U
	Yellow-billed Cuckoo *Coccyzus americanus*	R	R	R	R				
	Mangrove Cuckoo *Coccyzus minor*			U	R				U
	Black-billed Cuckoo *Coccyzus erythropthalmus*	R	R						
	Striped Cuckoo *Tapera naevia*	U	U		C	U			
	Pheasant Cuckoo *Dromococcyx phasianellus*								
	Lesser Ground-Cuckoo *Morococcyx erythropygus*			U					U

	Species	La Selva	Rancho Naturalista	Palo Verde	Bosque DRT	Las Cruces	Bosque DP	San Gerardo	Chomes
	Rufous-vented Ground-Cuckoo *Neomorphus geoffroyi*								
	Smooth-billed Ani *Crotophaga ani*				C	C			
	Groove-billed Ani *Crotophaga sulcirostris*	U	C	C					C
	Barn Owl—TYTONIDAE								
	Barn Owl *Tyto alba*			U		R	R		
	Typical Owls—STRIGIDAE								
	Pacific Screech-Owl *Megascops cooperi*			C					C
	Tropical Screech-Owl *Megascops choliba*					U	R		
	Vermiculated Screech-Owl *Megascops guatemalae*	C				C			
	Bare-shanked Screech-Owl *Megascops clarkii*						R	U	
	Crested Owl *Lophostrix cristata*	U	R		R	R			
	Spectacled Owl *Pulsatrix perspicillata*	U		U	U	R	R		
	Costa Rican Pygmy-Owl *Glaucidium costaricanum*							C	
	Central American Pygmy-Owl *Glaucidium griseiceps*	R							
	Ferruginous Pygmy-Owl *Glaucidium brasilianum*			C					C
	Mottled Owl *Ciccaba virgata*	U	C	U		C		R	
	Black-and-white Owl *Ciccaba nigrolineata*	U		R	R	U			
	Striped Owl *Pseudoscops clamator*			R		R			
	Unspotted Saw-whet Owl *Aegolius ridgwayi*						R	R	
	Nightjars—CAPRIMULGIDAE								
	Short-tailed Nighthawk *Lurocalis semitorquatus*	U	U			U			
	Lesser Nighthawk *Chordeiles acutipennis*			U					U
	Common Nighthawk *Chordeiles minor*			R					
	Common Pauraque *Nyctidromus albicollis*	C	C	C	C	C	C		C
	Chuck-will's-widow *Caprimulgus carolinensis*		R						

	Species	La Selva	Rancho Naturalista	Palo Verde	Bosque DRT	Las Cruces	Bosque DP	San Gerardo	Chomes
	Dusky Nightjar *Caprimulgus saturatus*						C	C	
	White-tailed Nightjar *Caprimulgus cayennensis*								

Potoos—NYCTIBIIDAE

	Species	La Selva	Rancho Naturalista	Palo Verde	Bosque DRT	Las Cruces	Bosque DP	San Gerardo	Chomes
	Great Potoo *Nyctibius grandis*	U							
	Common Potoo *Nyctibius griseus*		R			U			
	Northern Potoo *Nyctibius jamaicensis*			R					R

Swifts—APODIDAE

	Species	La Selva	Rancho Naturalista	Palo Verde	Bosque DRT	Las Cruces	Bosque DP	San Gerardo	Chomes
	Black Swift *Cypseloides niger*	R	R		R	R	C	U	
	White-chinned Swift *Cypseloides cryptus*								
	Spot-fronted Swift *Cypseloides cherriei*								
	Chestnut-collared Swift *Streptoprocne rutila*	R	U			R		R	
	White-collared Swift *Streptoprocne zonaris*	C	C	U	C	C	C	C	U
	Chimney Swift *Chaetura pelagica*	U							
	Vaux's Swift *Chaetura vauxi*	R	C			U	U	U	
	Costa Rican Swift *Chaetura fumosa*				C	C			
	Gray-rumped Swift *Chaetura cinereiventris*	C							
	Lesser Swallow-tailed Swift *Panyptila cayennensis*	R	R		U	R			

Hummingbirds—TROCHILIDAE

	Species	La Selva	Rancho Naturalista	Palo Verde	Bosque DRT	Las Cruces	Bosque DP	San Gerardo	Chomes
	Bronzy Hermit *Glaucis aeneus*	C			C	U			
	Band-tailed Barbthroat *Threnetes ruckeri*	C	R		C				
	Green Hermit *Phaethornis guy*		C			C	C		
	Long-billed Hermit *Phaethornis longirostris*	C			C	U			
	Stripe-throated Hermit *Phaethornis striigularis*	C	C	U	C	C	R		
	White-tipped Sicklebill *Eutoxeres aquila*	R	R		U	U			
	Green-fronted Lancebill *Doryfera ludovicae*						U	R	

	Species	La Selva	Rancho Naturalista	Palo Verde	Bosque DRT	Las Cruces	Bosque DP	San Gerardo	Chomes
	Scaly-breasted Hummingbird *Phaeochroa cuvierii*				C	C			
	Violet Sabrewing *Campylopterus hemileucurus*		C			U	C	R	
	White-necked Jacobin *Florisuga mellivora*	U	C		C	U			
	Brown Violetear *Colibri delphinae*		C				U		
	Green Violetear *Colibri thalassinus*		R			R	C	C	
	Green-breasted Mango *Anthracothorax prevostii*	R	C	U					R
	Veraguan Mango *Anthracothorax veraguensis*								
	Violet-headed Hummingbird *Klais guimeti*	C	U		U	R			
	Black-crested Coquette *Lophornis helenae*	R	U						
	White-crested Coquette *Lophornis adorabilis*				C	R			
	Green Thorntail *Discosura conversii*	R	C				R		
	Canivet's Emerald *Chlorostilbon canivetii*			U					R
	Garden Emerald *Chlorostilbon assimilis*		R		R	C			
	Violet-crowned Woodnymph *Thalurania colombica*	C	C		C	U			
	Fiery-throated Hummingbird *Panterpe insignis*						U	R	
	Blue-throated Goldentail *Hylocharis eliciae*	R		U	C	U			R
	Blue-chested Hummingbird *Amazilia amabilis*	U							
	Charming Hummingbird *Amazilia decora*				C	C			
	Mangrove Hummingbird *Amazilia boucardi*			R					U
	Blue-tailed Hummingbird *Amazilia cyanura*								
	Steely-vented Hummingbird *Amazilia saucerrottei*			C			R		U
	Snowy-bellied Hummingbird *Amazilia edward*					C			
	Rufous-tailed Hummingbird *Amazilia tzacatl*	C	C	C	C	C	C	R	U
	Cinnamon Hummingbird *Amazilia rutila*			C					C

		La Selva	Rancho Naturalista	Palo Verde	Bosque DRT	Las Cruces	Bosque DP	San Gerardo	Chomes
	Stripe-tailed Hummingbird *Eupherusa eximia*					R	R		
	Black-bellied Hummingbird *Eupherusa nigriventris*						C		
	White-tailed Emerald *Elvira chionura*					U			
	Coppery-headed Emerald *Elvira cupreiceps*						U		
	Snowcap *Microchera albocoronata*	R	C				R		
	Bronze-tailed Plumeleteer *Chalybura urochrysia*	C	U						
	White-bellied Mountain-gem *Lampornis hemileucus*								
	Purple-throated Mountain-gem *Lampornis calolaemus*						C		
	White-throated Mountain-gem *Lampornis castaneoventris*					R		C	
	Green-crowned Brilliant *Heliodoxa jacula*		C			U	C	U	
	Magnificent Hummingbird *Eugenes fulgens*						C	C	
	Purple-crowned Fairy *Heliothryx barroti*	U	C		C	C			
	Long-billed Starthroat *Heliomaster longirostris*	R	R		C	U	R		
	Plain-capped Starthroat *Heliomaster constantii*			U					U
	Magenta-throated Woodstar *Calliphlox bryantae*					R	U		
	Ruby-throated Hummingbird *Archilochus colubris*		R	U					R
	Volcano Hummingbird *Selasphorus flammula*						U	C	
	Scintillant Hummingbird *Selasphorus scintilla*					R	C	C	
	Trogons—TROGONIDAE								
	Black-headed Trogon *Trogon melanocephalus*			C					C
	Baird's Trogon *Trogon bairdii*				C	U			
	Violaceous Trogon *Trogon violaceus*	C	C	R	C	C			
	Elegant Trogon *Trogon elegans*			R					
	Collared Trogon *Trogon collaris*		C			C	C	C	

		La Selva	Rancho Naturalista	Palo Verde	Bosque DRT	Las Cruces	Bosque DP	San Gerardo	Chomes
	Orange-bellied Trogon *Trogon aurantiiventris*						C		
	Black-throated Trogon *Trogon rufus*	C	U		C	U			
	Slaty-tailed Trogon *Trogon massena*	C			C	U			
	Lattice-tailed Trogon *Trogon clathratus*		R						
	Resplendent Quetzal *Pharomachrus mocinno*						U	C	

Motmots—MOMOTIDAE

		La Selva	Rancho Naturalista	Palo Verde	Bosque DRT	Las Cruces	Bosque DP	San Gerardo	Chomes
	Tody Motmot *Hylomanes momotula*								
	Blue-crowned Motmot *Momotus momota*		C	U	C	C	U		
	Rufous Motmot *Baryphthengus martii*	C	C				U		
	Keel-billed Motmot *Electron carinatum*								
	Broad-billed Motmot *Electron platyrhynchum*	C	U						
	Turquoise-browed Motmot *Eumomota superciliosa*			C					C

Kingfishers—ALCEDINIDAE

		La Selva	Rancho Naturalista	Palo Verde	Bosque DRT	Las Cruces	Bosque DP	San Gerardo	Chomes
	Ringed Kingfisher *Megaceryle torquata*	U		C	C				C
	Belted Kingfisher *Megaceryle alcyon*	R		R	R				R
	Amazon Kingfisher *Chloroceryle amazona*	U	U	C	C				C
	Green Kingfisher *Chloroceryle americana*	C	C	C	C				U
	Green-and-rufous Kingfisher *Chloroceryle inda*	R							
	American Pygmy Kingfisher *Chloroceryle aenea*	R		U	U				U

Puffbirds—BUCCONIDAE

		La Selva	Rancho Naturalista	Palo Verde	Bosque DRT	Las Cruces	Bosque DP	San Gerardo	Chomes
	White-necked Puffbird *Notharchus macrorhynchos*	U		R	U				R
	Pied Puffbird *Notharchus tectus*	R							
	White-whiskered Puffbird *Malacoptila panamensis*	R			U	U			
	Lanceolated Monklet *Micromonacha lanceolata*		R						
	White-fronted Nunbird *Monasa morphoeus*								

	Jacamars—GALBULIDAE	La Selva	Rancho Naturalista	Palo Verde	Bosque DRT	Las Cruces	Bosque DP	San Gerardo	Chomes
	Rufous-tailed Jacamar *Galbula ruficauda*	C	U		C	C			
	Great Jacamar *Jacamerops aureus*								
	Barbets and Toucans—RAMPHASTIDAE								
	Red-headed Barbet *Eubucco bourcierii*		R			R	C		
	Prong-billed Barbet *Semnornis frantzii*						C		
	Emerald Toucanet *Aulacorhynchus prasinus*					U	C	C	
	Collared Aracari *Pteroglossus torquatus*	C	C	R					
	Fiery-billed Aracari *Pteroglossus frantzii*				C	C			
	Yellow-eared Toucanet *Selenidera spectabilis*		R						
	Keel-billed Toucan *Ramphastos sulfuratus*	C	C	R			C		
	Chestnut-mandibled Toucan *Ramphastos swainsonii*	C			C	C	C		
	Woodpeckers—PICIDAE								
	Olivaceous Piculet *Picumnus olivaceus*				R	U			
	Acorn Woodpecker *Melanerpes formicivorus*					R		C	
	Golden-naped Woodpecker *Melanerpes chrysauchen*				U	U			
	Black-cheeked Woodpecker *Melanerpes pucherani*	C	C						
	Red-crowned Woodpecker *Melanerpes rubricapillus*				C	C			
	Hoffmann's Woodpecker *Melanerpes hoffmannii*		C	C					C
	Yellow-bellied Sapsucker *Sphyrapicus varius*		R			R	U	R	
	Hairy Woodpecker *Picoides villosus*						C	C	
	Smoky-brown Woodpecker *Veniliornis fumigatus*	U	U			U	U		
	Red-rumped Woodpecker *Veniliornis kirkii*				U				
	Rufous-winged Woodpecker *Piculussimplex*	U	U		U	R			
	Golden-olive Woodpecker *Colaptes rubiginosus*		C			U	U	R	

		La Selva	Rancho Naturalista	Palo Verde	Bosque DRT	Las Cruces	Bosque DP	San Gerardo	Chomes
	Cinnamon Woodpecker *Celeusloricatus*	U							
	Chestnut-colored Woodpecker *Celeus castaneus*	U							
	Lineated Woodpecker *Dryocopus lineatus*	U	U	U	C	C			U
	Pale-billed Woodpecker *Campephilus guatemalensis*	C	R	U	U	C			U
	Ovenbirds and Woodcreepers—FURNARIIDAE								
	Tawny-throated Leaftosser *Sclerurus mexicanus*		U			R			
	Gray-throated Leaftosser *Sclerurus albigularis*						R		
	Scaly-throated Leaftosser *Sclerurus guatemalensis*				C	R			
	Pale-breasted Spinetail *Synallaxis albescens*				R	C			
	Slaty Spinetail *Synallaxis brachyura*	C	C		C	C	R		
	Red-faced Spinetail *Cranioleuca erythrops*					C	C		
	Spotted Barbtail *Premnoplex brunnescens*					U	C	R	
	Ruddy Treerunner *Margarornis rubiginosus*						R	C	
	Buffy Tuftedcheek *Pseudocolaptes lawrencii*						U	C	
	Striped Woodhaunter *Hyloctistes subulatus*	R			R	R			
	Lineated Foliage-gleaner *Syndactyla subalaris*					U	C	U	
	Scaly-throated Foliage-gleaner *Anabacerthia variegaticeps*					R			
	Buff-fronted Foliage-gleaner *Philydor rufum*								
	Buff-throated Foliage-gleaner *Automolus ochrolaemus*	U	U		C	C			
	Ruddy Foliage-gleaner *Automolus rubiginosus*					U			
	Streak-breasted Treehunter *Thripadectes rufobrunneus*						C	C	
	Plain Xenops *Xenops minutus*	U	C		C	C	R		
	Streaked Xenops *Xenops rutilans*					R	R	R	
	Plain-brown Woodcreeper *Dendrocincla fuliginosa*	U	R						

Species	La Selva	Rancho Naturalista	Palo Verde	Bosque DRT	Las Cruces	Bosque DP	San Gerardo	Chomes
Tawny-winged Woodcreeper *Dendrocincla anabatina*				U	U			
Ruddy Woodcreeper *Dendrocincla homochroa*			R		U			
Olivaceous Woodcreeper *Sittasomus griseicapillus*		U	U		C	R		
Long-tailed Woodcreeper *Deconychura longicauda*				U				
Wedge-billed Woodcreeper *Glyphorhynchus spirurus*	C	C	U	C	C			
Strong-billed Woodcreeper *Xiphocolaptes promeropirhynchus*						R		
Northern Barred-Woodcreeper *Dendrocolaptes sanctithomae*	C	U	U	C	R			
Black-banded Woodcreeper *Dendrocolaptes picumnus*						U		
Cocoa Woodcreeper *Xiphorhynchus susurrans*	C	U		C				
Ivory-billed Woodcreeper *Xiphorhynchus flavigaster*								
Black-striped Woodcreeper *Xiphorhynchus lachrymosus*	U			C				
Spotted Woodcreeper *Xiphorhynchus erythropygius*	U	C			U	R		
Streak-headed Woodcreeper *Lepidocolaptes souleyetii*	C	C	C		C	R		C
Spot-crowned Woodcreeper *Lepidocolaptes affinis*						C	C	
Brown-billed Scythebill *Campylorhamphus pusillus*		U			C	U		

Typical Antbirds—THAMNOPHILIDAE

Species	La Selva	Rancho Naturalista	Palo Verde	Bosque DRT	Las Cruces	Bosque DP	San Gerardo	Chomes
Fasciated Antshrike *Cymbilaimus lineatus*	C	R						
Great Antshrike *Taraba major*	U			C	R			
Barred Antshrike *Thamnophilus doliatus*	U	U	U					U
Black-hooded Antshrike *Thamnophilus bridgesi*				C	C			
Western Slaty-Antshrike *Thamnophilus atrinucha*	C							
Russet Antshrike *Thamnistes anabatinus*	R	C		C	U			
Plain Antvireo *Dysithamnus mentalis*		C		R	C	U		
Streak-crowned Antvireo *Dysithamnus striaticeps*	U							

		La Selva	Rancho Naturalista	Palo Verde	Bosque DRT	Las Cruces	Bosque DP	San Gerardo	Chomes
	Spot-crowned Antvireo *Dysithamnus puncticeps*								
	White-flanked Antwren *Myrmotherula axillaris*	U							
	Slaty Antwren *Myrmotherula schisticolor*		C		R	C	U		
	Checker-throated Antwren *Epinecrophylla fulviventris*	U	C						
	Dot-winged Antwren *Microrhopias quixensis*	U	U		C	U			
	Rufous-rumped Antwren *Terenura callinota*								
	Dusky Antbird *Cercomacra tyrannina*	C	U		U	R			
	Bare-crowned Antbird *Gymnocichla nudiceps*	R							
	Chestnut-backed Antbird *Myrmeciza exsul*	C			C				
	Dull-mantled Antbird *Myrmeciza laemosticta*		U						
	Immaculate Antbird *Myrmeciza immaculata*	R	U						
	Spotted Antbird *Hylophylax naevioides*	R	U						
	Bicolored Antbird *Gymnopithys leucaspis*	U			U	U			
	Ocellated Antbird *Phaenostictus mcleannani*	U							

Antthrushes—FORMICARIIDAE

		La Selva	Rancho Naturalista	Palo Verde	Bosque DRT	Las Cruces	Bosque DP	San Gerardo	Chomes
	Black-faced Antthrush *Formicarius analis*	C	R		C	C			
	Black-headed Antthrush *Formicarius nigricapillus*								
	Rufous-breasted Antthrush *Formicarius rufipectus*								

Gnateaters—CONOPOPHAGIDAE

		La Selva	Rancho Naturalista	Palo Verde	Bosque DRT	Las Cruces	Bosque DP	San Gerardo	Chomes
	Black-crowned Antpitta *Pittasoma michleri*								

Antpittas—GRALLARIIDAE

		La Selva	Rancho Naturalista	Palo Verde	Bosque DRT	Las Cruces	Bosque DP	San Gerardo	Chomes
	Scaled Antpitta *Grallaria guatimalensis*		U			U	U		
	Streak-chested Antpitta *Hylopezus perspicillatus*	R							
	Thicket Antpitta *Hylopezus dives*	R	U						
	Ochre-breasted Antpitta *Grallaricula flavirostris*		R			U	R		

	Species	La Selva	Rancho Naturalista	Palo Verde	Bosque DRT	Las Cruces	Bosque DP	San Gerardo	Chomes
	Tapaculo—RHINOCRYPTIDAE								
☐	Silvery-fronted Tapaculo *Scytalopus argentifrons*					U	C	C	
	Tyrant Flycatchers—TYRANNIDAE								
☐	Yellow-bellied Tyrannulet *Ornithion semiflavum*				U	R			
☐	Brown-capped Tyrannulet *Ornithion brunneicapillus*	U							
☐	Northern Beardless-Tyrannulet *Camptostoma imberbe*			U					U
☐	Southern Beardless-Tyrannulet *Camptostoma obsoletum*			R	U	R			
☐	Mouse-colored Tyrannulet *Phaeomyias murina*								
☐	Yellow Tyrannulet *Capsiempis flaveola*	C			C	U			
☐	Yellow-crowned Tyrannulet *Tyrannulus elatus*				R	U			
☐	Greenish Elaenia *Myiopagis viridicata*		R	U		U			
☐	Yellow-bellied Elaenia *Elaenia flavogaster*	C	C	C	C	C	C	R	C
☐	Lesser Elaenia *Elaenia chiriquensis*		R		R	U			
☐	Mountain Elaenia *Elaenia frantzii*					U	C	C	
☐	Torrent Tyrannulet *Serpophaga cinerea*		C				U	U	
☐	Olive-striped Flycatcher *Mionectes olivaceus*	R	C			C	C	R	
☐	Ochre-bellied Flycatcher *Mionectes oleagineus*	C	C	U	C	C			
☐	Sepia-capped Flycatcher *Leptopogon amaurocephalus*		R						
☐	Slaty-capped Flycatcher *Leptopogon superciliaris*		C			U			
☐	Rufous-browed Tyrannulet *Phylloscartes superciliaris*								
☐	Rough-legged Tyrannulet *Phyllomyias burmeisteri*					U		R	
☐	Paltry Tyrannulet *Zimmerius vilissimus*	C	C	R	C	C	C	C	R
☐	Northern Scrub-Flycatcher *Sublegatus arenarum*			U					C
☐	Black-capped Pygmy-Tyrant *Myiornis atricapillus*	C							
☐	Scale-crested Pygmy-Tyrant *Lophotriccus pileatus*		C		U	C	R		

		La Selva	Rancho Naturalista	Palo Verde	Bosque DRT	Las Cruces	Bosque DP	San Gerardo	Chomes
	Northern Bentbill *Oncostoma cinereigulare*	U		U	C	R			
	Slate-headed Tody-Flycatcher *Poecilotriccus sylvia*	R		R	U	R			
	Common Tody-Flycatcher *Todirostrum cinereum*	C	C	C	C	C	R		C
	Black-headed Tody-Flycatcher *Todirostrum nigriceps*	U	C						
	Eye-ringed Flatbill *Rhynchocyclus brevirostris*	U	U		C	C			
	Yellow-olive Flycatcher *Tolmomyias sulphurescens*	C	C	C	C	C			C
	Yellow-margined Flycatcher *Tolmomyias assimilis*	C							
	Stub-tailed Spadebill *Platyrinchus cancrominus*			U					
	White-throated Spadebill *Platyrinchus mystaceus*		U			U	U		
	Golden-crowned Spadebill *Platyrinchus coronatus*	R			C	U			
	Royal Flycatcher *Onychorhynchus coronatus*	R		R		R			
	Ruddy-tailed Flycatcher *Terenotriccus erythrurus*	R	R		R	C			
	Sulphur-rumped Flycatcher *Myiobius sulphureipygius*	U	U		C	C			
	Black-tailed Flycatcher *Myiobius atricaudus*				U				
	Bran-colored Flycatcher *Myiophobus fasciatus*				R	U			
	Tawny-chested Flycatcher *Aphanotriccus capitalis*	R	C						
	Tufted Flycatcher *Mitrephanes phaeocercus*					R	C	C	
	Olive-sided Flycatcher *Contopus cooperi*	R	U	R	U	U	U	R	
	Dark Pewee *Contopus lugubris*						C	C	
	Ochraceous Pewee *Contopus ochraceus*							U	
	Western Wood-Pewee *Contopus sordidulus*	R	C	U		C	R	U	U
	Eastern Wood-Pewee *Contopus virens*	C	C	U	U	C	R	U	U
	Tropical Pewee *Contopus cinereus*	C	C	U	C	U	R		U

	Species	La Selva	Rancho Naturalista	Palo Verde	Bosque DRT	Las Cruces	Bosque DP	San Gerardo	Chomes
	Yellow-bellied Flycatcher *Empidonax flaviventris*	C	C	U	C	C			
	Acadian Flycatcher *Empidonax virescens*	U	R	C					U
	Alder Flycatcher *Empidonax alnorum*		U			R	R		
	Willow Flycatcher *Empidonax traillii*	U	U	C		R	R		U
	White-throated Flycatcher *Empidonax albigularis*		R				R		
	Least Flycatcher *Empidonax minimus*								
	Yellowish Flycatcher *Empidonax flavescens*					U	C	C	
	Black-capped Flycatcher *Empidonax atriceps*							C	
	Black Phoebe *Sayornis nigricans*		C			R	C	C	
	Long-tailed Tyrant *Colonia colonus*	C							
	Bright-rumped Attila *Attila spadiceus*	C	U	C	C	C	C		R
	Rufous Mourner *Rhytipterna holerythra*	C	U		C	U			
	Dusky-capped Flycatcher *Myiarchus tuberculifer*	C	C	C	C	C	U		C
	Panama Flycatcher *Myiarchus panamensis*			U					C
	Nutting's Flycatcher *Myiarchus nuttingi*			U					
	Great Crested Flycatcher *Myiarchus crinitus*	C	U	C	C	U	R		U
	Brown-crested Flycatcher *Myiarchus tyrannulus*			C					C
	Great Kiskadee *Pitangus sulphuratus*	C	C	C	C	U	C	U	C
	Boat-billed Flycatcher *Megarhynchus pitangua*	C	C	C	C	C	R	U	C
	Rusty-margined Flycatcher *Myiozetetes cayanensis*								
	Social Flycatcher *Myiozetetes similis*	C	C	C	C	C	C	R	C
	Gray-capped Flycatcher *Myiozetetes granadensis*	C	U		C	C	R		
	White-ringed Flycatcher *Conopias albovittatus*	U							

	Species	La Selva	Rancho Naturalista	Palo Verde	Bosque DRT	Las Cruces	Bosque DP	San Gerardo	Chomes
	Golden-bellied Flycatcher *Myiodynastes hemichrysus*						C	C	
	Streaked Flycatcher *Myiodynastes maculatus*			C	U	U			C
	Sulphur-bellied Flycatcher *Myiodynastes luteiventris*	U	C	U	U	C	U		R
	Piratic Flycatcher *Legatus leucophaius*	C	C	C	C	C	R		C
	Tropical Kingbird *Tyrannus melancholicus*	C	C	C	C	C	U	C	C
	Western Kingbird *Tyrannus verticalis*			R					
	Eastern Kingbird *Tyrannus tyrannus*	R	U		R	R	R		
	Gray Kingbird *Tyrannus dominicensis*								
	Scissor-tailed Flycatcher *Tyrannus forficatus*			C					C
	Fork-tailed Flycatcher *Tyrannus savanna*								

Becards and Allies—Genera *Incertae sedis*

	Species	La Selva	Rancho Naturalista	Palo Verde	Bosque DRT	Las Cruces	Bosque DP	San Gerardo	Chomes
	Thrush-like Schiffornis *Schiffornis turdina*	R	R		R	U			
	Gray-headed Piprites *Piprites griseiceps*	R							
	Rufous Piha *Lipaugus unirufus*	U			C	C			
	Speckled Mourner *Laniocera rufescens*	R							
	Barred Becard *Pachyramphus versicolor*						U	C	
	Cinnamon Becard *Pachyramphus cinnamomeus*	C	C	R	R				C
	White-winged Becard *Pachyramphus polychopterus*	U	U	U	C	C			U
	Black-and-white Becard *Pachyramphus albogriseus*		R						
	Rose-throated Becard *Pachyramphus aglaiae*			C	U	C			U
	Masked Tityra *Tityra semifasciata*	C	C	C	C	C	U		U
	Black-crowned Tityra *Tityra inquisitor*	C	C	R	C	C			R

Cotingas—COTINGIDAE

	Species	La Selva	Rancho Naturalista	Palo Verde	Bosque DRT	Las Cruces	Bosque DP	San Gerardo	Chomes
	Lovely Cotinga *Cotinga amabilis*								
	Turquoise Cotinga *Cotinga ridgwayi*				C	U			

		La Selva	Rancho Naturalista	Palo Verde	Bosque DRT	Las Cruces	Bosque DP	San Gerardo	Chomes
	Yellow-billed Cotinga *Carpodectes antoniae*				R				
	Snowy Cotinga *Carpodectes nitidus*	U							
	Purple-throated Fruitcrow *Querula purpurata*	U							
	Bare-necked Umbrellabird *Cephalopterus glabricollis*	U					R		
	Three-wattled Bellbird *Procnias tricarunculatus*	R			R	R	R		
	Manakins—PIPRIDAE								
	White-collared Manakin *Manacus candei*	C	C						
	Orange-collared Manakin *Manacus aurantiacus*				C	U			
	White-ruffed Manakin *Corapipo altera*	U	C		R	C			
	Lance-tailed Manakin *Chiroxiphia lanceolata*								
	Long-tailed Manakin *Chiroxiphia linearis*			C					
	White-crowned Manakin *Pipra pipra*	R	U						
	Blue-crowned Manakin *Pipra coronata*				U	C			
	Red-capped Manakin *Pipra mentalis*	C			C	R			
	Sharpbill—OXYRUNCIDAE								
	Sharpbill *Oxyruncus cristatus*								
	Vireos—VIREONIDAE								
	White-eyed Vireo *Vireo griseus*								
	Mangrove Vireo *Vireo pallens*			R					U
	Yellow-throated Vireo *Vireo flavifrons*	C	C	C	C	C			C
	Blue-headed Vireo *Vireo solitarius*								
	Yellow-winged Vireo *Vireo carmioli*						U	C	
	Warbling Vireo *Vireo gilvus*								
	Brown-capped Vireo *Vireo leucophrys*						C	C	
	Philadelphia Vireo *Vireo philadelphicus*	U	U	U	U	C		R	R

		La Selva	Rancho Naturalista	Palo Verde	Bosque DRT	Las Cruces	Bosque DP	San Gerardo	Chomes
	Red-eyed Vireo *Vireo olivaceus*	C	C		U	U	U		
	Yellow-green Vireo *Vireo flavoviridis*	R	C	C	C	C	U		C
	Scrub Greenlet *Hylophilus flavipes*				C				
	Tawny-crowned Greenlet *Hylophilus ochraceiceps*	U	C		C	C			
	Lesser Greenlet *Hylophilus decurtatus*	C	C	C	C	C			U
	Green Shrike-Vireo *Vireolanius pulchellus*	U	R		C	U			
	Rufous-browed Peppershrike *Cyclarhis gujanensis*			U		U	U	C	U
	Jays—CORVIDAE								
	White-throated Magpie-Jay *Calocitta Formosa*			C					C
	Black-chested Jay *Cyanocorax affinis*								
	Brown Jay *Cyanocorax morio*	R	C				C		
	Azure-hooded Jay *Cyanolyca cucullata*						U		
	Silvery-throated Jay *Cyanolyca argentigula*							R	
	Swallows—HIRUNDINIDAE								
	Purple Martin *Progne subis*	R							
	Gray-breasted Martin *Progne chalybea*	U	U	U	C	C			C
	Brown-chested Martin *Progne tapera*								
	Tree Swallow *Tachycineta bicolor*								
	Mangrove Swallow *Tachycineta albilinea*	C		C	U				C
	Violet-green Swallow *Tachycineta thalassina*								
	Blue-and-white Swallow *Pygochelidon cyanoleuca*		C			C	C	C	
	Northern Rough-winged Swallow *Stelgidopteryx serripennis*	U	C	C	R				U
	Southern Rough-winged Swallow *Stelgidopteryx ruficollis*	C	C		C	C			U
	Bank Swallow *Riparia riparia*	R	R	R		R			U

	Species	La Selva	Rancho Naturalista	Palo Verde	Bosque DRT	Las Cruces	Bosque DP	San Gerardo	Chomes
☐	Cliff Swallow *Petrochelidon pyrrhonota*	U	R	U		U			U
☐	Cave Swallow *Petrochelidon fulva*								
☐	Barn Swallow *Hirundo rustica*	C	U	C		C			C
	Wrens—TROGLODYTIDAE								
☐	Band-backed Wren *Campylorhynchus zonatus*	C	C						
☐	Rufous-naped Wren *Campylorhynchus rufinucha*			C					C
☐	Rock Wren *Salpinctes obsoletus*								
☐	Black-throated Wren *Thryothorus atrogularis*	C	C						
☐	Black-bellied Wren *Thryothorus fasciatoventris*				C				
☐	Bay Wren *Thryothorus nigricapillus*	C	U						
☐	Riverside Wren *Thryothorus semibadius*				C	C			
☐	Stripe-breasted Wren *Thryothorus thoracicus*	C	C						
☐	Rufous-breasted Wren *Thryothorus rutilus*					C			
☐	Spot-breasted Wren *Thryothorus maculipectus*								
☐	Rufous-and-white Wren *Thryothorus rufalbus*			C					R
☐	Banded Wren *Thryothorus pleurostictus*			C					R
☐	Plain Wren *Thryothorus modestus*	C	C	U	C	C			U
☐	House Wren *Troglodytes aedon*	C	C	R	C	C	C	C	U
☐	Ochraceous Wren *Troglodytes ochraceus*					R	C	C	
☐	Sedge Wren *Cistothorus platensis*								
☐	Timberline Wren *Thryorchilus browni*								
☐	White-breasted Wood-Wren *Henicorhina leucosticta*	C	C			C	U		
☐	Gray-breasted Wood-Wren *Henicorhina leucophrys*					U	C	C	
☐	Nightingale Wren *Microcerculus philomela*	R							

Species	La Selva	Rancho Naturalista	Palo Verde	Bosque DRT	Las Cruces	Bosque DP	San Gerardo	Chomes
Scaly-breasted Wren *Microcerculus marginatus*		U		C	C			
Song Wren *Cyphorhinus phaeocephalus*	R							
Dipper—CINCLIDAE								
American Dipper *Cinclus mexicanus*		C				U	C	
Gnatwrens and Gnatcatchers—SYLVIIDAE								
Tawny-faced Gnatwren *Microbates cinereiventris*	R							
Long-billed Gnatwren *Ramphocaenus melanurus*	C	C	C	U	C			
White-lored Gnatcatcher *Polioptila albiloris*			C					
Tropical Gnatcatcher *Polioptila plumbea*	C	C	C	C	C			C
Thrushes—TURDIDAE								
Black-faced Solitaire *Myadestes melanops*		U			U	C	C	
Black-billed Nightingale-Thrush *Catharus gracilirostris*						U	U	
Orange-billed Nightingale-Thrush *Catharus aurantiirostris*		R			C	U		
Slaty-backed Nightingale-Thrush *Catharus fuscater*					R	U		
Ruddy-capped Nightingale-Thrush *Catharus frantzii*						C	C	
Black-headed Nightingale-Thrush *Catharus mexicanus*		C						
Veery *Catharus fuscescens*	R	R						
Gray-cheeked Thrush *Catharus minimus*	U	R		R				
Swainson's Thrush *Catharus ustulatus*	C	C		U	C	U	C	
Wood Thrush *Hylocichla mustelina*	C	C	U	U	R	U		
Sooty Thrush *Turdus nigrescens*						U	R	
Mountain Thrush *Turdus plebejus*					R	C	C	
Pale-vented Thrush *Turdus obsoletus*	U	R						
Clay-colored Thrush *Turdus grayi*	C	C	C	C	C	C	C	C
White-throated Thrush *Turdus assimilis*	R			U	C	R		

	La Selva	Rancho Naturalista	Palo Verde	Bosque DRT	Las Cruces	Bosque DP	San Gerardo	Chomes
Mimic Thrushes—MIMIDAE								
Gray Catbird *Dumetella carolinensis*	R	R						
Tropical Mockingbird *Mimus gilvus*								
Waxwing—BOMBYCILLIDAE								
Cedar Waxwing *Bombycilla cedrorum*		R	R	R	R	R		R
Silky-Flycatchers—PTILOGONATIDAE								
Black-and-yellow Silky-Flycatcher *Phainoptila melanoxantha*							R	
Long-tailed Silky-Flycatcher *Ptilogonys caudatus*					R	C	C	
New World Warblers—PARULIDAE								
Blue-winged Warbler *Vermivora pinus*	R	U			R	R		
Golden-winged Warbler *Vermivora chrysoptera*	U	C		R	C	C	R	
Tennessee Warbler *Vermivora peregrina*	C	C	C	C	C	U	C	C
Flame-throated Warbler *Parula gutturalis*						C	C	
Northern Parula *Parula americana*								
Tropical Parula *Parula pitiayumi*		C		R	C	C		
Yellow Warbler *Dendroica petechia*	C	R	C	C	C	U		C
Chestnut-sided Warbler *Dendroica pensylvanica*	C	C	C	C	C	C		U
Magnolia Warbler *Dendroica magnolia*	R							
Black-throated Blue Warbler *Dendroica caerulescens*								
Yellow-rumped Warbler *Dendroica coronata*	R	R			R			
Black-throated Green Warbler *Dendroica virens*	R	U			U	C	C	
Townsend's Warbler *Dendroica townsendi*							R	
Hermit Warbler *Dendroica occidentalis*								
Blackburnian Warbler *Dendroica fusca*	U	C			C	C	R	
Yellow-throated Warbler *Dendroica dominica*		R			R			
Prairie Warbler *Dendroica discolor*								

		La Selva	Rancho Naturalista	Palo Verde	Bosque DRT	Las Cruces	Bosque DP	San Gerardo	Chomes
	Palm Warbler *Dendroica palmarum*								
	Bay-breasted Warbler *Dendroica castanea*	U	R		R	U	R		
	Blackpoll Warbler *Dendroica striata*								
	Cerulean Warbler *Dendroica cerulea*	R	R			R			
	Black-and-white Warbler *Mniotilta varia*	U	C	U	U	C	C	U	U
	American Redstart *Setophaga ruticilla*	U	C	U	U	U	R		U
	Prothonotary Warbler *Protonotaria citrea*	R		C		U	U		C
	Worm-eating Warbler *Helmitheros vermivorum*	R	U			U			
	Ovenbird *Seiurus aurocapilla*	U	R	R		U			
	Northern Waterthrush *Seiurus noveboracensis*	C		C	C	C			C
	Louisiana Waterthrush *Seiurus motacilla*	U				U	C	C	
	Kentucky Warbler *Oporornis formosus*	C	C	U	R	C	R		U
	Mourning Warbler *Oporornis philadelphia*	C	C	R	C	C			
	MacGillivray's Warbler *Oporornis tolmiei*	R	R						
	Common Yellowthroat *Geothlypis trichas*			R	U	R			
	Olive-crowned Yellowthroat *Geothlypis semiflava*	C	C				R		
	Masked Yellowthroat *Geothlypis aequinoctialis*					U			
	Gray-crowned Yellowthroat *Geothlypis poliocephala*	C	U	U	R	C	U		U
	Hooded Warbler *Wilsonia citrina*	R	R						
	Wilson's Warbler *Wilsonia pusilla*	R	C	R		C	C	C	
	Canada Warbler *Wilsonia canadensis*	U	U			U	R		
	Slate-throated Redstart *Myioborus miniatus*		C			C	C		
	Collared Redstart *Myioborus torquatus*						C	C	

	Species	La Selva	Rancho Naturalista	Palo Verde	Bosque DRT	Las Cruces	Bosque DP	San Gerardo	Chomes
☐	Golden-crowned Warbler *Basileuterus culicivorus*		C			U	C		
☐	Rufous-capped Warbler *Basileuterus rufifrons*		C	C		C			R
☐	Black-cheeked Warbler *Basileuterus melanogenys*						C	C	
☐	Three-striped Warbler *Basileuterus tristriatus*					R	C		
☐	Buff-rumped Warbler *Paeothlypis fulvicauda*	C	U		C	C			
☐	Wrenthrush *Zeledonia coronata*						U	C	
☐	Yellow-breasted Chat *Icteria virens*	R		R					

Bananaquit—Genus *Incertae sedis*

	Species	La Selva	Rancho Naturalista	Palo Verde	Bosque DRT	Las Cruces	Bosque DP	San Gerardo	Chomes
☐	Bananaquit *Coereba flaveola*	C	C		C	C			

Tanagers and Allies—THRAUPIDAE

	Species	La Selva	Rancho Naturalista	Palo Verde	Bosque DRT	Las Cruces	Bosque DP	San Gerardo	Chomes
☐	Common Bush-Tanager *Chlorospingus ophthalmicus*		U			C	C	C	
☐	Sooty-capped Bush-Tanager *Chlorospingus pileatus*							C	
☐	Ashy-throated Bush-Tanager *Clorospingus canigularis*		U						
☐	Black-and-yellow Tanager *Chrysothlypis chrysomelas*		U						
☐	Rosy Thrush-Tanager *Rhodinocichla rosea*								
☐	Dusky-faced Tanager *Mitrospingus cassinii*	C							
☐	Carmiol's Tanager *Chlorothraupis carmioli*	R	C						
☐	Gray-headed Tanager *Eucometis penicillata*			U	U	C			
☐	White-throated Shrike-Tanager *Lanio leucothorax*	R			U				
☐	Sulphur-rumped Tanager *Heterospingus rubrifrons*								
☐	White-shouldered Tanager *Tachyphonus luctuosus*	C	C		C	C			
☐	Tawny-crested Tanager *Tachyphonus delatrii*	R							
☐	White-lined Tanager *Tachyphonus rufus*	C	C		R	U			
☐	Red-crowned Ant-Tanager *Habia rubica*					C			

		La Selva	Rancho Naturalista	Palo Verde	Bosque DRT	Las Cruces	Bosque DP	San Gerardo	Chomes
	Red-throated Ant-Tanager *Habia fuscicauda*	C	C						
	Black-cheeked Ant-Tanager *Habia atrimaxillaris*				C				
	Hepatic Tanager *Piranga flava*		R			R	R	R	
	Summer Tanager *Piranga rubra*	C	C	C	C	C	C	R	C
	Scarlet Tanager *Piranga olivacea*	U	C	R	U	C			
	Western Tanager *Piranga ludoviciana*			R					U
	Flame-colored Tanager *Piranga bidentata*					R	R	C	
	White-winged Tanager *Piranga leucoptera*					U	R	U	
	Crimson-collared Tanager *Ramphocelus sanguinolentus*	U	C						
	Passerini's Tanager *Ramphocelus passerinii*	C	C						
	Cherrie's Tanager *Ramphocelus costaricensis*				C	C			
	Blue-gray Tanager *Thraupis episcopus*	C	C	U	C	C	C	C	C
	Palm Tanager *Thraupis palmarum*	C	C	R	C	C			R
	Blue-and-gold Tanager *Bangsia arcaei*								
	Golden-hooded Tanager *Tangara larvata*	C	C		C	C			
	Speckled Tanager *Tangara guttata*		U			C			
	Spangle-cheeked Tanager *Tangara dowii*		R				U	C	
	Plain-colored Tanager *Tangara inornata*	U							
	Rufous-winged Tanager *Tangara lavinia*	R							
	Bay-headed Tanager *Tangara gyrola*	R	C		C	C			
	Emerald Tanager *Tangara florida*		U						
	Silver-throated Tanager *Tangara icterocephala*	C	C		U	C	U	C	
	Scarlet-thighed Dacnis *Dacnis venusta*	U	U		R	C	U		

		La Selva	Rancho Naturalista	Palo Verde	Bosque DRT	Las Cruces	Bosque DP	San Gerardo	Chomes
	Blue Dacnis *Dacnis cayana*	U			C	U			
	Green Honeycreeper *Chlorophanes spiza*	C	C		C	C			
	Shining Honeycreeper *Cyanerpes lucidus*	C			C	C			
	Red-legged Honeycreeper *Cyanerpes cyaneus*	U		R	C	C		R	

Sparrows, Seedeaters, and Allies—EMBERIZIDAE

		La Selva	Rancho Naturalista	Palo Verde	Bosque DRT	Las Cruces	Bosque DP	San Gerardo	Chomes
	Blue-black Grassquit *Volatinia jacarina*	U	C	U	C	C	R		C
	Slate-colored Seedeater *Sporophila schistacea*								
	Variable Seedeater *Sporophila americana*	C	C	R	C	C	R		U
	White-collared Seedeater *Sporophila torqueola*	R		C	U	C			C
	Yellow-bellied Seedeater *Sporophila nigricollis*				R				
	Ruddy-breasted Seedeater *Sporophila minuta*				R				
	Nicaraguan Seed-Finch *Oryzoborus nuttingi*	U							
	Thick-billed Seed-Finch *Oryzoborus funereus*	U	U		U	C			
	Blue Seedeater *Amaurospiza concolor*						R		
	Yellow-faced Grassquit *Tiaris olivaceus*	U	C			C	C	C	
	Slaty Finch *Haplospiza rustica*						R	R	
	Peg-billed Finch *Acanthidops bairdii*						R	U	
	Slaty Flowerpiercer *Diglossa plumbea*						C	C	
	Wedge-tailed Grass-Finch *Emberizoides herbicola*								
	Yellow-thighed Finch *Pselliophorus tibialis*						C	C	
	Large-footed Finch *Pezopetes capitalis*							C	
	White-naped Brush-Finch *Atlapetes albinucha*		R			C	C		
	Orange-billed Sparrow *Arremon aurantiirostris*	C	C		C	C			
	Sooty-faced Finch *Arremon crassirostris*								

	Species	La Selva	Rancho Naturalista	Palo Verde	Bosque DRT	Las Cruces	Bosque DP	San Gerardo	Chomes
	Chestnut-capped Brush-Finch *Arremon brunneinucha*		U			C	C	R	
	Stripe-headed Brush-Finch *Arremon torquatus*					U			
	Olive Sparrow *Arremonops rufivirgatus*			C					C
	Black-striped Sparrow *Arremonops conirostris*	C	C		C	C			
	Prevost's Ground-Sparrow *Melozone biarcuata*								
	White-eared Ground-Sparrow *Melozone leucotis*								
	Stripe-headed Sparrow *Aimophila ruficauda*			C					C
	Botteri's Sparrow *Aimophila botterii*								
	Rusty Sparrow *Aimophila rufescens*								
	Grasshopper Sparrow *Ammodramus savannarum*								
	Rufous-collared Sparrow *Zonotrichia capensis*		U			U	C	C	
	Volcano Junco *Junco vulcani*								

Saltators, Grosbeaks, and Buntings—CARDINALIDAE

	Species	La Selva	Rancho Naturalista	Palo Verde	Bosque DRT	Las Cruces	Bosque DP	San Gerardo	Chomes
	Streaked Saltator *Saltator striatipectus*				R	C			
	Grayish Saltator *Saltator coerulescens*	C	R						
	Buff-throated Saltator *Saltator maximus*	C	C		C	C			
	Black-headed Saltator *Saltator atriceps*	C	C						
	Slate-colored Grosbeak *Saltator grossus*	U							
	Black-faced Grosbeak *Caryothraustes poliogaster*	C	R						
	Black-thighed Grosbeak *Pheucticus tibialis*	R					U	C	
	Rose-breasted Grosbeak *Pheucticus ludovicianus*	U	C	U	U	C	U	C	U
	Blue-black Grosbeak *Cyanocompsa cyanoides*	C	R		C	C			
	Blue Grosbeak *Passerina caerulea*			U					U
	Indigo Bunting *Passerina cyanea*	R	R	C	R	R			C

Species	La Selva	Rancho Naturalista	Palo Verde	Bosque DRT	Las Cruces	Bosque DP	San Gerardo	Chomes
Painted Bunting *Passerina ciris*			U					U
Dickcissel *Spiza americana*			R					R

Blackbirds, Orioles, and Allies—ICTERIDAE

Species	La Selva	Rancho Naturalista	Palo Verde	Bosque DRT	Las Cruces	Bosque DP	San Gerardo	Chomes
Bobolink *Dolichonyx oryzivorus*								
Red-winged Blackbird *Agelaius phoeniceus*			C					
Red-breasted Blackbird *Sturnella militaris*								
Eastern Meadowlark *Sturnella magna*		R	R					C
Melodious Blackbird *Dives dives*		U						C
Great-tailed Grackle *Quiscalus mexicanus*		R	U	C	C			C
Nicaraguan Grackle *Quiscalus nicaraguensis*								
Shiny Cowbird *Molothrus bonariensis*								
Bronzed Cowbird *Molothrus aeneus*	R	U	U	U	C	C	R	C
Giant Cowbird *Molothrus oryzivorus*	R	U			R			
Black-cowled Oriole *Icterus prosthemelas*	C	U						
Orchard Oriole *Icterus spurius*	R	R			R			
Yellow-tailed Oriole *Icterus mesomelas*	U							
Streak-backed Oriole *Icterus pustulatus*			C					C
Spot-breasted Oriole *Icterus pectoralis*			U					U
Baltimore Oriole *Icterus galbula*	C	C	C	C	C	C	C	C
Yellow-billed Cacique *Amblycercus holosericeus*	C	U		C	U		R	
Scarlet-rumped Cacique *Cacicus uropygialis*	C	C		C				
Crested Oropendola *Psarocolius decumanus*					U			
Chestnut-headed Oropendola *Psarocolius wagleri*	C	C				U		
Montezuma Oropendola *Psarocolius montezuma*	C	C	R			U		

Species	La Selva	Rancho Naturalista	Palo Verde	Bosque DRT	Las Cruces	Bosque DP	San Gerardo	Chomes
Euphonias and Goldfinches—FRINGILLIDAE								
Scrub Euphonia *Euphonia affinis*			C					C
Yellow-crowned Euphonia *Euphonia luteicapilla*	C	U		C	C			
Thick-billed Euphonia *Euphonia laniirostris*				C	C			
Yellow-throated Euphonia *Euphonia hirundinacea*		U	R					
Elegant Euphonia *Euphonia elegantissima*		R			U	R	C	
Spot-crowned Euphonia *Euphonia imitans*				C	C			
Olive-backed Euphonia *Euphonia gouldi*	C	U						
White-vented Euphonia *Euphonia minuta*	U	U		U	R			
Tawny-capped Euphonia *Euphonia anneae*		C						
Golden-browed Chlorophonia *Chlorophonia callophrys*		U			U	C	C	
Yellow-bellied Siskin *Carduelis xanthogastra*					R	C	C	
Lesser Goldfinch *Carduelis psaltria*					C			
Old World Sparrow—PASSERIDAE								
House Sparrow *Passer domesticus*								
Munia—ESTRILDIDAE								
Tricolored Munia *Lonchura malacca*								
Cocos Island Specialties								
Masked Booby *Sula dactylatra*								
Red-footed Booby *Sula sula*								
Great Frigatebird *Fregata minor*								
Swallow-tailed Gull *Creagrus furcatus*								
Black Noddy *Anous minutus*								
White Tern *Gygis alba*								
Sooty Tern *Onychoprion fuscatus*								
Long-tailed Jaeger *Stercorarius longicaudus*								

		La Selva	Rancho Naturalista	Palo Verde	Bosque DRT	Las Cruces	Bosque DP	San Gerardo	Chomes
	Cocos Cuckoo *Coccyzus ferrugineus*								
	Cocos Flycatcher *Nesotriccus ridgwayi*								
	Cocos Finch *Pinaroloxias inornata*								
	Additional Species								

Bibliography

Baker, Christopher P. 2004. *Moon Handbooks: Costa Rica.* Emeryville, California: Avalon Travel Publishing.

Fogden, Michael. 1993. *An Annotated Checklist of the Birds of Monteverde and Peñas Blancas.* Monte Verde, Costa Rica: Self-published.

Fogden, Michael, and Patricia Fogden. 2005. *Hummingbirds of Costa Rica.* San José, Costa Rica: Zona Tropical Publications.

Fogden, Susan C. L. 2005. *A Photographic Guide to Birds of Costa Rica.* London: New Holland Publishers.

Franke, Joseph. 1993. *Costa Rica's National Parks and Preserves.* Seattle, Washington: The Mountaineers.

Garrigues, Richard, and Robert Dean. 2007. *The Birds of Costa Rica: A Field Guide.* Ithaca, New York: Cornell University Press/San José, Costa Rica: Zona Tropical Publications.

Gill, Frank B. 1990. *Ornithology.* New York: W. H. Freeman and Company.

Henderson, Carrol L. (2002). *A Field Guide to the Wildlife of Costa Rica.* Austin: University of Texas Press.

Hilty, Steven. 1994. *Birds of Tropical America.* Austin: University of Texas Press.

Howell, Steve N. G. 1999. *A Bird-Finding Guide to Mexico.* Ithaca, New York: Cornell University Press.

Janzen, Daniel H. 1983. *Costa Rican Natural History.* Chicago: University of Chicago Press.

Kricher, John. 1997. *A Neotropical Companion.* Princeton, New Jersey: Princeton University Press.

Ridgely, Robert S., and John A. Gwynne Jr. 1989. *A Guide to the Birds of Panama.* Princeton, New Jersey: Princeton University Press.

Rogers, Dennis W. 1996. *Costa Rica and Panama, the Best Birding Locations.* Portland, Oregon: Cinclus Publications.

Ross, David L., Jr. 1998. *Costa Rican Bird Song Sampler.* Ithaca, New York: Cornell Laboratory of Ornithology.

Ross, David L., Jr., and Bret M. Whitney. 1995. *Voices of Costa Rican Birds: Caribbean Slope.* Ithaca, New York: Cornell Laboratory of Ornithology.

Sánchez, Julio E. 2002. *Birds of Tapantí National Park.* San José, Costa Rica: Instituto Nacional de Biodiversidad.

Schram, Brad. 1998. *A Birder's Guide to Southern California.* Colorado Springs: American Birding Association.

Sekerak, Aaron D. 1996. *A Travel and Site Guide to the Birds of Costa Rica*. Edmonton, Alberta: Lone Pine Publishing.

Stiles, Gary F., and Alexander F. Skutch. 1989. *A Guide to the Birds of Costa Rica*. Ithaca, New York: Cornell University Press.

Taylor, Keith. 1990. *A Birder's Guide to Costa Rica*. Victoria, British Colombia: Self-published.

Wheatley, Nigel, and David Brewer. 2001. *Where to Watch Birds in Central America, Mexico and the Caribbean*. Princeton, New Jersey: Princeton University Press.

Zuchowski, Willow. 2007. *Tropical Plants of Costa Rica: A Guide to Native and Exotic Flora*. Ithaca, New York: Cornell University Press/San José, Costa Rica: Zona Tropical Publications.

Index